GUIDELINES FOR

Investigating Chemical Process Incidents

Second Edition

This book is one of a series of titles published by the Center for Chemical Process Safety of the American Institute of Chemical Engineers. A complete list of available titles appears at the end of this book.

GUIDELINES FOR

Investigating Chemical Process Incidents

Second Edition

Center for Chemical Process Safety
of the
American Institute of Chemical Engineers
3 Park Avenue, New York, NY 10016-5991

American Institute of Chemical Engineers
3 Park Avenue
New York, New York 10016-5991

Library of Congress Cataloging-in-Publication Data:

Guidelines for investigating chemical process incidents.— 2nd ed.
p. cm.
Includes bibliographical references and index.
ISBN 0-8169-0897-4
1. Chemical plants—Safety measures. I. American Institute of Chemical Engineers. Center for Chemical Process Safety.

TP155.5.G775 2003
660'.2804—dc21

2003009470

It is sincerely hoped that the information presented in this document will lead to an even more impressive safety record for the entire industry; however, neither the American Institute of Chemical Engineers, its consultants, CCPS Technical Steering Committee and Subcommittee members, their employers, their employers' officers and directors, nor AntiEntropics, Inc. .and its employees warrant or represent, expressly or by implication, the correctness or accuracy of the content of the information presented in this document. As between (1) American Institute of Chemical Engineers, its consultants, CCPS Technical Steering Committee and Subcommittee members, their employers, their employers' officers and directors, and AntiEntropics, Inc. and its employees, and (2) the user of this document, the user accepts any legal liability or responsibility whatsoever for the consequence of its use or misuse.

This book is available at a special discount when ordered in bulk quantities. For information, contact the Center for Chemical Process Safety at the address shown above.

PRINTED IN THE UNITED STATES OF AMERICA
10 9 8 7 6 5 4 3 2 1

Contents

Preface

The American Institute of Chemical Engineers (AIChE) has helped chemical plants, petrochemical plants, and refineries address the issues of process safety and loss control for over 30 years. Through its ties with process designers, plant constructors, facility operators, safety professionals, and academia, the AIChE has enhanced communication and fostered improvement in the high safety standards of the industry. AIChE's publications and symposia have become an information resource for the chemical engineering profession on the causes of incidents and the means of prevention.

The Center for Chemical Process Safety (CCPS), a directorate of AIChE, was established in 1985 to develop and disseminate technical information for use in the prevention of major chemical accidents. CCPS is supported by a diverse group of industrial sponsors in the chemical process industry and related industries who provide the necessary funding and professional guidance for its projects. The CCPS Technical Steering Committee and the technical subcommittees oversee individual projects selected by the CCPS. Professional representatives from sponsoring companies staff the subcommittees and a member of the CCPS staff coordinates their activities.

Since its founding, CCPS has published many volumes in its "Guidelines" series and in smaller "Concept" texts. Although most CCPS books are written for engineers in plant design and operations and address scientific techniques and engineering practices, several guidelines cover subjects related to chemical process safety management. A successful process safety program relies upon committed managers at all levels of a company who view process safety as an integral part of overall business management and act accordingly.

Incident investigation is an essential element of every process safety management program. This book presents underlying principles, management system considerations, investigation tools, and specific methodologies for investigating incidents in a way that will support implementation of a rigorous process safety program at any facility.

A team of incident investigation experts from the chemical industry drafted the chapters for this guideline and provided real-world examples to illustrate some of the tools and methods used in their profession. The subcommittee members reviewed the content extensively and industry peers evaluated this book to help ensure it represents a factual accounting of industry best practices. This second edition of the guideline provides updated information on many facets of the investigative process as well as additional details on important considerations such as human factors, forensics, legalities surrounding incident investigation, and near miss reporting.

Acknowledgments

The American Institute of Chemical Engineers wishes to thank the Center for Chemical Process Safety (CCPS) and those involved in its operation, including its many sponsors whose funding made this project possible; the members of its Technical Steering Committee who conceived of and supported this Guidelines project; and the members of its Incident Investigation Subcommittee. The Incident Investigation Subcommittee of the Center for Chemical Process Safety authored this second edition of the Guidelines for Investigating Chemical Process Incidents.

The members of the CCPS Incident Investigation Subcommittee were:

Michael Broadribb, Chair, *BP America, Inc.*
Curtis Clements, *DuPont*
Jim Bartlett, *Proctor and Gamble (retired)*
Dennis Blowers, *BP Solvay Polyethylene NA*
Bill Bridges, *ABS Consulting*
Marty Clancy, *AIChE*
Don Connolley, *Akzo Nobel Chemicals, Inc.*
Christy Franklyn, *Risk, Reliability, and Safety Engineering, Inc.*
Brian Kelly, *Syncrude Canada Ltd.*
Jack McCavit, *Celanese Chemicals*
Lisa Morrison, *NOVA Chemicals, Inc.*
Mickey Norsworthy, *Arch Chemicals*
Henry Ozog, *ioMosaic*
Mark Paradies, *System Improvements, Inc.*
Katherine Pearson, *Rohm and Haas Company*
Jack Philley, *Baker Engineering and Risk Consultants*
Pat Ragan, *Aventis*
Adrian Sepeda, *Occidental Chemical Corporation (retired)*

David Tabar, *Sherwin Williams*
Lee Vanden Heuvel, *ABS Consulting*

Dan Sliva was the CCPS staff liaison and was responsible for overall administration of the project. AntiEntropics, Inc. of New Market, Maryland, was contracted to provide editing services for this book. Sandra A. Baker and Robert Walter were the principal technical editors. CCPS would like to thank Mr. Ludwig Benner for providing historical perspective on MES information. CCPS would also like to thank Ms. Angella Lewis of Rohm and Haas Company for her timely graphics support.

CCPS also gratefully acknowledges the comments and suggestions received from the following peer reviewers:

Arthur M. Dowell, III, *Rohm and Haas Company*
Andrew Hart, *NOVA Chemicals (Canada) Ltd*
David Heller, *US Chemical Safety and Hazard Investigation Board*
Alistair D. McNab, *UK Health and Safety Executive*
Mike Marshall, *USDOL - OSHA*
David A. Moore, *AcuTech Consulting Group*
C. Robert Nelms, *Failsafe Network, Inc.*
William M. Olsen, *Merck & Company, Inc.*
Robert Ormsby, *Air Products and Chemicals, Inc.*
Ed Perz, *BP plc*
Armando L. Santiago, *U.S. Environmental Protection Agency*

Their insights, comments, and suggestions helped ensure a balanced perspective to this Guideline.

The members of the CCPS Incident Investigation Subcommittee wish to thank their employers for allowing them to participate in this project and lastly, we wish to thank Scott Berger and Les Wittenberg of the CCPS staff for their support and guidance.

Acronyms and Abbreviations

ACC	American Chemistry Council
AIChE	American Institute of Chemical Engineers
ALARP	As Low as Reasonably Practicable
ANSI	American National Standards Institute
API	American Petroleum Institute
ARIP	Accidental Release Information Program
ASME	American Society of Mechanical Engineers
B.P.	Boiling Point
BI	Business Interruption
BLEVE	Boiling Liquid Expanding Vapor Explosion
BPCS	Basic Process Control System
C	Consequence factor, related to magnitude of severity
CCF	Common Cause Failure
CCPS	Center for Chemical Process Safety,
CE/A	Change Evaluation/Analysis
CEI	Dow Chemical Exposure Index
CELD	Cause and Effect Logic Diagram
CIRC	Chemical Incidents Report Center
CLC	Comprehensive List of Causes
CPQRA	Chemical Process Quantitative Risk Assessment
CSB	Chemical Safety and Hazards Investigation Board (US)
CTM	Causal Tree Method
CW	Cooling Water
D	Number of times a component or system is challenged (hr^{-1} or $year^{-1}$)
DCS	Distributed Control System
DIERS	Design Institute for Emergency Relief Systems,
DOT	Department of Transportation
E&CF	Events & Causal Factor Charting
EBV	Emergency Block Valve

EPA	United States Environmental Protection Agency
ERPG	Emergency Response Planning Guideline
ETA	Event Tree Analysis
F	Failure Rate (hr^{-1} or $year^{-1}$)
f	Frequency (hr^{-1} or $year^{-1}$)
F&EI	Dow Fire and Explosion Index
F/N	Fatality Frequency versus Cumulative Number
FCE	Final Control Element
FMEA	Failure Modes and Effect Analysis
FTA	Fault Tree Analysis
HAZMAT	Hazardous Materials
HAZOP	Hazard and Operability Study
HAZWOPER	Hazardous Waste Operations and Emergency Response
HBT	Hazard–Barrier–Target
HE	Hazard Evaluation
HRA	Human Reliability Analysis
IChemE	Institution of Chemical Engineers
IEC	International Electrotechnical Commission
IEEE	Institute of Electrical and Electronic Engineers
IPL	Independent Protection Layer
ISA	The Instrumentation, Systems, and Automation Society (formerly, Instrument Society of America)
JSA	Job Safety Analysis
LAH	Level Alarm—High
LEL	Lower Explosive Limit
LFL	Lower Flammability Limit
LI	Level Indicator
LIC	Level Indicator—Control
LNG	Liquefied Natural Gas
LOPA	Layer of Protection Analysis
LOTO	Lockout/Tagout
LT	Level Transmitter
MARS	Major Accident Reporting System
MAWP	Maximum Allowable Working Pressure
MCSOII	Multiple-Cause, Systems-Oriented Incident Investigation
MES	Multilinear Event Sequencing
MHIDAS	Major Hazard Incident Data System
MOC	Management of Change
MORT	Management Oversight Risk Tree
MSDS	Material Safety Data Sheet
N_2	Nitrogen
NTSB	National Transportation Safety Board
OREDA	The Offshore Reliability Data project

ORPS	Occurrence Reporting and Processing System
OSBL	Outside Battery Limits
OSHA	United States Occupational Safety and Health Administration
P_{fatality}	Probability of Fatality
P_{ignition}	Probability of Ignition
$P_{\text{person present}}$	Probability of Person Present
P	Probability
P&ID	Piping and Instrumentation Diagram
PFD	Probability of Failure on Demand
PHA	Process Hazard Analysis
PI	Pressure Indicator
PL	Protection Layer
PM	Preventive Maintenance
PSID	Process Safety Incident Database
PSM	Process Safety Management
PSV	Pressure Safety Valve (Relief Valve)
R	Risk
RMP	Risk Management Program
RV	Relief Valve
SCAT	Systematic Cause Analysis Technique
SCE	Safety Critical Equipment
SIF	Safety Instrumented Function
SIS	Safety Instrumented System
SOP	Standard Operating Procedure
SOURCE	Seeking Out the Underlying Root Causes of Events
SSDC	System Safety Development Center
STEP	Sequentially Timed Events Plot
T	Test Interval for the Component or System (hours or years)
T_0	starting time
T_n	ending time
VCE	Vapor Cloud Explosion
VLE	Vapor Liquid Equilibrium
XV	Remote Activated/Controlled Valve

1

Introduction

1.1. Building on the Past

Flixborough, Bhopal, Piper Alpha—All three are now synonyms for catastrophe. These names are inextricably linked with images of death and disastrous loss tied to the production of chemicals or oil. An objective review of the world's industrial history reveals a story punctuated with infrequent yet similarly tragic incidents. Invariably, in the wake of such tragedy, companies, industries, and governments work to learn the causes. Their ultimate goal is that the knowledge acquired through diligent investigation can help prevent recurrence.

However, these investigations have revealed something of more significance—the key to preventing disaster first lies in recognizing the leading indicators. These leading indicators exist in incidents that are less than catastrophic. They can even be seen in so-called near misses that may have no discernable impact on routine operation. By examining lower-consequence, higher-frequency occurrences, companies may avoid those rare incidents that cause major consequences. The two most significant roles incident investigations can play in comprehensive process safety programs are:

1. Preventing disasters by consistently examining and learning from near misses and
2. Preventing disasters by consistently examining and learning from major consequence accidents.

The Center for Chemical Process Safety (CCPS) of the American Institute of Chemical Engineers (AIChE) recognized the role of incident investigation when it published the original *Guidelines for Investigating Chemical*

Process Incidents in 1992. The first edition provided a timely treatment of incident investigation including:

- a detailed examination of incident investigation's role in a process safety management system,
- guidance on implementing an incident investigation system, and
- in-depth information on conducting incident investigations, including the tools and techniques most useful in understanding the underlying causes.

This second edition builds on the first text's solid foundation. The goal is to retain the knowledge base provided in the original book while simultaneously updating and expanding upon it to reflect the latest thinking. This edition presents techniques used by the world's leading practitioners in the science of process safety incident investigation.

Successful investigations are dependent on preplanning and appropriate training. Preplanning allows organizations to respond properly and promptly. The first step in conducting a successful incident investigation is to recognize when an incident has occurred so that it can be investigated appropriately. To enhance effective recognition and communication during an investigation, the following definitions for key terms will apply throughout this book. Some investigators may define the terms presented below slightly differently or use other descriptive terms that mean the same things. The heart of the issue is that members of an operating investigation team all share a common language that supports their investigation objectives efficiently and accurately.

Incident*—an unusual or unexpected occurrence, which either resulted in, or had the potential to result in:*

- *serious injury to personnel,*
- *significant damage to property,*
- *adverse environmental impact, or*
- *a major interruption of process operations.*

This definition implies three categories of incidents:

1. Accidents
2. Near misses
3. Operational interruptions

An ***accident*** *is an occurrence in which property damage, material loss, detrimental environmental impact, or human loss (either injury or death) occurs.*

A ***near miss*** *is an occurrence in which an accident (that is, property damage, material loss, environmental impact, or human loss) or an operational interruption could have plausibly resulted if circumstances had been slightly different.*

*An **operational interruption** is an occurrence in which production rates or product quality is seriously impacted.*

The second step in conducting a thorough investigation is to assemble a qualified team to determine and analyze the facts of the incident. This team's charter, using appropriate investigative techniques and methodologies, is to reveal the true underlying root causes. The terms causal factor and root cause help investigators analyze the facts and communicate with each other during the investigation phase.

*A **causal factor**, also known as a critical factor or contributing cause, is a major unplanned, unintended contributor to the incident (a negative occurrence or undesirable condition), that if eliminated would have either prevented the occurrence, or reduced its severity or frequency.*

*A **root cause** is a fundamental, underlying, system-related reason why an incident occurred that identifies a correctable failure(s) in management systems. There is typically more than one root cause for every process safety incident.*

The third step in incident investigation is to generate a report detailing facts, findings, and recommendations. Typically, recommendations are written to reduce risk by:

- improving the process technology,
- upgrading the operating or maintenance procedures or practices, and
- upgrading the management systems. (When indicated in a recommendation, this is often the most critical area.)

After the investigation is completed and the findings and recommendations are issued in the report, a system must be in place to implement those recommendations. This is not part of the investigation itself, but rather the follow-up related to it. It is not enough to put a technological, procedural, or administrative response into effect. The action should be monitored periodically for effectiveness and, where appropriate, modified to meet the intent of the original recommendation.

These four steps will result in the greatest positive effect when they are performed in an atmosphere of openness and trust. Management must demonstrate by both word and deed that the primary objective is not to assign blame, but to understand what happened for the sake of preventing future incidents. This book helps organizations define and refine their incident investigation systems to achieve positive results effectively and efficiently.

1.2. Who Should Read This Book?

This book assists three target groups:

1. Incident investigation team leaders
2. Incident investigation team members
3. Corporate and site process safety managers and coordinators

For anyone directly involved in leading or participating on incident investigation teams, the book provides a valuable reference tool. It presents knowledge, techniques, and examples to support successful investigations. For persons in technical and management roles responsible for implementing the incident investigation element of an integrated process safety system, it offers a model for success in building or upgrading their program.

Like the previous edition, the book remains focused primarily on investigating process-related incidents that present realized or potential catastrophic consequences (that is, accidents as well as near misses). However, readers will find that the methodologies, tools, and techniques described in the following chapters may also be applied when investigating other types of occurrences such as reliability, quality, and occupational health and safety incidents.

1.3. The Guideline's Objectives

Readers should be able to achieve the following objectives.

- Describe the basic principles behind successful incident investigations.
- Identify the essential features of a management system designed to foster and support high quality incident investigations.
- List detailed information for planning and conducting incident investigations including investigative tools, techniques, and methodologies for determining causes.
- Use the findings of an investigation to make effective recommendations that can reduce the likelihood of recurrence or mitigate the consequences of similar incidents (or even dissimilar incidents with common root causes).
- Plan an effective system for documenting, communicating, and resolving investigation findings and recommendations including a method to track closure of incident recommendations.

The summaries below provide an overview of the content and organization of the book chapter-by-chapter and assist in quickly locating a particular area of interest.

Chapter 2—Designing an Incident Investigation Management System
This chapter provides an overview of a management system for investigating process safety incidents. It opens with a review of management responsibilities and presents the important features that a management system must address to be effective. It examines systematic approaches that help implement incident investigation teams, root cause determinations, recommendations, follow-up, and documentation.

Chapter 3—An Overview of Incident Causation Theories
This section discusses the basics of determining incident causation and describes the general categories of incidents—from near miss to major catastrophe. It examines the anatomy of process incidents as related to theoretical models of incident causation.

Chapter 4—An Overview of Incident Investigation Tools and Methodologies
This chapter provides a brief overview of investigation tools in simple, generic terms, and demonstrates the benefits of using a more structured approach. It describes both public and proprietary methodologies.

Chapter 5—Reporting and Investigating Near Misses
Many major process safety incidents were preceded by precursor occurrences. These occurrences were unrecognized or ignored because "nothing bad" actually happened. The lessons learned from such occurrences, typically referred to as near misses, can be extremely valuable in averting disaster. However, this benefit is only realized when they are recognized, reported, and investigation techniques are properly applied. This chapter describes near misses, discusses their importance, and presents the latest methods for helping ensure appropriate near misses are reported.

Chapter 6—The Impact of Human Factors
This chapter describes human factor considerations in incident causation. It provides insight and tools to identify and address applicable human factors issues during an investigation.

Chapter 7—Building and Leading an Investigation Team
Personnel with proper training, skills, and experience are critical to the successful outcome of an incident investigation. This chapter describes team composition as a function of incident type, complexity, and severity as well as suggested training topics. It also provides team leaders with a high-level overview of the basic team activities typically required in the course of conducting an investigation.

Chapter 8—Gathering and Analyzing Evidence
Facts are the fuel an investigation needs to reach a successful conclusion. This chapter addresses the practical considerations of data-gathering activities. It describes types of data, sources of data, data-gathering tools, and techniques.

Chapter 9—Determining Root Causes—Structured Approaches
This chapter addresses methods and tools used successfully to identify multiple root causes. Process safety incidents are usually the result of more than one root cause. This chapter provides a structured approach for determining root causes. It details some powerful, widely used tools and techniques available to incident investigation teams including timelines, logic trees, predefined trees, checklists, and fact/hypothesis. Examples are included to demonstrate how they apply to the types of incidents readers are likely to encounter.

Chapter 10—Developing Effective Recommendations
Once the likely causes of an incident have been identified, investigation teams evaluate what can be done to help prevent recurrence. The incident investigation recommendations are the product of this evaluation. This chapter addresses types of recommendations, some attributes of good recommendations, methods to document and present recommendations, and management's responsibilities.

Chapter 11—Communication Issues and Preparing the Final Report
In the case of incident investigation, a major milestone is completed when the final incident investigation report is submitted. The incident report documents the investigation team's findings, conclusions, and recommendations. This chapter describes practical considerations for writing formal incident reports, a discussion of the attributes of quality reports, and the issue of communicating the report findings to affected persons, both internally and externally.

Chapter 12—Considering Legal Issues
The work products of incident investigations are subject to the legal process of discovery. The incident investigation team must keep two purposes in mind. First, the ultimate purpose of the incident investigation is to determine what happened, why it happened, and how to prevent future occurrences. Second, there are important legal issues associated with the conduct, documentation, and follow-up of incident investigations. This chapter provides insight into legal issues and is written for a lay audience.

Chapter 13—Implementing the Team's Recommendations
The recommendations generated from an incident investigation should be properly implemented in a timely fashion to decrease the probability of

recurrence. This chapter focuses on the critical aspects of implementing recommendations. It addresses initial resolution of the recommendations, their implementation, and sharing lessons learned from the investigation.

Chapter 14—Seeking Continuous Improvement for Your System
The adage "if it ain't broke, don't fix it" does not apply to process safety management systems. Continuous improvement should be an integral part of the management system. This chapter describes techniques that can help the incident investigation element of process safety remain strong and viable in an ever-changing technical, business, and regulatory environment. It includes considerations for assessing existing incident investigation programs as well as approaches for implementing continuous improvement.

Chapter 15—Lessons Learned
This chapter focuses on the value of critically analyzing incident information and discusses the benefits of using databases to evaluate lessons learned. The power of both internal databases and industry-wide incident databases can help organizations improve their performance in process safety and environmental responsibility.

Appendices
The appendices provide a wealth of supplemental information on the subject of incident investigation. Topics include:

A. Relevant Organizations
B. Professional Assistance Directory
C. Photography Guidelines for Maximum Results
D. Example Case Study—Fictitious NDF Company Incident
E. Example Case Study—True Incident
F. Selected US OSHA and US EPA Incident Investigation Regulations
G. Quick Checklist for Investigators
H. Additional References
I. CD-ROM Contents

CCPS Incident Investigator's Companion CD ROM
This useful companion disk contains root cause analysis examples, predefined tree examples, practical checklists that can be customized, and incident evidence photograph examples. It includes a quick checklist for investigators traveling to an incident, examples of methodologies that may be useful in training the onsite team, and checklists and samples from the text that can be printed out at the incident site to help organize the team's work.

1.4. The Continuing Evolution of Incident Investigation

Like all of the elements of process safety management, the incident investigation element continues to evolve. The AIChE Center for Chemical Process Safety assists this evolution by providing interested parties with information to help them in safely operating process facilities. To this purpose, CCPS and the contributing authors offer this second edition of the *Guidelines for Investigating Chemical Process Incidents*.

2

Designing an Incident Investigation Management System

Process safety incidents are the result of management system failures.

—*Guidelines for Technical Management of Chemical Process Safety*, CCPS

This chapter describes how to build and implement a practical management system for investigating process safety incidents. The primary objective of incident investigation is to prevent recurrence by applying our knowledge and experience. This can best be accomplished by establishing a management system for investigation that assists in achieving the following four goals.

1. Encouraging employees to report all incidents including near misses
2. Ensuring investigations identify root causes
3. Ensuring investigations identify recommended preventive measures that reduce the probability of recurrence or mitigate potential consequences
4. Ensuring follow-up action to resolve all recommendations effectively

The items in this list are essential to maintaining a well-designed incident investigation program. A high priority should be to promote reporting to learn from near miss incidents *before* a substantial loss occurs.

The ***incident investigation management system*** *should be described in a written document that defines the roles, responsibilities, protocols, and specific activities to be carried out by personnel performing an incident investigation.*

This chapter highlights management's responsibilities and the importance of leadership as well as management system content and proven methods for implementing a management system. Figure 2-1 depicts a typical view of the management system model used throughout this book.

Note that near misses (Chapter 5), human factors (Chapter 6), and legal issues (Chapter 12) are special considerations in both preplanning and deploying the incident investigation management system.

2.1. Preplanning Considerations

2.1.1. An Organization's Responsibilities

Establishing a high quality incident investigation program begins with management's support, commitment, and action. To demonstrate support, it is common practice to establish a written policy regarding incident reporting and investigation, to communicate this policy to the workforce, and to sustain it over time. This is often expressed in a formal statement written to achieve the following goals.

- Communicate management's commitment to prevent recurrences by determining root causes, recommending preventive measures, and taking follow-up action.
- Recognize the importance of investigation as a primary hazard control mechanism
- Strongly support reporting and investigating near misses.
- Clearly focus on finding causes and management system weaknesses, and avoid assigning blame.
- Endorse sustained commitment of resources for the investigation program, through training team members and managers. This supports employee participation in the investigation program and the appropriate and timely implementation of recommendations.
- Emphasize the value and necessity of communicating and sharing the lessons learned from the investigation to all that could reasonably benefit.
- Support a system to ensure that all recommendations and findings are resolved and that decisions and actions are documented.

Management demonstrates support for this policy by nurturing an atmosphere of trust and respect. This encourages openness in reporting incidents throughout the organization. Failure to achieve this positive atmosphere may result in hidden incidents or low or no reporting of near misses, potentially leading to an avoidable catastrophic incident.

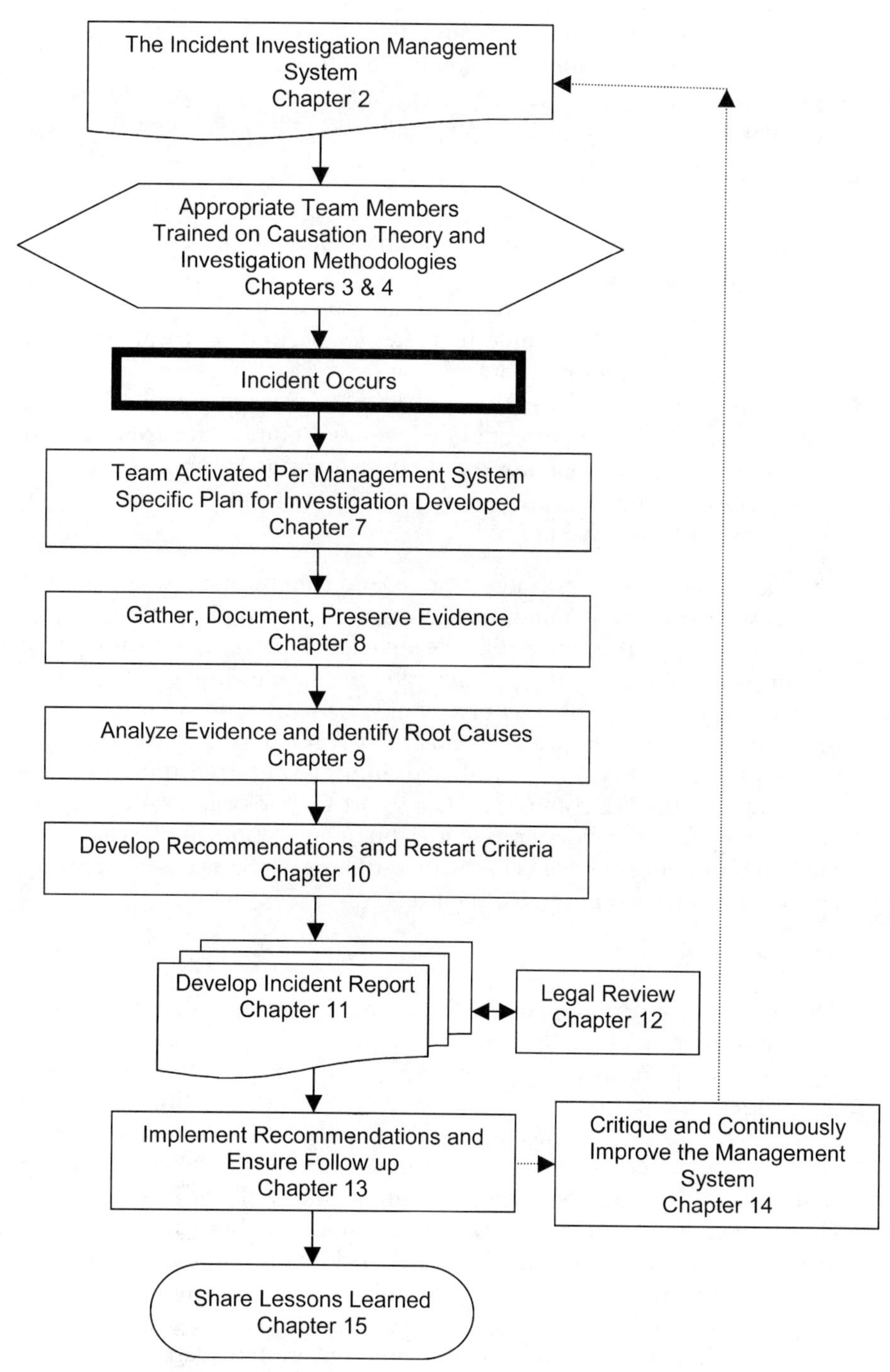

FIGURE 2-1. **Management system for process safety investigation.**

Management demonstrates commitment by recognizing that periodic reviews and reevaluations of the incident investigation management system are necessary to ensure it continues to function as originally intended and achieves the desired results. Periodic reviews provide four obvious benefits.

1. They provide verification that all action items resulting from recommendations have been resolved to completion, are documented, and are effective.
2. They ensure documentation exists explaining why a recommendation was rejected or modified after its original approval in an incident investigation report.
3. They allow early detection of both encouraging and disturbing trends. An example would be increased number (or frequency) of near misses (actual, not reported) in a particular area or process.
4. They provide an opportunity for continuous improvement for the investigation system itself.

Management demonstrates support and commitment by action, such as the establishment of a high-quality incident investigation training program. This helps to ensure that the management system is understood and implemented as designed. Each job position's training on the system will vary in level of detail and objectives. Persons assigned to lead roles on incident investigation teams should be targeted to receive the most focused training. Periodic refresher training is an opportunity for management to reinforce commitment, support for the organization's policy and philosophy on incident reporting and investigation, and discuss modifications and improvements in the investigation process based on lessons learned from performing investigations.

2.1.1.1. First Notification

All employees, including contractors, involved in or learning of an incident, whether accident, near miss, or operational interruption, should be required to report details of the incident immediately to their supervisor. The supervisor would customarily be responsible for initiating further action to investigate the incident, and take required action. First notification may also need to follow company protocol to report details of the incident to specific individuals or organizations internally or externally. The circumstances of the incident and the results of the investigation should also be communicated via the company's incident reporting system, which should be described in the management system.

The term *report* can have several meanings. Sometimes the term could mean a verbal initial notification or communication to alert the organization that an incident has occurred. The term also refers to the final, formal written incident investigation report. Even after 30 years, the term *report* still causes confusion when discussing the U.S. Department of Labor, Occupational Health and Safety Administration (US OSHA) regulations. US OSHA requires a verbal notification *report* for certain incidents such as a fatality or hospitalization of three or more employees. US OSHA requires verbal notification within 48 hours following the incident, but does not require the use of a specific form. The reporter, however, should make a written documentation of this verbal communication, noting the time, person involved, extent of information disclosed, and any special instructions or requests made by US OSHA at the time of notice.

US OSHA further requires a written *report* for each work-related injury or illness that is severe enough to be recordable.

Many regulatory agencies require *immediate notification*. However, the application of the term is inconsistent. For example, the State of California requires formal notification for dismemberment or disfigurement injury incidents within hours. The State of New Jersey requires immediate notice when certain quantities of hazardous materials are released. Historically, nothing longer than 15 minutes has been considered immediate by New Jersey regulators. The U.S. Coast Guard and the Environmental Protection Agency also have specific notification requirements.

The format and timing of all external notifications should be identified and incorporated into the incident investigation management system before an incident. With this information readily at hand, the proper notifications may be made quickly and accurately when an incident occurs.

Internal notifications, sometimes called alerts or *flash reports,* are trigger mechanisms for starting specific portions of the incident investigation management system and for decision making. Obviously, medical treatment of injured personnel and stabilization of the incident site always takes priority over other activities if there is a conflict regarding the use of available resources during the early stages of an incident. These notification alerts may be initiated for near misses with serious implications, serious incidents, or high potential serious incidents such as those which result in serious injury or illness, spill or release with consequential damage, or public exposures.

Initial notifications, which should be part of the company's emergency response plan, may include the following.

Internal

- **Within the facility to summon emergency responders**
- Within the site to start administrative response
- To the company headquarters and administrative departments

External
- Resources for mutual aid emergency response
- Regulatory agencies, as required
- Family members or next-of-kin
- Neighboring facilities
- News media where appropriate
- Insurance carriers
- Neighboring community

Making initial notification in a timely manner can present serious challenges immediately following an incident. The incident investigation management system should address how to handle these communications and how to coordinate with facility emergency response plans. A checklist with prearranged names, titles, and phone numbers should be developed and kept up to date for this use.

2.1.2. The Benefits of Management's Commitment

Management's commitment to a systematic incident investigation system results in benefits such as:

- fewer worker injuries and illnesses,
- fewer environmental issues,
- greater return on investment capital,
- increases in process capability and uptime,
- improved product quality,
- reduced costs, and
- an enhanced image in the eyes of employees, industry, and the public.

These benefits will be realized with a dedicated and sustained commitment of resources.

The incident investigation management system should help managers:

- understand the specific responsibilities of each level of the organization in regard to the management system,
- develop a clear understanding of the organization's commitment,
- recognize, accept, and address root causes,
- proactively address near misses, and
- persistently follow up on recommendations to ensure their effective resolution.

Incident investigation is only one of the many elements of a process safety management program, but it plays an essential role in identifying overall management system weaknesses on a continuous basis.

Some upper-level managers may not intuitively understand the structure, function, benefits, and requirements of multiroot cause incident investigations. Educating this audience is critical because management participation helps promote a sense of sponsorship and assists in establishing investigations as a normal task within a manager's or supervisor's duties. Promoting sponsorship reduces the tendency to perceive investigations as primarily the domain of a narrowly focused group of full-time investigators. For example, one company requires that a senior business unit manager (a refinery manager or chemical plant manager) lead any fatality investigation with the support of a specially trained root cause analyst. All business unit managers receive a minimum of 8 hours training in the conduct of these investigations—*no exceptions*. No one in the organization should believe that "incident investigation is the Safety Department's responsibility."

Management's continuing endorsement and approval for the program is essential. It is beneficial to reaffirm that management understands the concepts of incident investigation on a periodic basis since changes in company leadership may affect the level of awareness and emphasis on incident investigation.

2.1.3. The Role of the Developers

One way to achieve the support of management is to include managers in development activities. Developers lead teams to establish an entire incident investigation management system or to upgrade one that is already in place. In either case, developers need top management's support. System developers prepare their team by researching the basic incident investigation principles and priorities. This book is a good resource for orienting a development team. Developers can provide leadership to help the team determine which investigation methodologies best fit the particular culture of their organization.

2.1.4. Integration with Other Functions and Teams

An active incident investigation will touch other functions within the organization. Preplanning for this interaction begins during the development stages of the management system by identifying known areas of mutual interest. The management system developers should review other existing management systems such as those listed below to identify opportunities for integration and communication.

- Emergency response
- Environmental protection
- Employee safety

- Security
- Regulatory compliance
- Insurance interactions
- External media communications
- Corporate legal policies and procedures
- Engineering design and risk reviews (such as process hazard analyses or management of change reviews)
- Accounting and purchasing practices
- Quality assurance

One approach is to mesh all investigation and root cause analysis activities under one management system for investigation. Such a system must address all four business drivers: (1) process and personnel safety, (2) environmental responsibility, (3) quality, and (4) profitability. This approach works well since techniques used for data collection, causal factor analysis, and root cause analysis can be the same regardless of the type of incident. Many companies realize that root causes of a quality or reliability incident may become the root cause of a safety or process safety incident in the future and vice versa.

This approach also helps to avoid redundancy regarding assignment of responsibility, authority, or priorities. It also makes it easier to report occurrences, as the reporter does not need to know the occurrence classification in order to determine who to notify.

2.1.5. Regulatory and Legal Issues

Regulatory and legal considerations should be addressed in the management system. Regulatory aspects of the system need to be monitored for changing demands. In the United States and Europe, government agency attention to industrial incidents has steadily increased due to new and revised safety and environmental regulations. Some have very specific requirements for reporting, documentation, and investigation.

Legal considerations continue to evolve and continue to affect cause determinations, remedy implementation, communication, and documentation aspects of process incident investigations. As a minimum, the legal department should be consulted at the beginning and the end of the investigation process for all major incidents. One issue common to both the regulatory and legal concerns is the additional responsibility a corporation assumes once it has increased knowledge of a hazard or remedy. Failure to act on this knowledge may result in significantly larger legal and regulatory consequences. Many chemical processing facilities use nonproprietary technologies that present common hazards. This allows for meaningful sharing of incident investigation findings throughout the industry. Public awareness of chemical processing incidents is high and its tolerance is low. Public knowledge of detailed chemical hazards and specific process characteristics is limited. Therefore, one company's poor safety performance can affect the entire industry through regulation by public outcry.

The management system should address methods for sharing incident causes and lessons learned through appropriate channels so that others can benefit. It is often a challenge for a company's management to share the details of investigations due to litigation concerns. However, when similar facilities might benefit, finding a way to share displays a company's concern for the public welfare and the entire industry's performance. In addition to litigation issues, practical logistics sometimes make it difficult to communicate lessons learned within and between companies. Determining which people or companies have a potential interest can sometimes be problematic. Despite these challenges, broad communication of investigation findings is a recognized good practice. Chapter 12 explores these issues in detail.

2.2. Typical Management System Topics

As stated in the introduction, *the incident investigation management system is a written document that defines the roles, responsibilities, protocols, and specific activities to be carried out by personnel performing an incident investigation.* The management system may include a purpose statement, definitions, classifications, and responsibilities. It provides the structure for activities such as evidence gathering, witness interviewing, and data control as well as standard practices for notification, reporting, and follow up. The following sections summarize the recommended elements of a management system for incident investigation.

2.2.1. Classifying Incidents

When developing an incident investigation management system, it is important to define common terms and classifications. Several incident categories can be used to develop a classification system. Classification has two main purposes:

1. Determining importance of the accident and resulting consequences. This often dictates team leadership, size, and composition
2. Determining how investigation results will be routed and to whom (including regulatory-required routing).

The system must describe specific mechanisms for deciding to activate an investigation team and the team composition for each incident category. There should also be a mechanism that describes required internal and external notification. The incident investigation management system should specify:

- Who will make the notification
- Who is to be notified
- How and when they must be notified

This is usually captured in a procedure and associated routing forms. Chapter 11 provides descriptions of the required notifications.

Classification systems can vary depending on company and site organization. There is no perfect one-size-fits-all system of classification. Traditionally, classification systems fit an incident into severity (actual or potential) categories. However, an alternate method is to assign lead investigators and team members based on the nature and complexity of the incident, rather than its severity. In practice, there will be gray areas in every system. Honest differences of technical opinion or significant changes in perspective during the initial stages of an investigation may lead the team to change the initial classification during the course of the investigation. This is common when the team is investigating a near miss and reaches a point where they can determine the reasonable probabilities of a consequence. The management system should provide guidance for resolving these issues. Table 2-1 shows various classification schemes.

From many companies' perspectives, classification by severity is the most common classification system used to establish when to investigate an incident and who should be on a team. The main disadvantage with using severity alone to establish team membership is that it does not consider the potential loss from the occurrence, such as that with a high potential incident. In addition, the complexity of the system involved should also be used to determine team composition as more complex systems might require a larger team to understand the data.

Regardless, the classification system should achieve a specific outcome and add value. To achieve this goal, companies should consider developing a classification scheme that helps establish the proper team composition based on the complexity, nature, *and* severity of the occurrence. Chapter 7 describes considerations for building a team based on classification of an incident. The chapter defines terms such as minor incidents, limited impact incidents, significant incidents, high-potential incidents (HIPO), and catastrophic incidents.

A classification scheme should display the following characteristics:

- It should be easily understood.
- It should include clear examples.
- It should detail specific mechanisms to authorize an investigation and who may do so.
- It should help determine team composition.

Table 2-1
Common Classification Schemes

Classification by System Complexity	Classification by Type of Incident	Classification by Severity	Classification by Applicable Regulation
• **High** –nuclear materials –high pressure (>50 psig) –high temperature (>2000°F) –exothermic reactions –explosive environment –several relief devices –highly automated –several operators • **Moderate** –10–50 psig –100–2000°F –minor reactivity –low probability of explosions –single relief device –1–3 operators • **Simple** –ambient conditions –little/no reactions –nonexplosive environment –single/no relief valve –1–2 operators	• **Accident** –major releases –minor releases –explosion –fire –personnel harm –high potential incidents • **Near Miss** –small release –safety permit violation –failure of critical safeguard –challenge last line of defense –serious process excursion • **Other** –process upsets –quality variations –downtime –offsite consequences	• **Multiple fatalities/ serious injuries** • **Fatality** • **Injury** –hospitalization –lost work day –recordable –first aid • **Evacuation** • **Shelter-in-place** • **Reportable (EPA)** • **Levels of business interruption/ product losses** • **Levels of equipment damage**	US OSHA PSM US EPA RMP US OSHA General Duty Clause US EPA General Duty Clause US Coast Guard US DOE US DoD US DOT NRC Permit Violation None

One alternative classification scheme simply specifies the experience level of the lead investigator and then leaves the team composition to the leader. This approach depends upon the leader's experience and training.

2.2.2. Other Options for Establishing Classification Criteria

The amount of direct monetary value of the loss, interruption, harm, or damage is sometimes used as a sorting category. This is often an internally established value related to insurance coverage deductibles or management financial authorization structure. The loss of production is another classification criterion. These criteria could be expressed in units of hours, days, or weeks of expected downtime. A further improvement is to estimate both the actual and *potential* severity of the impact of such incidents. Potential frequency of recurrence is also a useful indicator in deciding on the severity ranking. Making such a determination is an imprecise effort, and organizations are best served when a decision is made quickly with the evidence at hand rather than waiting for more perfect data.

If a particular regulatory agency must be notified or becomes involved, there may be certain corresponding internal action and notifications. In the United States, some incidents reported to the Environmental Protection Agency (EPA) are an example of classification that can cause specific actions beyond simple notification. The approval loop for the notification report may expand to include legal representatives and the investigation team composition may change to meet specific regulatory requirements or to provide stewardhip of the company's interests.

2.2.3. Specifying Documentation

The management system should specify documentation requirements for interim data and work products of the investigation. The company's legal staff may have a valuable opinion on this decision or they may offer case-by-case opinions. Several issues are obvious. Witness interviews and physical evidence are examples of notable issues with which the legal department may wish to be involved. Other important documentation issues include:

- the minutes of team deliberations,
- official notifications to external agencies,
- the method for tracking documents and evidence requested, received, or issued by the team, and
- the final report.

Certain documents or evidence may need special attention due to potential litigation. A chain-of-custody record may be necessary.

2.2.4. Describing Team Organization and Functions

The incident investigation management system may include a description of how a team is organized and how it functions. The team organization, composition, and functions must be structured to provide flexibility based on the particular incident and the management system should emphasize that fact. The system may describe an investigation team's basic objectives and priorities. When establishing the charter for a major investigation, it is important to remember that the team members are not full time, professional investigators. They may only serve on such a team once during their entire work career.

The team leader's responsibilities need to be explicit. Normally a team leader chosen for more serious or more complex incident investigations will be independent from the operation or facility where the incident occurred. Actual team composition may vary significantly based on the nature of the process and the degree of technical sophistication. This flexibility of team composition is an important feature of a well-designed incident investigation management system.

The investigation process generally follows a problem solving process sequence. Once the team has developed the specific investigation plan, evidence is gathered. These two activities consume much of the team's time. A typical checklist for the plan is included in Chapter 7.

The management system may define some specific team functions and responsibilities. Some examples are listed below.

- Selecting and developing an incident investigation plan defining the scope of the investigation
- Identifying support and supplies
- Developing evidence handling procedures
- Establishing communication channels both within the company and with outside groups
- Conducting witness interviews
- Summarizing findings and recommendations in a report

Implementation and associated follow-up on resolution of all recommendations is an essential component of a management system. As written, it should specifically address the assignment of responsibility for follow-up. In rare cases, the incident investigation team will retain responsibility and authority for the final resolution of the recommendations. However, in most cases the primary responsibility will shift to a designated member of management who is not a member of the incident investigation team. If management rejects or significantly modifies the recommendations from an incident investigation team, management has the responsibility to discuss these with the team to determine if the team needs to clarify their stance.

The management system should promote continuous improvement by including a process for feedback. To ensure continuous improvement, the team should perform a self-evaluation after each investigation. This self-evaluation should include:

- Team thoroughness in the investigation

- Team effectiveness in applying the techniques
- Team preparedness in advance of the investigation
- Equipment performance during the investigation
- Supply logistics and quality

2.2.5. Setting Training Requirements

The management system should describe minimum initial training and refresher training for four major groups. The level of detail contained in the management system may vary. For example, it may provide a brief summary and then refer to a training management system document or position curricula for the detailed training information. A summary of training topics for each group is provided below.

- **Management**
  This group must be familiar with the concepts, policies, extent of commitment from upper-level management, and specific assignments of responsibility associated with process safety incident investigation.

- **All employees in a position to notice and report a near miss or accident**
  This group includes operators, mechanics, first line supervisors, auxiliary staff groups such as technicians and engineers, and middle-level management. They should be trained on how to differentiate an incident from a nonincident and what to do once an incident is identified.

- **Incident investigation team members**
  This group has an additional need to be trained in the support functions of an investigation, particularly how to effectively gather data. For instance, team members should be trained on how to preserve evidence, interview peers, develop test plans, and develop sampling procedures. Depending on their role in the investigation, some team members may need training in data analysis and the use of specific investigative tools.

- **Investigation leaders**
  Some organizations break this training into two or more levels with team leaders given more training if they will lead investigations of higher level or complex incidents. Leaders learn how to determine the appropriate investigation methodology, how to gather data, how to analyze data for causal factors, how to determine root causes

of causal factors, and how to develop effective recommendations and reports.

Leader training deserves special attention. Training for leaders, and others, could include role-playing for witness interviews, conflict resolution, the applicable laws, regulator powers, and confidentiality issues. They must feel free to request help or training when needed, especially at the early stages of an investigation. Lower level investigators may handle low to moderate complexity incidents. This training usually consists of classroom training plus experienced coaching during their first few investigations. They can benefit from participating as team members on an incident investigation led by an experienced leader. Low complexity incidents may require one helper (team member) to support the leader in data gathering and analysis. The higher-level investigation leader should be able to handle almost any incident within the company. The training for this level usually consists of additional classroom training and coaching by a more experienced investigator during their first few major investigations.

In some cases, employees, rather than supervisors, lead investigations for lower level incidents. Companies have found it beneficial for employees to feel ownership of the investigation results. This philosophy helps encourage workers to report more near misses by reducing the fear caused when a supervisor leads the investigation. Most incidents are low complexity. Many of these are near misses and benefit from investigation by persons closest to the process.

Chapter 7 provides details on the training, selection, and organization of incident investigation teams.

2.2.6. *Emphasizing Root Causes*

Identifying causes is a major objective of the entire investigation process and should be specified in the management system. Initial selection or custom development of the root cause determination process will require special attention to the concept of multiple causes and to underlying system-related causes. The approach should emphasize finding management system weaknesses and failures versus placing blame on individuals. (Refer to Chapter 6.) Some employees may need to adjust to this approach particularly if past methods did not previously encourage it. Everyone involved in the resolution process for recommendations needs to understand the concept of multiple root causes of an incident.

Chapter 9 describes this important aspect of the management system in detail.

2.2.7. *Developing Recommendations*

Identifying and evaluating practical recommendations are critical team activities. The management system should include attention to evaluating proposed recommendations. For example, recommendations should eliminate the causes of the incident or near miss while being practical, cost effective, and within the control of the organization. Ineffective recommendations may only serve to transfer the hazard or even create a new hazard that was not present before the initial incident. The management system for incident investigation needs a built-in mechanism to require safety analyses of the proposed recommendations. A tie should exist between the facility's incident investigation management system and their management of change (MOC) program. The investigation team needs to think whether its recommendations are practical and will adequately address the root causes. Additionally, site management should ensure that any changes to equipment or procedures as a result of recommendations are properly evaluated before implementation.

The space shuttle *Challenger* disaster is a classic example for the need to evaluate proposed recommendations. Before the *Challenger* incident, NASA was aware of the poor performance of the ring joint seal systems from previous near-miss incident investigations. In a well-meant effort to improve the safety margin, it was decided to increase the pressure test from 100 to 200 psig (6.8 to 13.6 atmospheres) after the ring joints were reassembled. In reality, this recommendation actually decreased the integrity and reliability of the ring joint seals by increasing the deformation of the sealing putty. A MOC analysis might have uncovered this increased risk.

Chapter 10 provides guidance for formulating effective responses to investigation findings.

2.2.8. *Fostering a Blame-Free Policy*

Disciplinary action is not part of the investigation. The management system for investigation should ensure that a blame-free policy precluding disciplinary action for honest mistakes is clearly stated and enforced.

Disciplinary action may be appropriate if malicious or criminal intent is positively identified as a root cause. An example would be when an investigation reveals horseplay, practical jokes, fights, or even sabotage was among the root causes. These activities have no place in any workplace and are especially undesirable in the chemical processing industry. It is most likely that a company's employee handbook, human resources documents, or union contract addresses these situations and communicates the policy in advance of an incident. In short, the investigators determine the facts, analyze for cause, and make recommendations. Managers then must react to those recommendations, and when human actions are called into question, discipline might be appropriate if there have been violations.

2.2.9. Implementing the Recommendations and Follow-up Activities

Resolving the recommendations and following-up on their effectiveness is a cornerstone of management systems. To reduce the probability of a repeat incident, the recommendations must be implemented and sustained. For lasting results, it is wise to audit all implemented recommendations periodically to ensure that they are continuing to achieve the intended objectives.

Another requirement is documenting resolution of all recommendations. The concept of an auditable trail is increasingly mentioned in regulatory and legal activities. If a team recommendation is rejected or modified, the basis for the rejection or change must be thoroughly documented after review with the investigation team. These requirements should be reflected in the incident investigation management system. The management review and approval process for the system encourages adoption of this important feature. It should be emphasized when personnel at all levels are trained.

The management system should indicate the importance (priority) of the recommendation, assignment of responsibility, and method for verifying and documenting its resolution. Management should acknowledge findings and observations expressed in the team's written report.

Chapter 13 examines various approaches to help ensure closure.

2.2.10. Resuming Normal Operation and Establishing Restart Criteria

If the incident has resulted in shutdown of a process, the decision to restart becomes a very important consideration of the investigation process. Although the actual decision to restart is a line management decision, one of the most important responsibilities of the team is to identify recommended criteria and conditions for resuming operations. It is desirable to postpone restart until the incident investigation team specifies minimum criteria for action by line management. Occasionally restart must be coordinated with other groups such as US OSHA or another regulatory agency. Communications and notifications need to be precise, clear, and verified. The management system may benefit by including descriptions of the following items.

- Responsibility for setting restart criteria and conditions

- Coordination with internal groups or outside parties
- Communications required to occur before restart
- Notifications of progress and status of restart approval

Restart criteria should focus on short-term prevention of one or more causal factors while the other causal factors and root causes are determined. One best practice is to have a task force quickly evaluate potential similar underlying causes elsewhere in the plant. Limitations imposed for restart are often undesirable for long-term operation as severe operational limits may be set. These limits, however, allow the system to run with minimal risk of recurrence of the incident. It is the duty of a company to demonstrate it is safe to restart. It is the duty of the regulator to prohibit restart if there is evidence of significant deficiencies.

2.2.11. Providing a Template for Formal Reports

Reports that document incident investigations are different from most business and technical reports. Business reports traditionally only address financial considerations. Process safety incident reports, however, can contain a full range of human elements: serious injury, fatality, flawed management systems, financial aspects, as well as complex technical issues. Although most business documents could become legal documents, the incident report has a higher likelihood of subpoena. This section of the management system benefits when the developers seek a critique by the legal department before deciding on a final template format and rules for writing.

Root causes and other findings contained in the report should be captured in a database.

A topic of increasing attention is the desire to spot detrimental trends in management systems as soon as possible. Advanced computer software programs have opened new opportunities for tracking and recognizing trends and comparing performance against a larger group. The management system for incident investigation should include provisions for computer coding of data.

The intended distribution and required approval levels should be addressed in the preplanning stages, and should be clearly identified in the written management system description.

Topics covered include the following:

- Recording the results of the investigation
- Categorizing the occurrences according to location, material, and other characteristics
- Tracking and closing recommendations

- Performing queries of the data across many investigations
- Trending against types of occurrences, categories, and root causes

2.2.12. *Review and Approval*

The management system should be reviewed, approved, and fully implemented by the appropriate company personnel. Investigations can have significant interaction with several other company functions. Each of these groups needs the opportunity to participate in the development of the initial management system through review and comment.

2.2.13. *Planning for Continuous Improvement*

Continuous improvement is a necessary element of a successful management system. Each investigation or implementation of the management system provides an opportunity to evaluate its effectiveness. The lessons learned strengthen and refine the management system. It is also valuable to recognize and share the positive aspects of those investigation activities that were especially successful.

To ensure the management system continues to provide the intended results, periodic reviews and updates are necessary. This action recognizes that organizations are dynamic, ever changing, and evolving.

Consider the following critique questions.

- Were the investigation techniques applied correctly and fully?
- Did the teams find the management system failures that led to the incident (that is, did they get to root causes)?
- Is the internal (nonpublished) team documentation adequate?
- Were the right skills available within the team?
- What other resources could be used next time?
- What should be changed next time?
- Are near misses being reported?

Chapter 14 details proven methods for enhancing your incident investigation system.

2.3. Implementing the Management System

Implementing a new or upgraded management system normally begins with training employees, supervision, and management in their respective roles in the investigation program. Implementation also includes development and refinement of the incident data management systems. The data management system should allow users to easily develop consistent reports and perform queries of incident data to spot systemic trends. Additionally, the management team's endorsement of the incident investigation management system is important when introducing a new or revised system. Management may find it useful to formalize this support. A sample program endorsement letter is included in the *CCPS Investigator's Companion CD ROM*.

2.3.1. Initial Implementation—Training

Implementation of a new or revised management system often begins with presenting training for the four groups described earlier in this chapter.

- Management
- All employees in a position to notice and report a near miss or accident
- Incident investigation team members
- Incident investigation team leaders

Typical training agendas for management and employees who may report an incident can be brief. Special training may be indicated for those employees and functions that will interface with the incident investigation team during an investigation. These would include, for example, emergency response teams, fire brigade, maintenance, security, site safety, site industrial hygiene, public relations, legal, and environmental. Table 2-2 describes general guidelines for the content of training sessions for various functions.

2.3.2. Initial Implementation—Data Management System

An incident data management system will need to consider the following four items.

1. The approach for investigation
2. The definition table for root causes
3. The needs of the investigation for data input and report output
4. The needs of management for data trending

Chapter 13 provides details on database systems for investigation data.

An example of a typical approach for writing an incident investigation management system is presented on pages 30–31. It addresses all of the items described previously.

Table 2-2
Suggested Training for Effective Implementation

Complex Incidents Investigation Team Leader Training	Moderate/Minor Incident Investigation Team Leader Training	Incident and Near Miss Reporting/ Notification	Awareness Training
These leaders will handle the most complex incidents (top 10% or less)	These leaders will handle low to moderate complexity incidents (90% or more or the incidents)	All operations and maintenance staff; appropriate purchasing, accounting, and other staff Individuals who are expected to identify and report all incidents, including near misses. Some of these individuals may become team leaders or members or may be interviewed during an investigation	All staff Individuals may fill any role in the system; this is the starting module of training
Training Agenda • Investigation planning • Data protection • Data collection • Causal factor determination • How to fill gaps in data • Root cause identification • Writing recommendations • Using the incident database • Programmatic issues such as reporting, communication, legal issues	**Training Agenda** • Data collection • Causal factor determination • Root cause identification • Writing recommendations • Using the incident database	**Training Agenda** • Near miss definitions and examples • The learning value of incidents • No blame approach • Root causes are management system failures • Incident reporting system	**Training Agenda** • What is changing in how you approach incidents? • What can each person do to help the system work? • Expected impact to most jobs

Process Safety Management System
Sample Administrative Procedure
Title: Incident Investigation

1. PURPOSE

This section provides a summary description of what this management system document contains. Some organizations choose to place their incident investigation policy statement here.

2. REFERENCES

This section lists sources used to develop the incident investigation management system: corporate guidance, regulatory requirements, and other reference materials such as this book.

3. DEFINITIONS

This section identifies and defines special terms related to the facility's preferred methods of incident investigation.

4. PROCEDURE

The procedure section serves as both a job-aid for employees to use while investigating incidents and a training tool for the facility personnel whose job descriptions or special assignments include leading or participating on an investigation team.

4.1 Facility Incident Investigation Statement
If the site or corporation has a written policy statement for incident investigation, it typically appears here.

4.2 What To Do After an Incident Occurs
This section identifies the accountabilities, responsibilities, information flow, and specific tasks to complete as soon as any incident is identified. It describes the initial incident report form and minimum requirements for satisfactory completion.

4.3 Level of Effort and Team Selection Guidelines
This section helps the managers on duty determine the type of investigation to perform, and guidance for selecting the make-up and the size of the initial team.

4.4 Performing the Investigation

This section addresses team leader and team member conduct during the investigation. It may be as detailed as desired or it can refer to other resources such as this book. It may include guidance on writing effective recommendations and how to write the report.

4.5 Communicating the Report

This section describes the reporting requirements, both internal and external, based upon regulations, corporate guidance, and the type of incident. This section may describe how the information will be added to the company incident database.

4.6 Recommendation Follow-up

This section describes how the facility will follow-up on the recommendations. It can include responsibility assignment guidelines and other information to help ensure the recommendations are addressed. A clear tie to the management of change system should be included here.

4.7 Report Retention and Document Control

This section describes the formal approval and acceptance method for the report and where the final report and any supporting documentation will be maintained. It may state the company document retention policy for incident investigation or refer to another resource to determine how information must be kept.

4.8 Follow-up Report

Some companies choose to issue a follow-up report summarizing the status of action items related to incident investigations. It can be a useful tool for keeping the entire plant management staff up-to-date on what has been done and what is left to do to prevent recurrence.

Typical attachments to the management system document are:

Attachment 1—Initial Incident Report Form

This is the form used as soon as possible after any incident to capture reporting data and classify the incident for team assignment. It can even be an electronic template ready for a supervisor to fill out on-line.

Attachment 2—Flow Chart of Initial Incident Report Occurrences

This optional attachment provides a visual way to view the flow of the major steps and decision-making points of an incident investigation. It can be a helpful overview for management.

Attachment 3—Sample Incident Investigation Report

This is a sample report. It is useful during team training and can provide the template for the team to use when preparing their report.

Attachment 4—Incident Investigation–Audit Checklist

This optional attachment is used to promote continuous improvement of the entire safety management system. It is a checklist for anyone to use as a guide when auditing incident investigation activity. The facility from which this example was derived includes a similar checklist for each safe work practice at the site.

Example incident investigation management system courtesy of AntiEntropics, Inc.

References

29 CFR 1904, *Recording and Reporting Occupational Injuries and Illnesses*. Effective January 1, 2002; The US OSHA website for recordkeeping revisions is http://www.osha.gov/recordkeeping/index.html

Bridges, W. G. "Get Near Misses Reported," *Proceedings of the International Conference and Workshop on Process Industry Incidents, Orlando, FL*. New York: Center for Chemical Process Safety (CCPS), AIChE, October 2000.

Dictionary of Terms Used in the Safety Profession, 3rd ed, American Society of Safety Engineers, Des Plains, IA, 1988.

Center for Chemical Process Safety. *Guidelines for Technical Management of Chemical Process Safety*, New York: American Institute of Chemical Engineers, 1989.

Winsor, D. A. "*Challenger:* A Case of Failure to Communicate," *Chemtech Magazine*, American Chemical Society, September 1989.

3

An Overview of Incident Causation Theories

Every incident has one or more root causes. To understand what these are and how they interact, an investigator must use a systematic approach. As a rule, the benefits of this systematic approach result from:

- implementing sound process safety management principles and
- applying consistent and accurate investigative effort.

To be effective the investigation must apply an approach which is based on basic incident causation theories and use tested data analysis techniques. Investigating incidents to determine root causes and make recommendations can be as much an art as a science. Within the industry, best practices in incident investigation have evolved substantially in the last 20 years. This chapter provides a brief overview of some of the more relevant causation theories.

Several theories of incident causation exist and each has associated investigation techniques. Incident investigators use their judgment to make adaptations to selected techniques based on the size and complexity of the investigation effort. Judgment based on knowledge and experience is important in determining how and why an incident occurred.

3.1. Stages of a Process-Related Incident

Investigators can systematically analyze data from past incidents to identify lessons learned and develop incident stereotypes. This makes it possible to develop a model displaying the anatomy of a process-related incident using a conceptual framework. Figure 3-1 provides a tool to help us understand incident causation.

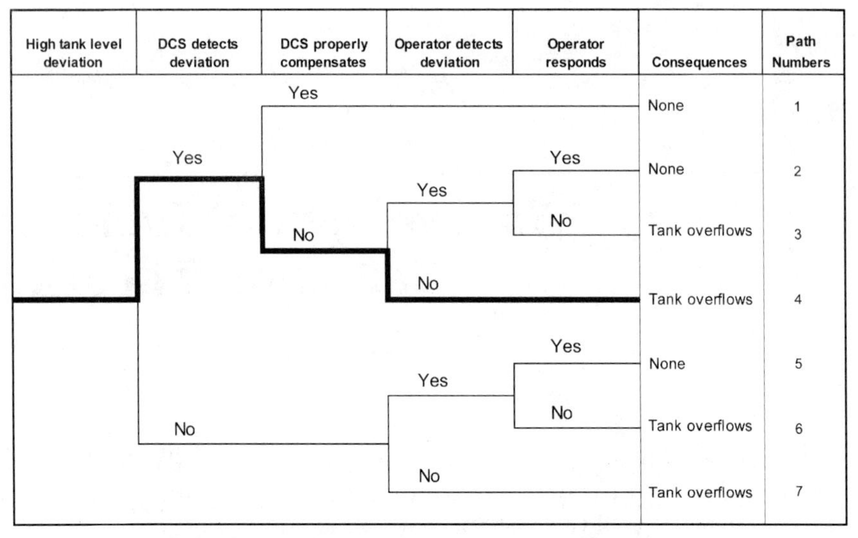

FIGURE 3-1. **Event tree for a process-related incident.**

In this example, there are two detection systems and two reaction opportunities. These yield three paths that lead to no adverse consequences and four paths that lead to failure with overflow as the consequence. The point is that sometimes there are more opportunities for things to go wrong than to go right. When a system or process fails, it may be difficult to trace the reasons for its failure. Based on available historic incident data, the anatomy of a major incident is rarely simple and rarely results from a single root cause. Serious incidents typically involve a complex sequence of occurrences and conditions. This sequence can include:

- equipment faults,
- latent unsafe conditions,
- environmental circumstances, and most importantly,
- human errors.

3.1.1. Three Phases of Process-Related Incidents

The progression of any process-related incident could be described as occurring in three different phases: [1]

1. Change from normal operating state into a state of abnormal (or disturbed) operation. An example is the tank level deviation in Figure 3-1.

2. Breakdown of the control of the abnormal operating phase. An example is the distributive control system (DCS) not compensating properly in Figure 3-1. Another example is the operator not detecting the deviation in Figure 3-1.
3. Loss of control of energy accumulations. An example is the operator not responding in Figure 3-1.

The four potential contributors to the incident causes in all three phases are:

1. Equipment
2. Process systems
3. Humans
4. The organization

The second phase may involve a breakdown of a barrier function. A barrier function is a safety feature such as a shutdown valve or containment system, a procedure, or the communication system. When these safety systems fail, the incident then evolves from an undesirable occurrence to a near miss and, if enough barriers fail, the incident could finally progress to a minor or major accident or operational interruption depending upon the consequences or circumstances.

The potential consequence of an incident is a function of the following five factors:

1. *Inventory of hazardous material*: type and amount
2. *Energy factor*: energy of chemical reaction or of material state
3. *Time factor:* the rate of release, its duration, and the warning time
4. *Intensity-distance relation*: the distance over which the hazard may cause injury or damage
5. *Exposure factor*: a factor that mitigates the potential effects of an incident

3.1.2. The Importance of Latent Failures

Historic incident data show that latent failures, also called latent conditions, have played an important role in incident causation. The term latent failure implies the condition is dormant or hidden. Normally the latent failure can be revealed before an incident through testing or auditing during typical operations within the process as shown in Figure 3-2.

There is always a possibility, however, that a latent failure may remain hidden during testing. There are several reasons a latent failure may not be detected.

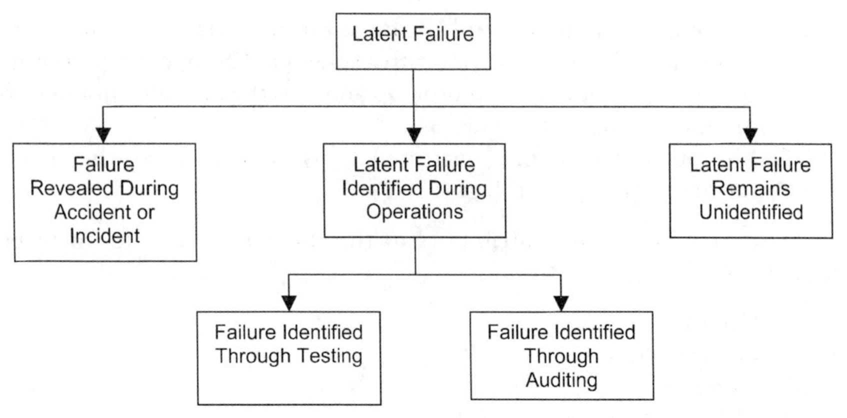

FIGURE 3-2. **Latent (hidden) failure.**

- It was not activated by the test used.
- The test was deficient, gave wrong results, or did not test the system properly.
- The test activity itself activates failure upon the next use of the process
- The deficiency was communicated poorly.

Latent component failures, human errors, and related unsafe acts and errors are all results of weaknesses in our management systems. This is why the terms *root cause* and *management system weaknesses* are used interchangeably. The term latent failure or latent error is still used in some academic settings.

3.2. Theories of Incident Causation

Theoretical incident concepts and associated models have evolved from investigations into the how and why of case histories. Resulting insights have made it possible to better explain and understand incident causation. There are many other incident causation theories besides the ones presented in this chapter, such as the Process Theory. (See the additional references for this chapter.) Key theories on incident causation discussed in this overview are:

1. Domino Theory of Causation
2. System Theory or Multiple-Causation Theory
3. Hazard–Barrier–Target Theory

These theories have encouraged development of techniques that support systematic incident investigation.

3.2.1. Domino Theory of Causation

A classic incident theory is H.W. Heinrich's domino theory of causation, which has had a significant influence on practical incident investigation. [2] Many adaptations of Heinrich's original proposal have been developed by later researchers. Heinrich labeled his five dominoes as follows:

1. ancestry and social environment,
2. fault or person,
3. unsafe act,
4. unsafe condition, and
5. injury.

Heinrich's approach is to identify, evaluate, and work on the middle dominoes, not just the last one or two dominoes in the line. The domino theory has significant limitations. The basic assumption is that there is a linear relationship between causation and progression. In other words, one occurrence follows another and ends in an incident. In the context of process-related incidents, this assumption is not always valid. Often parallel occurrences coincide to result in an incident rather than occurring as purely sequential occurrences. Nevertheless, the domino theory can provide a useful conceptual framework for simple incidents.

This theory led to the Updated Domino Theory by Kuhlmann, Seven-Domino Sequence by Marcum, 'Relabeled' Five-Domino Sequence by Bird, Modified Domino by Weaver, and 'Relabeled' Five-Domino Sequence by Adams.

3.2.2. System Theory

Today one of the most widely accepted and adapted incident theory relies on the system theory developed by Recht. [3] According to this theory, an incident is seen as an abnormal effect or result of the technological or management system. System theory analyzes the structure and state of a physical system for its elements and their interdependencies. A physical system is either a technological system or a human factors system. The theory provides:

- a framework for analyzing system requirements and constraints,
- detailed descriptions of component processes, and
- detailed descriptions of operational and task event sequences including environmental conditions.

It allows for the development of models of complex engineering systems and management structures. These models can be analyzed for inter-relationships between individual elements and the overall system function. Theoretically, there could be as many causes of an incident as there are system components. The term *multiple-cause theory,*[4] coined by Peterson, is often used instead of system theory.

3.2.3. Hazard–Barrier–Target Theory

The Hazard–Barrier–Target (HBT) theory, developed by Skiba, provides an interesting view of the multiple-cause or system theory. In HBT, an investigator starts with the understanding that a process has one or more inherent hazards. The hazard is a property of the process such as toxicity of a chemical, stored energy such as pressure much higher or lower than ambient, electrical hazards, etc. The target can be a person or the environment, and in an abstract sense, some interpret the target to be any loss impact. For example, the target could be *product* and *lower quality* could be the impact. The barriers are actually layers of protection and prevent the hazard from having a negative impact on the target. One important concept that is stressed in HBT is that **all** barriers have weaknesses, therefore each barrier has a probability of not working when needed. For example, any process aspect that has a probability of not working when needed is a hole in the barrier. The most important concept for any investigator to learn may be the following statement:

No layer of protection is perfect.

In fact, all layers of protection are fully dependent on management system implementation to ensure a reasonable probability of working when needed.

A hazard must get past all barriers to realize a negative impact on the target. This is always theoretically possible. Therefore, incidents occur when all barriers fail to prevent harm and a near miss occurs when one or more barriers fail. HBT is an excellent teaching tool for incident mechanisms and for describing the probabilistic nature of incidents, even for protected systems. Initially, investigators expanded HBT into an investigative technique. However, after much experimentation it was found to be a poor investigation technique, but an excellent model for describing the occurrence *after* the investigation is complete. This was because it provides little useful methods or rules for helping the investigator determine a specific sequence of positive and negative occurrences that led to an incident. Other techniques, such as logic tree analysis and causal factor charting are superior incident analysis tools and are discussed in detail in Chapter 9.

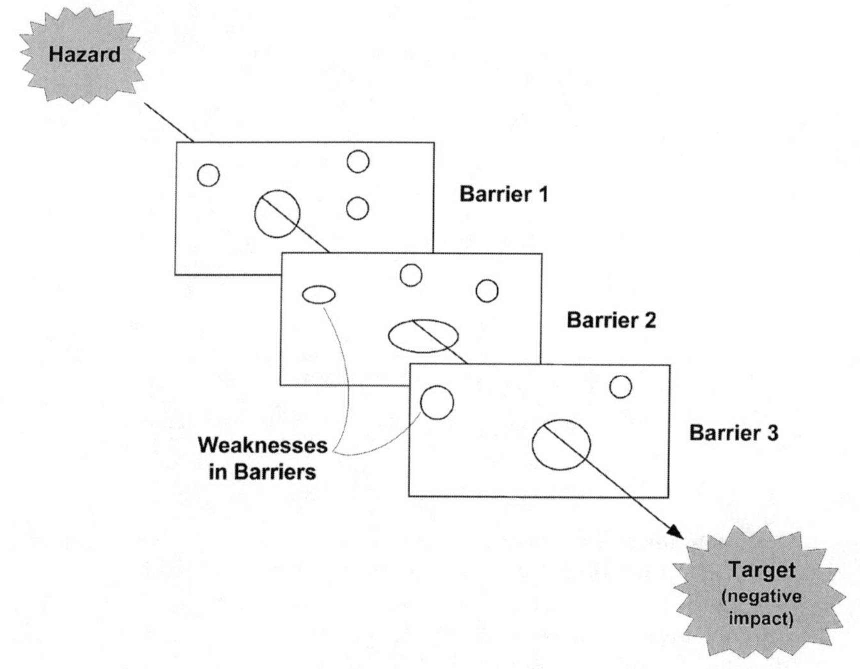

FIGURE 3-3. **Hazard–Barrier–Target concept.**

3.3. Investigation's Place in Controlling Risk

System theory can be applied to incident investigation, reliability problems, quality problems, and other business losses. One of several reasons why system theory has received broad recognition relative to incident investigation is that it builds directly on current, verified process safety principles. In process safety, as in all other systems used to control risk to a business, there are three basic keys to controlling the risk (see Figure 3-4):

1. **Understanding Risk:** To predict incidents, it is important to understand the risk associated with the process or system. In process safety, this is accomplished by identifying the potential incident or loss scenarios, then predicting the magnitude and likelihood of the occurrence. This is often what is done during a process hazard analysis or during management of change hazard reviews. The result is an understanding of the specific barriers, also called layers of protection, necessary to control the risk to a tolerable level.

FIGURE 3-4. **Universal concept for controlling risk (Kletz).**

2. **Management Systems:** To manage risk, appropriate management systems must be in place to ensure the barriers against incidents remain intact. These preventive, error detection, and mitigation management systems make up the bulk of process safety efforts and include written operating and maintenance procedures, effective training, control of up-to-date process safety information, management of change protocols, performance measurement, auditing, and others.
3. **Analyze Weaknesses:** To learn from incidents, the final step is to recognize that the incident prediction and management systems are not perfect. Implementing practices to learn from mistakes and allowing continuous improvement to the systems to prevent incidents is essential. These practices are incident reporting and investigation processes.

This book focuses on learning lessons from incidents to lower the risk of future major incidents. It is important to use a structured approach to incident investigation that builds on proven and recognized techniques; this makes it easier to develop consistent understanding from incidents and to communicate insights and results from investigations effectively.

3.4. Relationship between Near Misses and Incidents

From the domino theory onward, it has become apparent that there are always less severe precursors to an incident. These can be called near hits, near misses, or close calls. For every incident labeled a near miss, more subtle precursors exist that, if uncovered and resolved earlier, would have prevented the near miss and therefore a subsequent incident.

Uncovering and analyzing the precursors to incidents is more cost effective than only investigating losses. Chapter 5 discusses the definition of a near miss and how to get these precursors reported and investigated.

Endnotes

1. US Department of Energy, *Accident/Incident Investigation Manual*, Second Edition. Idaho Falls, ID: System Safety Development Center, Idaho National Engineering Laboratory 1985. (DOE/SSDC 76-45/27)
2. Heinrich, H.W. *Industrial Accident Prevention*. New York: McGraw-Hill, 1936.
3. Recht, I.L. "System Safety Analysis - A Modern Approach to Safety Problems," *National Safety News*, December, February, April, June, 1965–66.
4. Peterson, D. *Human-Error Reduction and Safety Management*. Goshen, NY: Aloray Inc. Professional & Academic Publisher, 1984.

Additional References

29 CFR 1904, *Recording and Reporting Occupational Injuries and Illnesses*. Effective January 1, 2002; The US OSHA website for recordkeeping revisions is http://www.osha.gov/recordkeeping/index.html.

American Society of Safety Engineers. *Dictionary of Terms Used in the Safety Profession*, 3rd ed. Des Plains, IA: American Society of Safety Engineers, 1988.

Bridges, W. G. "Get Near Misses Reported." International Conference and Workshop on Process Industry Incidents, Center for Chemical Process Safety (CCPS)/AIChE, Orlando, FL, October 2000.

Center for Chemical Process Safety. *Guidelines for Hazard Evaluation Procedures, Second Edition with Worked Examples*. New York: American Institute of Chemical Engineers, 1992.

Center for Chemical Process Safety. *Guidelines for Technical Management of Chemical Process Safety*. New York: American Institute of Chemical Engineers, 1989.

Greenwood, M., and Woods, H. M. "The Incidence of Industrial Accidents with Special Reference to Multiple Accidents," *Ind. Fatigue Res. Board*, Report 4, HMSO, London, England, 1919.

Kepner, C. H., and Tregoe, B. B. *The Rational Manager*. 2nd ed. Princeton, NJ: Kepner-Tregoe, Inc., 1976.

Petersen, D. *Techniques of Safety Management*, 2nd ed. New York: McGraw-Hill, 1978.

4

An Overview of Investigation Methodologies

Best practices in incident investigation have evolved substantially in the last 20 years. Investigators recognize that every incident is caused by one or more root causes. To identify and understand what these root causes are and how they interact, an investigator must collect evidence and conduct an analysis of that evidence. Today, organizations use a variety of methodologies to investigate incidents. These methodologies use combinations of various investigation tools.

This chapter provides a brief overview of investigation tools in simple, generic terms and demonstrates the benefits of using a more structured approach. There are a number of public and proprietary methodologies that employ generic tools, and some are referenced in the appendices and accompanying CD-ROM.

The following terminology is used throughout this chapter:

Tool*—A device or means used at a discrete stage of the incident investigation to facilitate understanding of event chronology, causal factors, and/or root causes.*

Technique*—The manner in which an incident investigation tool is developed or used.*

Methodology*—The use of a combination of two or more incident investigation tools to analyze the evidence and determine the root causes of the incident.*

4.1. Historical Approach

The historical approach to investigating incidents has been an informal, one-on-one interview typically comprising the person involved in the incident with his or her immediate supervisor. This tool when used alone has generally been discredited for process safety incidents, especially complex incidents resulting in, or having the potential to result in, serious or catastrophic consequences. The tool is still often used for investigating low severity incidents, including minor occupational injuries.

The focus of informal, one-on-one investigations was often limited to determining the immediate remedies that would prevent an exact repeat of the incident circumstances. For example, a common finding may have been that *an operator failed to follow an established procedure*. Based on the finding, the investigator might proceed to evaluate how best to get this *specific* operator to follow the procedure as a recommendation to prevent recurrence. This informal type of investigation required little time or training, but the disadvantages of the approach were grave. The resulting investigations did not provide a comprehensive understanding of:

- Why the incident really occurred
- What was needed to prevent recurrence of similar incidents elsewhere
- How the management systems failed and should be corrected

The investigation committee method is another unsuccessful approach. This unstructured approach is historically significant and was judged inadequate for investigating process safety incidents because it produced incomplete and inconsistent results. It often did not find the root cause level or all the root causes.

4.2. Modern Structured Approach

Progressive companies use a more structured and comprehensive team approach to identify root causes. Scientific principles and concepts are applied to determine root causes and make recommendations to prevent recurrence. Effective investigations should use tested data analysis tools and methodologies to seek the identification of multiple causes. To be repeatable, the investigation should use a systematic approach, which may also be prescriptive. As a rule, the benefits of this systematic approach result from two actions:

1. Implementing sound process safety management principles, and
2. Applying consistent and accurate investigative effort.

The disciplines of engineering and quality control have long recognized the principles of root cause analysis. Some process safety tools for root cause analysis have been borrowed from these disciplines. For example, fault tree analysis was developed as an engineering tool, but its "logic tree" structure has been adapted to meet process safety requirements.

The overall investigation approach within the process safety field is similar across many of the available methodologies. However, differences arise in the particular emphasis. Some methodologies focus on management and organizational oversights and omissions, while others consider human performance issues in more depth. Busy personnel working within the organization where the incident occurred often perform investigations. Investigative tools, therefore, need to be practical and relatively easy to use. Sometimes investigators may use judgment to make adaptations to selected tools based on the size and complexity of the investigation effort. Judgment based on knowledge and experience is important in determining how and why an incident occurred.

4.3. Methodologies Used by CCPS Members

The Center for Chemical Process Safety (CCPS) conducted a survey of its membership and other processing companies in preparation for this revised edition. Based on the responses, some general observations can be made about incident investigations:

- Companies reported using an average of two or three different methodologies for both major and minor incidents. The surveyed companies used both public domain and proprietary tools and methodologies.
- The most popular methodologies use different combinations of the tools described in Figure 4-1.

Appendix B provides a list of resources and a summary table showing some of the tools associated with various methodologies.

The methodologies used today provide improved results over simplified techniques such as informal, one-on-one interviewing. Most methodologies have adopted a battery of tools for application at particular stages of the investigation to determine root causes. None of the individual tools adequately addresses every stage of the incident investigation process. As a minimum, a tool representing the incident sequence is used prior to identifying causal factors (also known as critical factors), to which root cause analysis is subsequently applied.

In general, the companies surveyed use one of two main methodologies to determine root causes. The first involves timeline construction followed by logic tree development. The second involves timeline construction, identification of causal factors, followed by the use of predefined trees or checklists. These two approaches are discussed in detail in Chapter 9.

Overview of Investigation Tools

Logic Trees
Logic trees are committee-based investigation tools that use a multiple cause, system-oriented approach to determine root causes integrated with process safety management program. Examples: fault tree, event tree, causal tree, and why tree.

Pre-Defined Trees
The team uses ready-made, off-the-shelf tree tools. The investigators do not have to build the tree, but rather apply the causal factors to each branch in turn, and discard those branches that are not relevant to the specific incident.

Checklists
The team reviews causal factors against investigative checklists to determine why that factor existed at the time of the incident. A combined *what if/checklist* approach may be used.

Causal Factor Identification
The team identifies negative events, conditions, and actions that made major contributions to the incident. Tools such as Barrier Analysis or Change Analysis may be used.

Sequence Diagram
The team constructs a graphical depiction of a timeline that allows investigators to exhibit related events and conditions in parallel branches.

Timeline
Investigation teams make a chronological listing of events using a variety of formats from a simple sequential list to diagrams showing events and conditions along a straight axis.

Brainstorming
The team uses its judgment and experience to find credible causes. Structured brainstorming may employ tools such as *what if* and *five whys.*

Informal, One-on-One
Traditional, informal investigation usually performed by immediate supervision.

FIGURE 4-1. **Overview of investigation tools.**

4.4. Description of Tools

When choosing a root cause analysis methodology, it is important to recognize that no single tool does everything. Good methodologies use combinations of tools to counteract their various individual weaknesses. Considerable care should be given in choosing a methodology, depending on the existing culture within the organization, the specific investigation leaders, the level of training resources available, and the complexity of the incident.

It is important to understand that the various tools use different types of logic to arrive at the result. These types of logic are intuitive, deductive, inductive, or a combination. Most of the tools described in this guideline are deductive or intuitive.

Deductive techniques look backward in time to examine preceding occurrences necessary to produce a specified result. Deduction is reasoning from the general to the specific. In the deductive analysis, it is postulated that a system or process has failed in a certain way. Next, an attempt is made to find out what modes of system, component, operator, or organizational behavior contributed to the failure. A typical general application of deductive reasoning to the incident investigation might be: What instrumental or human failures contributed to the overpressurization of the process reactor? Most of the logic trees are deductive.

Intuitive techniques rely on the experience and knowledge of the people involved to identify causes. Brainstorming utilizes intuitive techniques, while structured brainstorming utilizes a combination of intuitive and deductive techniques.

Inductive tools are characterized as "forward search strategies" for identifying the impact of potential process deviations. Inductive tools can support incident investigation. They are especially useful when the evidence and facts of an incident have been exhausted or are not attainable. The team must then rely on inductive reasoning to point to where to search for more information to fill in gaps to understand the causes and occurrences of the incident. A fact hypothesis matrix is described in Chapter 9.

Other examples of inductive tools that have limited application in incident investigation include failure mode and effects analysis (FMEA), hazard and operability study (HAZOP), and event tree analysis (ETA). These are detailed in the CCPS book, *Guidelines for Hazard Evaluation Procedures.*[1]

4.4.1. Brainstorming

Brainstorming brings together a group of people from diverse backgrounds to discuss the incident and intuitively determine the causes of the incident. It is essentially an unstructured approach, but can provide more perspective and experience than one-on-one investigations. The group will typically have an understanding of the sequence of occurrences that led up to the incident through a timeline or sequence diagram. The group may also have identified causal factors, and typically focus on establishing barriers to reduce the risk (probability or consequences) of recurrence.

The disadvantage to unstructured group brainstorming is that the discussion may be dominated by certain vocal individuals, who are not shy about stating their opinions, and who may or may not be experts but believe that they are. Each person may enter the discussion with his or her own bias, thereby directing the thinking toward incorrect conclusions. The results of group brainstorming are very dependent on the collective experiences of the group, which may be incomplete, lacking critical knowledge or a competency skill set. Two different groups may reach two different conclusion on the cause of an incident.

A slightly more structured approach uses What-If Analysis,[1] which involves the team asking "What if?" questions that usually concern equipment failures, human errors, or external occurrences. Some examples are: *What if the procedure was wrong? What if the steps were performed out of order?* The questions can be generic in nature or highly specific to the process or activity where the incident occurred. Sometimes these questions are preprepared by one or two individuals, which may also potentially bias the discussion.

The 5-Whys tool [2] is another approach used to add some structure to group brainstorming. The tool utilizes a logic tree approach without actually drawing the logic tree diagram. The group questions why unplanned, unintended, or adverse occurrences occurred or conditions existed. Typically the group needs to ask "why?" five times to reach root causes; hence the name. Judgment and experience are required to use the 5-Whys tool effectively to reach management system failures. The level of analysis is up to the group and does not always ensure reaching root causes.

4.4.2. Timelines

The first phase of incident investigation involves developing a preliminary chronological description of the sequence of occurrences that led to the failure. This requires collecting evidence through interviewing key witnesses and examining all relevant evidence (equipment and documents) in order to piece together the circumstances of the incident in chronological order. This timeline development can range from a simple list of occurrences in sequence to diagrams showing occurrences and conditions along a straight axis.

Development of the timeline should start as soon as facts emerge about the incident. By starting early, the investigator will become aware of gaps in the sequence of occurrences that may be used as a focus for subsequent stages of the investigation in order to resolve the gaps. Construction of the timeline is an iterative activity. The timeline is refined and adjusted as the team gains a more complete and accurate understanding of the actual incident scenario and sequence of occurrences.

Timelines alone do not identify the root causes of an incident. They should be used in conjunction with other tools, described in the following sections.

For a detailed description of how to develop timelines along with several examples, refer to Chapter 9.

4.4.3. Sequence Diagrams

Although investigators had previously used diagrams and charts, The Bureau of Surface Transportation Safety of the National Transportation Safety Board (NTSB) introduced Multilinear Event Sequencing (MES) concepts in the early 1970s to analyze and describe accidents. Ludwig Benner, Jr., and his colleagues at NTSB pioneered the development of MES concepts and many implementing procedures such as Sequentially Timed Events Plot (STEP).[3,4] W. G. Johnson adapted the basic MES concepts to create a form of sequence diagrams known as Events & Causal Factors Charting (E&CF)[5] first introduced as part of the MORT program. The term *causal factor chart* is commonly used for a sequence diagram in this industry.

Techniques for developing sequence diagrams encompass a number of fundamental principles. Before developing a sequence diagram, it is important to define the end of the incident sequence. The starting point of the incident must also be defined but may not be immediately apparent until partway through the investigation. Typically, sequence diagrams start at the end and work backward, identifying the immediate contributing events first.

Like timelines, construction of the diagram may start as soon as facts emerge about the incident in order to identify gaps for resolution. It is important to choose a format that may be easily updated and revised as new evidence is gathered, such as Post-It® notes on which a single event or condition is written.

Like timelines, sequence diagrams do not identify root causes, and therefore they should be used in conjunction with other tools. The mechanics of these tools are relatively easy to learn, but the investigator must exercise care to avoid locking into a preconceived scenario. For more information on sequence diagram tools refer to Chapter 9.

4.4.4. Causal Factor Identification

Once the evidence has been collected and a timeline or sequence diagram developed, the next phase of the investigation involves identifying the causal factors. These causal factors are the negative occurrences and actions that made a major contribution to the incident. Causal factors involve human errors and equipment failures that led to the incident, but can also be undesirable conditions, failed barriers (layers of protection, such as process controls or operating procedures), and energy flows. Causal factors point to the key areas that need to be examined to determine what caused that factor to exist.

There are a number of tools, such as Barrier Analysis [6, 7] and Change Analysis,[8] that can assist with the identification of causal factors. The concepts of incident causation encompassed in these tools are fundamental to the majority of investigation methodologies. (See Chapter 3 for information about the Domino Theory, System Theory, and HBT Theory.) The simplest approach involves reviewing each unplanned, unintended, or adverse item (negative event or undesirable condition) on the timeline and asking, "Would the incident have been prevented or mitigated if the item had not existed?" If the answer is yes, then the item is a causal factor. Generally, process safety incidents involve multiple causal factors.

Causal factor identification tools are relatively easy to learn and easy to apply to simple incidents. For more complex incidents with complicated timelines, one or more causal factors can be overlooked, ultimately leading to missed root causes. Another disadvantage is that an inexperienced investigator could potentially assume that suppositions are causal factors, when in reality the supposed event or condition did not occur.

Once the causal factors have been identified, the factors are analyzed using a root cause analysis tool, such as 5-Whys or predefined trees. See Chapter 9 for a more detailed discussion of Barrier Analysis (sometimes called hazard–barrier–target analysis or HBTA) and Change Analysis (also referred to as Change Evaluation/Analysis or CE/A). In essence, these tools act as a filter to limit the number of factors, which are subjected to further analysis to determine root causes.

4.4.5. Checklists

Checklist analysis tools can be a user-friendly means to assist investigation teams as they conduct root cause analysis.[1] Each causal factor is reviewed against the checklist to determine why that factor existed at the time of the incident. The Systematic Cause Analysis Technique (SCAT)[9, 10] is an example of a proprietary checklist tool.

The comprehensiveness of the various checklists varies greatly. Some are very detailed with numerous categories and subcategories, whereas others do not reach down to the level of root causes.

The advantage of checklists is that they are simple to use and the investigation team does not require a lot of training. The checklist essentially directs the investigation team and keeps them on track. Another advantage is consistency; the standard categories (and subcategories) on the checklist can be easily trended to identify recurring problems at a facility.

A disadvantage is that a checklist may allow an investigation team to jump to conclusions, and does not provide the opportunity to think "outside the box." This is especially important if the checklist is one of the less comprehensive types. It is also tempting to use the checklist too early, before all causal factors have been identified. Be sure to determine *what* happened and *how* it happened **before** determining *why* it happened. Otherwise, the team will think it has identified the right root causes, when in reality not all of the root causes have been determined.

Some checklists must be used carefully because to the casual observer they can imply blame, contrary to the intended policy of discouraging blame seeking.

Checklists may also be used to supplement other tools; for example, checklists on human factors may be used in conjunction with logic trees. Similarly, checklists may be used in combination with structured brainstorming tools such as What If/Checklist and Hazard and Operability (HAZOP) Analysis.(1) It is also a good practice to apply a tool like the 5-Whys to the root causes identified from the checklist to verify whether they are truly root causes.

The use of checklists, such as SCAT, is discussed in Chapter 9. See Chapter 6 for an example of a human factors checklist.

4.4.6. Predefined Trees

Owing largely to a scarcity of investigation methods available 30 years ago, attempts were made to apply engineering fault tree analysis (FTA) logic to incident reports. This approach appeared to demonstrate value. It was observed that many of the incidents displayed similar patterns of causal factors. This commonality led to the development of the Management Oversight Risk Tree (MORT)(11, 12, 13) that was based loosely on fault tree conventions. MORT was modeled on safety management system concepts that were considered best practice at that time. A simpler Mini-MORT variation has been developed to reduce complexity. (14)

Today there are several predefined tree tools available from public and proprietary sources, including TapRooT™,(15, 16) Seeking Out the

Underlying Root Causes of Events (SOURCE™),[17] and Comprehensive List of Causes (CLC).[18] The majority of the predefined tree tools are prescriptive and list potential root causes for consideration among their branches. This offers the investigator a systematic method of considering the possible root causes associated with an incident. This approach encourages investigators to contemplate a wide range of causal factors and not just those that come to mind through brainstorming. The investigator does not have to build the tree, but rather apply the causal factors to each branch in turn, and discard those branches that are not relevant to the specific incident.

Like checklists, the comprehensiveness of the various predefined trees varies. Some are very detailed with numerous categories and subcategories, whereas others may not fully reach root causes. This is hardly surprising, as the predefined trees are essentially a graphical representation of numerous checklists, organized by subject matter, such as human error, equipment failure, or other topics. The more comprehensive techniques were developed from many years of incident experience and management system experience across the chemical and allied industries.

The advantages of predefined trees are that they may bring expertise into the investigation that the team does not have, and by presenting all investigators with the same classification system, greater consistency is encouraged among investigators. Largely, the technique ensures a comprehensive analysis and simplifies statistical trend analysis of the collected data. A disadvantage of predefined trees, as with a checklist, may be a tendency to discourage lateral thinking if the incident involves novel factors not previously experienced by those who developed the original tree.

The use of predefined trees, overall, requires less resources and prior training than the nonprescriptive techniques involving team-developed trees discussed below. Some organizations have taken a generic, predefined tree and structured it along the lines of the company's management system. The effectiveness of a predefined tree is dependent on how well the tree models the data and system of dealing with the incident. When choosing a predefined tree, the user should confirm that the tree models the technology and system of the user.

The use of predefined trees is also discussed in detail in Chapter 9. See the accompanying CD-ROM for examples of predefined trees, including Comprehensive List of Causes (CLC) and SOURCE/Root Cause Map.

4.4.7. Team-Developed Logic Trees

A number of deductive techniques require that the investigation team develop a tree. This is accomplished by reasoning to organize causal factors into a diagram (tree) and define their interrelationship. These logic trees can vary over a wide range from simple trees to complex fault trees. Most start at the end occurrence and work backward until a point is reached at which the team agrees it would be unproductive to go further.

The earliest logic trees were based on engineering fault tree analysis methods. Today, companies use a number of variations or combinations of logic trees and call them by different names, such as Why Tree,[19] Causal Tree,[20, 21] Cause and Effect Logic Diagram (CELD),[22] and Multiple-Cause, Systems-Oriented Incident Investigation (MCSOII).[23, 24] The tools have more similarities than differences.

Logic trees are best developed using a multidiscipline team. Starting at the end event, the discussion is guided by asking, "why?" and recording the results in a tree format. The general approach encourages investigators to contemplate a wide range of causal factors, but relies on group discussion. This makes its success dependent on the experience and knowledge of the team. These tools recognize that incidents have multiple, underlying causes, and the investigations attempt to identify and implement system changes that will eliminate recurrence not only of the exact incident, but of similar occurrences as well.

There are five main strengths to a logic tree approach.

1. It provides the ability to separate a complex incident into discrete smaller events (segments) and then to examine each piece individually.
2. It allows the investigation team to understand how the causes worked together to allow the incident to occur.
3. It improves the quality of investigations by directing the focus past the immediate surface causes to the underlying root causes and management system failures, and mandating a search for multiple causes.
4. It offers a clear record that the investigation team understands the incident through the logic diagram/tree.
5. It provides an opportunity to include human factors in the incident investigation process.

The disadvantages of logic trees center on their dependency on the cumulative expertise within the assembled investigation team and their ability to compile sound fact finding related to the incident. No technique can be a substitute if the team does not have the requisite knowledge and experience. Logic trees can also be somewhat time consuming to develop and may not use a consistent set of categories and subcategories, making trend analysis of recurring problems difficult.

Examples of logic trees—fault, event, causal, and why—are discussed below in order of increasing rigor. Chapter 9 contains detailed information on developing logic trees.

Why Tree

The Why Tree provides a simple method for depicting the logical relationship between causes and effects of an incident.[19] The process starts by displaying all direct losses and associated consequences in separate boxes. A drill down question that asks "why?" challenges each box. Plausible explanations are entered into new boxes attached by straight lines to the subject or receptor box above. Ultimately, the page will fill up with several boxes attached by straight lines.

Unlike a formal fault tree, this method is empirical and does not require logic gates to be established. All boxes are scrutinized to determine their validity. If the content of a box is refuted by hard evidence, it is crossed off with an appropriate explanation. Otherwise, the boxes are left connected to show the logical progression upward toward an incident.

At the base of the why tree where fundamental management systems are implicated, the process ceases. The investigation team should then focus its efforts on the rigor and quality of the management systems that could have prevented the incident. Recommendations are developed to address system deficiencies and these are tested against the why tree. An example is included on the accompanying CD ROM.

Causal Tree

Causal Trees were developed in an effort to use the principles of deductive logic found in Fault Tree but make it more user-friendly. Originally, private companies developed the Causal Tree Method (CTM) for safety, process safety, and environmental incident investigations applications. Rhone-Poulenc, for example, was an early user.[20, 21] Multiple-Cause Systems Oriented Incident Investigation (MCSOII) is another name for the CTM. At this time, most companies use simplified versions of fault trees for complex incident investigations.

Causal tree methods rely on group discussion among experts from different fields, including workers, witnesses, supervisors, and process safety specialists. Starting at the end event, and working one level of the tree at a time, the group asks three questions:

1. What was the cause of this result?
2. What was directly *necessary* to cause the end result?
3. Are these factors (identified from question 2 above) *sufficient* to have caused the result?

In recognition that most incidents have multiple root causes, the team is generally required to identify a minimum of three factors; one from each of the following categories: organizational, human, and material factors.

Causal trees may be drawn from top to bottom, left to right, or right to left. Connectors such as AND- and OR-gates are often omitted. Some methods use only AND-gates.

Event Tree

Another type of logic tree, the event tree, is an inductive technique. Event Tree Analysis (ETA) also provides a structured method to aid in understanding and determining the causes of an incident.[1] While the fault tree starts at the undesired event and works backward to identify root causes, the event tree looks forward to display the progression of various combinations of equipment failures and human errors that result in the incident graphically.

Each event, such as equipment failure, process deviation, control function, or administrative control, is considered in turn by asking a simple yes/no question. Each is then illustrated by a node where the tree branches into parallel paths. Each relevant event is addressed on each parallel path until all combinations are exhausted. This can result in a number of paths that lead to no adverse consequences and some that lead to the incident as the consequence. The investigator then needs to determine which path represents the actual scenario. Generally, a qualitative event tree is developed when used for incident investigation purposes.

Chapter 3 contains an example of an event tree in Figure 3-1 on page 34.

Fault Tree

Fault Tree Analysis (FTA) provides a structured method for determining the causes of an incident.[1, 25, 26, 27] The fault tree itself is a graphic model that displays the various combinations of equipment failures and human errors that can result in an incident.

The undesired event appears as the top event and the trees are drawn from top to bottom. Two basic logic gates connect event blocks: the AND-gate and the OR-gate. The facts dictate the structure of the incident diagram and limit the influence of presupposed conclusions invariably drawn by team members before all of the facts are identified and logically matched. Logic rules are used to test the tree structure.

The term *fault tree* means different things to different people. Some people use the term to describe trees that have frequency terms included. These quantitative trees can be solved mathematically to provide a frequency of the incident. However, for incident investigation, the term commonly refers to a qualitative tree.

4.5. Selecting an Appropriate Methodology

As mentioned previously, no single tool does everything. Good methodologies use combinations of tools to counteract the various individual weaknesses of the tools involved. Methodologies are chosen based on the

- organizational culture,
- experience and expertise of the incident investigators, and
- nature and complexity of the incident.

There are a number of common generic features shared by most root cause methodologies, which are presented in Figure 4-2.

Three key components of the overall methodology that need to be selected to ensure effective incident investigation and identification of root causes are:

1. ***"What" happened?***
   A component for describing and schematically representing the incident sequence and its contributing events and conditions.

2. ***"How" it happened?***
   A component for identifying the critical events and conditions (causal factors) in the incident sequence.

3. ***"Why" it happened?***
   A component for systematically investigating the management and organizational factors that allowed the critical events and conditions to occur, that is, root causes identification.

Finally, in selecting an appropriate incident investigation methodology, consideration needs to be given to whether the method facilitates the identification of management system and organizational inadequacies and oversights. The methodology should specifically single out factors that influence and control an organization's risk management practices and procedures.

As previously discussed, CCPS member companies generally use one of two main approaches:

- Timeline construction followed by logic tree development.
- Timeline construction, identification of causal factors, and predefined trees or checklists.

These two approaches are discussed in detail in Chapter 9.

- ✓ **SYSTEMATIC** – Provides a systematic and thorough approach that is applied in an organized and logical fashion.
- ✓ **CONSISTENT RESULTS** – Similar teams working with similar information will be able to produce similar results, not overly influenced by team composition and individual team experience factors.
- ✓ **GRAPHIC DIAGRAM** – Most root cause methodologies apply some type of graphic illustration/diagram to record the specific events, conditions, and relevant facts. This may be a logic tree, cause diagram, or other graphic. In some instances, the logic diagram is combined with the chronology tool (for example, Causal Factor Charting). Causal relationships are applied in constructing and examining the diagram, such as the *necessary and sufficient* tests.
- ✓ **CHRONOLOGY** – Most root cause methodologies call for arranging the facts, evidence, and events in chronological order to be able to understand the scenario.
- ✓ **ROOT CAUSE TEST** – Most root cause methodologies will provide a specific definition of the term root cause and will require examination for confirmation that the causes that have been identified as root causes are indeed root causes.
- ✓ **MULTIPLE CAUSES** – Most current methods recognize the concept of multiple root causes.
- ✓ **HUMAN FACTORS** – Most current root cause methodologies will address the human reliability and human performance aspects that are involved in the occurrence.
- ✓ **MANAGEMENT SYSTEMS** – Most current root cause investigation techniques emphasize the need for examining the administrative management systems that were involved in the occurrence, and will evaluate these systems for any inherent weaknesses or defects.
- ✓ **CHECKLISTS** – Root cause investigation methods often include a set of integrated and comprehensive checklists (or accompanying reference documents) to ensure that common generic causes and issues are considered.
- ✓ **QUALITY ASSURANCE** – A comprehensive root cause methodology will often include a component to ensure that the method is being applied properly.

FIGURE 4-2. **Common features of root cause methodologies.**

Endnotes

1. Center for Chemical Process Safety. *Guidelines for Hazard Evaluation Procedures, Second Edition with Worked Examples*. New York: American Institute of Chemical Engineers, 1992.
2. ABS Consulting. *Introduction to Reliability Management: An Overview*. Knoxville, TN: ABS Group Inc. 2000.
3. Benner, L., Jr. *10 MES Investigation Guides*. Oakton, VA: Starline Software Ltd. 2000.

4. Hendrick, K. and Benner, L., Jr. *"Investigating Accidents with S-T-E-P."* New York: Marcel Dekker, 1987.
5. Buys, R. J., and Clark, J. L. *"Events and Causal Factors Charting."* Revision 1, Idaho Falls, ID: System Safety Development Center, Idaho National Engineering Laboratory, 1978. (DOE 76-45/14 SSDC-14)
6. Dew, J. R. "In Search of the Root Cause." *Quality Progress*. **23**(3): 97–102, 1991.
7. Trost, W. A., and Nertney, R. J. *"Barrier Analysis."* Idaho Falls, ID: System Safety Development Center. Idaho National Engineering Laboratory 1985. (DOE/SSDC 76-45/29)
8. Kepner, C. H., and Tregoe, B. B. *The Rational Manager,* 2nd ed. Princeton, NJ: Kepner-Tregoe, Inc., 1976.
9. Bird, F. E., Jr., and Germain, G .L. *Practical Loss Control Leadership.* Loganville, GA: International Loss Control Institute (ILCI), 1985.
10. International Loss Control Institute. *SCAT—Systematic Cause Analysis Technique.* Loganville, GA: Det Norske Veritas, 1990.
11. Johnson, W. G. *MORT, Safety Assurances Systems.* New York: Marcel Dekker, 1980.
12. Department of Energy, *Accident/Incident Investigation Manual.* 2nd ed., Idaho Falls, ID: System Safety Development Center. Idaho National Engineering Laboratory, 1985. (DOE/SSDC 76-45/27)
13. Buys, R. J. *Standardization Guide for Construction and Use of MORT-Type Analytical Trees.* Idaho Falls, ID: System Safety Development Center. Idaho National Engineering Laboratory 1977. (ERDA 76-45/8)
14. Ferry, T.S. *Modern Accident Investigation, and Analysis* 2nd ed. New York: Wiley, 1988.
15. Unger, L.. and Paradise, M. "TapRooT™—A Systematic Approach for Investigating Incidents." *Paper, Process Plant Safety Symposium*. Houston, TX, 1992.
16. Paradise, M. "Root Cause Analysis and Human Factors." *Human Factors Bulletin*. **34**(8): 1–5, 1991.
17. ABS Consulting. *Root Cause Analysis Handbook: A guide to effective Incident Investigation*. Knoxville, TN: ABS Group Inc., 1999.
18. BP (formerly BP Amoco). *Incident Investigation. Root Cause Analysis Training. Comprehensive List of Causes.* London, 1999.
19. Nelms, Robert, C. *The Go Book*. C. Robert Nelms publisher.
20. Leplat, J. "Accident Analyses and Work Analyses." *Journal of Occupational Accidents*. **1**: 331–340, 1978.
21. Boissieras, J. *Causal Tree, Description of the Method.* Princeton, NJ: Rhone-Poulenc, 1983.
22. Mosleh, A. et al. *Procedures for Treating Common Cause Failures in Safety and Reliability Studies*. Palo Alto, CA: Electric Power Research Institute. 1988. (EPRI NP-5613)
23. Dowell, A. M. *Guidelines for Systems Oriented Multiple Cause Incident Investigations*. Deer Park, TX: Rohm and Haas Texas Inc. Risk Analysis Department, 1990.

24. Anderson, S. E., and Skloss, R. W. "More Bang for the Buck: Getting the Most from an Accident Investigation." Paper, *Loss Prevention Symposium*. New York: AIChE, 1991.
25. Browning, R. L. "Analyze Losses by Diagram." *Hydrocarbon Processing*, **54**:253–257,1975.
26. Arendt, J. S. "A Chemical Plant Accident Investigation Using Fault Tree Analysis." *Proceedings of 17th Annual Loss Prevention Symposium*. Paper 11a, New York: AIChE, 1983.
27. Vesely, W.E. et al. *"Fault Tree Handbook."* Washington, DC: US Government Printing Office, 1981. (NUREG-0492)

Additional Reference

Nelms, C. Robert. *What You Can Learn from Things that Go Wrong*. C. Robert Nelms Publisher, 1996.

5

Reporting and Investigating Near Misses

Major chemical process incidents are often preceded by multiple warning symptoms in the months, days, or hours before the incident. These symptoms may be occurrences that are called near misses, but what exactly is a near miss? A near miss is an incident that presents an opportunity to learn valuable information that may prevent future accidents.

This chapter describes near misses, discusses their importance, and presents the latest methods for getting near misses reported and investigated. The term *near miss* has a long history of use, but the terms *near hit* or *close call* more closely describe what actually happens. The term *near miss* is used throughout this book since it is so widely accepted as the term for incidents that are direct precursors to accidents.

5.1. Defining a Near Miss

A ***near miss*** *is an occurrence in which an accident (that is, property damage, environmental impact, or human loss) or an operational interruption could have plausibly resulted if circumstances had been slightly different.*

This definition may be difficult to use in practice without examples. Different companies may interpret it in different ways. To develop a common understanding of the term, a company may choose to develop a list of examples it believes to be near misses. Some stated common examples of near miss incidents might include:

- Excursions of process parameters beyond preestablished critical control limits
- Releases of less than threshold quantities of materials that are not classified as environmentally reportable

- Activation of layers of protection such as relief valves, interlocks, rupture disks, blowdown systems, halon systems, vapor release alarms, and fixed water spray systems
- Activation of emergency shutdowns (in some instances)

Near misses must be reported so that they can be investigated. Investigation provides valuable information that can help prevent future near misses, accidents, and operational interruptions. The first step is reporting the near miss and then investigating it to determine the causes and underlying reasons why it occurred. A thorough investigation to discover root causes helps investigators identify management system weaknesses that lead to near misses and potential incidents. Investigating near misses is a high value activity. Learning from near misses is much less expensive than learning from accidents.

It helps to define two terms to understand the hierarchy of sequencing for the occurrences that make up an incident. The most widely accepted terms are *causal factor* and *root cause*:

A ***Causal factor,*** *also known as a critical factor, is the major contributor to the incident (a negative occurrence or undesirable condition), that if eliminated would have either prevented the occurrence, or reduced its severity or frequency.*

A process safety incident typically has multiple causal factors. The term *direct cause* is often used interchangeably with the term causal factor, but this can be confusing because it is also used to refer to just the *last* causal factor in an incident sequence.

Root cause*—A fundamental, underlying, system-related reason why an incident occurred that identifies a correctable failure(s) in management systems. There is typically more than one root cause for every process safety incident.*

Given an understanding of the definition of a near miss, enhanced by specific examples for a facility, it may be possible to estimate how many near misses one might expect to be reported compared with the number of accidents that occur. A greater number of erroneous acts or undesirable conditions may occur compared to the number of near misses. Figure 5-1 illustrates the relationships among accidents, near misses, and nonincidents.

The ratios shown above depend heavily on the definition of a near miss and depend on the type of loss. For example, regarding quality-related incidents, there appear to be far fewer near misses and errors or failure conditions per operational interruption. The reason for the lower ratio is that quality-related incidents typically result from excursions that are less severe than personnel safety or process safety occurrences. Chemical processes are

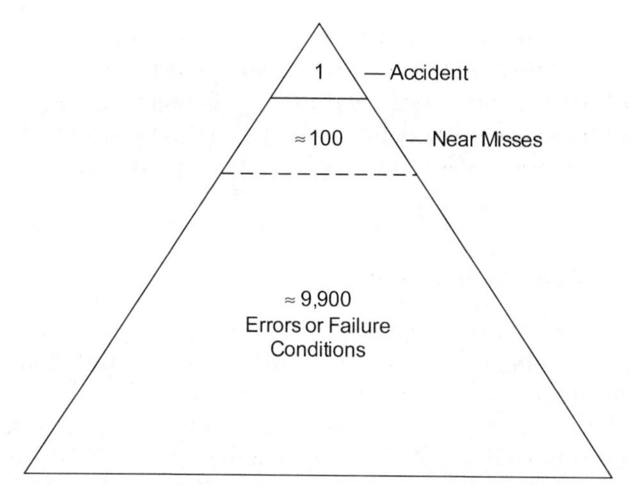

FIGURE 5-1. **Typical relationship between errors and potential or actual impacts.**

designed to prevent the safety occurrence, which usually means a far greater number of failures (causal factors) are required to result in harm to people. A general corollary is *as the process gets simpler, and the path of occurrences to reach harm becomes shorter, there are fewer near misses and errors per accident.*

If two processes handle the same materials, the simpler one is typically inherently safer. Simplicity usually implies that operating conditions are closer to ambient and less layers of protection are needed. However, the simpler process may also produce fewer near misses.

Many chemical companies indicate that they have one or fewer near misses reported for every accident. Although some companies achieve a ratio of 20 or higher, many see fewer near misses reported than accidents, which raises the following general question.

Why are so few near misses reported?

The next sections of this chapter describe the industry surveys and results that help answer this question.

5.2. Obstacles to Near Miss Reporting and Recommended Solutions

Many companies recognize that there are obstacles to reporting near misses. The following is a list of the obstacles gleaned from surveys and from the experience of the surveyors. The more critical obstacles appear first in the list, but some of the lower ranked obstacles can still keep the reporting rates low. Solutions are presented for each obstacle. These have been tested and have worked, but not every company will achieve the same level of results. Some solutions may fit one company's culture better than others may. [1], [2]

5.2.1. Fear of Disciplinary Action

This obstacle easily ranks highest on the list. Basic human nature is to not act if the consequences are negative or punishing. Individuals will not report a near miss if they believe their bosses will hold the near miss against them or their peers. If this obstacle is not overcome, near misses will not be reported regularly. This has been proven by an extensive survey[1] and research projects[2] and is acknowledged by regulatory authorities worldwide. [3]

Overcoming this obstacle requires management to do two things:

1. Recognize that all incidents, including near misses, are the result of management system weaknesses. The goal is to find the reasons why a human made a mistake. These reasons will identify the management system weaknesses. When they are improved, other humans are less prone to repeat the mistake.
2. Implement a policy that does NOT punish individuals when their errors lead to incidents and near misses (except for malfeasance or acts of malicious intent, such as fights and sabotage).

These two concepts, according to all survey respondents with high near miss reporting ratios, provide the best approach for overcoming this obstacle.

The two concepts apply to different levels of the error pyramid presented in Figure 5-1. Individual accountability should be enforced in the nonincident portion at the base of the pyramid using one or more known successful practices. Three examples are:

1. Self-directed work teams
2. Behavior-based management
3. Discipline by the supervisors

When a sequence of errors and failures propagate to the incident level, enough precursor occurrences have occurred to indicate that the current chain of occurrences represents a systemic problem. Systemic problems should not be blamed on an individual. Instead, management must find the system weaknesses and fix them before the next individual runs into a similar situation or problem. Therefore, another important step in overcoming the blame obstacle is to:

Find the root causes (management system weaknesses) of each causal factor and write recommendations that address the root causes, not the causal factors.

A causal factor may be a mistake someone makes, but finding the reasons why the individual made the mistake is more productive in preventing recurrence than punishing the individual for his or her mistakes. The focus should be on finding the root causes and ensuring appropriate recommendations are written and implemented. This solution is closely related to the first solution of establishing a blame-free culture for incidents. By the way, remember not to blame the managers either. Fix the system instead. However, managers should be held accountable for getting near misses reported.

A related fallacy is if a company trains enough investigators, near misses will be reported so there is no need to establish a blame-free system. This assumption has been proven false when fear of blame is not addressed.

At one facility, approximately ten percent of the operating and maintenance staff was trained in how to lead investigations. Achieving that percentage is recognized as a good practice. However, management still used the incidents to assign blame to the individuals involved in the chain of occurrences and in some instances used an incident as a reason to fire employees. The near miss reporting ratio at that facility has not increased past one. A ratio of one is better than zero but statistically a larger sample of incident data is needed to prevent future incidents related to the same management system weaknesses.

Another fallacy somewhat related to the fear of discipline is that getting rid of the incident-prone individuals will prevent future incidents. Studies have shown that fewer than 20% of the incidents involved a repeater.(4) The *incident proneness theory* is generally discredited as a flawed incident causation theory. It is probably more likely that repeaters are just less adept at hiding near misses and incidents; or perhaps they are more proactive or open about fixing the problems when they are involved.

Management must enforce a no blame policy once it is implemented. Once enforced, the system may need months or years to show results. Tremendous results can appear in just one year when management proves that they will not assign blame due to an incident. Building trust is the key.

Of course, criminal action, incompetence, or malfeasance warrants discipline. These include sabotage, horseplay, fights, and other acts of malicious intent. Discipline should be addressed by company policies, not the incident investigation.

Some solutions that will enhance near miss reporting include:

- Having peers help investigate incidents involving peers.
- Making the employees owners of the incident reporting and investigation system.
- Telling the employees that the new policy is not to assign blame. Hold management accountable to this commitment.
- Beginning with a system for anonymous reporting of incidents.

5.2.2. Fear of Embarrassment

A recognized motivator for humans to achieve job satisfaction is pride in their expertise, whatever their job level may be. Some employees are reluctant to report incidents because they are too embarrassed about their mistake or they think their peers will never let them hear the end of it. This is a negative motivator. For example, in some plant environments, the senior operators may name a piece of equipment after the person responsible for breaking it. In some cases, an employee will have several pumps and reactor lids named after them. No one wants this dubious honor. This is merely an example of social correction within a peer group. If morale is generally good and the inherent humor is accepted, the phenomenon is not harmful. However, there is anecdotal evidence that this natural human tendency can lead to one crew of employees avoiding letting the other crews find out what mistakes they made.

This obstacle is probably the most difficult one to overcome and one should expect somewhat limited success. The solutions to this obstacle include the following:

- Ensure that all employees understand the importance of near miss reporting.
- Demonstrate, through feedback of lessons learned, the importance of near miss reporting.

5.2.3. Lack of Understanding: Near Miss versus Nonincident

The definition of a near miss is often only partially understood. During basic incident investigation training, it is common for many class participants to believe that one example occurrence—a relief valve opening on demand—is never an incident, while the rest of the class believes it is. The individuals who believe it is a nonincident hold the position that the valve worked as designed. However, the rest of the class believes that the relief valve opening is a near miss (or possibly an accident or operational interruption in some facilities). Which group is correct? The second group based its logic on the assumption that if the valve had not opened, there could have been a catastrophic loss of containment. This is the preferred approach. Categorizing the relief valve episode as a near miss provides a learning opportunity. Companies with a good track record of near-miss reporting have established a management system that clearly defines (with an appropriate list of examples) near misses in the organization. They have also communicated clear expectations for reporting by all.

The real issue is which occurrences provide high learning value and which provide low learning value. Accidents and near misses are high-learning-value occurrences. Some companies encourage reporting of all undesirable occurrences and then decide which to investigate. If all occurrences are entered into a database, then the set of nonincidents can be analyzed to see which are chronic, and these can be investigated. Their learning value has now been determined to be higher than first thought because of their frequency. Figure 5-2 illustrates the concept of reporting all occurrences and then deciding which ones have high learning value and should be investigated immediately.

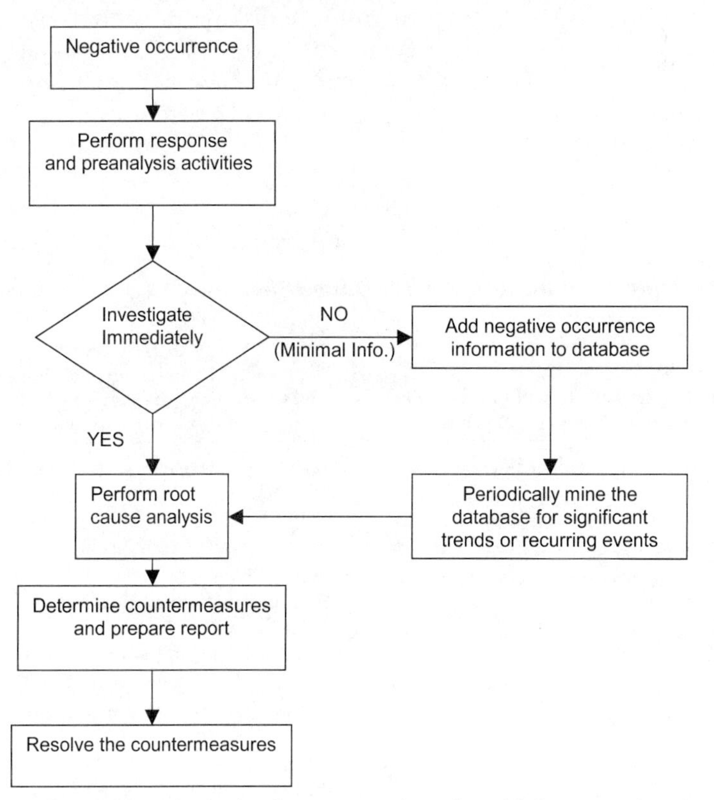

FIGURE 5-2. **Incident investigation process based on high- versus low-learning-value concept.**

Several solutions may be necessary to overcome lack of understanding about what constitutes a near miss. First:

Develop a list of in-context examples that illustrate high-learning-value incidents, particularly near misses.

This list should be created with input from various disciplines in the facility. Start the list by reviewing emergency work orders, process excursions, trouble reports in operating logbooks, and other similar sources.

The list can be used as a training tool for all personnel who work in or near the process. Listing these in a two-column format, with examples of incidents listed in one column and an example of nonincidents listed in the other is recommended. The examples should be as parallel as possible so that everyone can clearly see the differences as shown in Table 5-1.

Process excursions that reach or exceed the specified safety or quality limits of the process are important near misses that should be reported. However, the employees may not know that reaching the high-high pressure alarm point or reaching the rupture disk set pressure constituted a near miss. They checked the system to make sure the disk was intact, made sure the pressure returned to normal, and then continued operating. These may have been the warning symptoms of a pending major chemical process incident.

TABLE 5-1
An Example Training Tool for Teaching Differences between Incidents and Nonincidents

Incident (resources will be expended to promptly investigate these)	Nonincident (report but do not investigate immediately; may be trended and investigated later)
Safety relief device opens	Safety relief device found to be outside of tolerances during routine scheduled inspection
Pressure reaches relief valve set pressure, but relief valve apparently does not open	Pressure excursion occurs but remains within the process safety limits
High-high pressure trip/shutdown (one layer of defense against overpressure of the system)	High pressure alarm (possible quality impact)
Toxic gas detector in the area tripped/alarms	Toxic gas detector found to be defective during routine inspection/testing
Walking under a suspended crane load	Not wearing a hard hat in a designated area
Suspended crane load slips	Crane wire rope found to be defective during prelift check

Note: This is an example. It should be reworked for use with specific facility processes.

The types of questions to ask when developing the list of near miss examples include the following:

- What could the consequences reasonably be if the circumstances were a little different?
- How likely is it for the near miss to be spotted before it continues to an incident?
- How complex is the process or operation and how many layers of defense are there against the incident?
- Is the near miss one step away from disaster? Is it challenging the last line of defense?
- Is the near miss two steps away, which may be a near miss for a high hazard/high complexity system?
- Is the risk associated with potential incidents well understood?
- Is there high learning value in this near miss?

Once the starting list of examples is complete:

Train personnel on the examples.

This will paint the picture of what the company means by the term near miss. Over time, expect the list to change and grow as unanticipated occurrences present themselves along the way.

Differentiate between a near miss and a behavior-based management observation.

Many companies have implemented a system to have peers observe and try to correct the behavior of peers by coaching or other means. This is part of a behavior based safety management system. This system should operate in the nonincident portion of the error pyramid. Include examples in a listing, such as Table 5-1, to illustrate the differences.

Use safety meetings to capture and communicate near misses that were not previously identified.

This will keep the topic of near misses high on everyone's mind and will continually improve the understanding of what a near miss is. This system works best when a dedicated scribe is in the meeting. Consider including trade union safety representatives.

5.2.4. Lack of Management Commitment and Follow-through

Management must demonstrate commitment. What is one measure of commitment? Funding. Management must provide in-depth training for investigators and appropriate training for interviewers. All operations and maintenance staff should be trained on how to recognize and report near misses. Selected staff should be trained on how to check quality of the results of investigations, tabulate, and query the data for systemic trends. Management should allow employees the time necessary to investigate incidents and generate reports. Management should communicate incidents and associated lessons learned to all affected employees and other sites where the lessons would be important. Finally, management should show an interest in the results and enforce follow-through and documentation of the resolution of recommendations. The solutions to this obstacle are rather straightforward, but can take many forms. It begins with the following steps:

- Provide training to an appropriate number of operations and maintenance personnel on a consistent approach to investigation, which includes causal factors and root causes determination.
- Hold regular meetings with employees to discuss the successes of and obstacles to near miss reporting. Praise employees for submitting near misses.
- Emphasize to employees how important it is for them to invest the time to investigate near misses even if overtime labor is necessary.
- Hold management accountable for achieving realistic near miss reporting.

5.2.5. High Level of Effort to Report and Investigate

An apparently high level of effort is required to report and investigate near misses. The costs of this effort are quantifiable. The benefits of these investigations are not as easy to tabulate. The actual number of accidents that have been prevented by improved near miss reporting may never be known. However, organizations that have seen dramatic increases in near miss reporting have also seen dramatic reductions in losses. The root causes of near misses of safety consequences may be the same management system weaknesses that adversely affect operability, quality, and profitability.

- Share both subjective and tangible benefits that are expected from increased near miss reporting with the entire workforce.
- Ensure that the data are entered in a database and queried regularly. Also, ensure that the results of the query are shared with employees so they can see the value of the near misses they are reporting.

One company increased its near miss reporting ratio from 1 to roughly 80 in just a year. The company entered all the data in a simple database developed in house and then queried the data regularly. They found the most frequent near miss was suspended crane loads slipping. The second most common near miss was employees walking under suspended crane loads. Based on this data, it is clear that an accident is likely to occur soon unless positive action is taken. Management shared these findings with the employees and let them draw their own conclusions. Two great benefits were achieved. People stopped walking under crane loads because they recognized the increased probability that a crane load might slip and smash them. Second, the employees saw immediate benefit to reporting near misses.

- Track the benefits of near miss reporting and trend these versus the near miss reporting rate or the near miss ratio.
- Implement user-friendly tools such as forms, software, or database applications that ease the burden of documenting and disseminating incident results.

5.2.6. Disincentives for Reporting Near Misses

Those who do not understand the process can view near miss reporting negatively. This obstacle has stopped near miss reporting in several instances. One plant manager was called to headquarters to explain why his incident rate climbed so suddenly after implementing some of the recommendations in this chapter. What his superiors failed to understand was that this was an expected and positive outcome of implementing an effective near miss reporting system. The company culture was focused on enforcing standards, and the company had a history of disciplining employees who caused an incident. Many individuals in this company still do not believe that giving up the freedom to punish employees when an incident occurs is a good business decision.

Disincentive occurs when department goals are tied to lower incident rates. The solution here is obvious and necessary:

Ensure that goals affecting profit sharing incentives are not tied to lower overall incident reporting rates. This discourages near miss reporting.

The company will learn that reporting and investigating near misses will enhance overall business performance, particularly because the near misses of a safety incident or environmental release have the same root causes as incidents that detract from quality and productivity. Safety personnel can assist in defining an appropriate near-miss reporting culture. All managers intuitively understand the return on investment from preventing incidents. The effort pays for itself directly through improvements in productivity.

5.2.7. Not Knowing Which Investigation System to Use

Some companies have one investigation system for occupational safety incidents, another one for process safety incidents, another for environmental releases, another for reliability issues, and yet another for quality and customer services issues. The same investigative approach and training may work well for incidents in any facet of a business. There is merit in combining the systems and, in particular, in combining the incident databases. Combining the incident systems will require more work on defining near misses and in determining success in reporting near misses.

A related consideration is that incidents can affect more than one aspect of a business. Table 5-2 illustrates this point for an incident involving a 1000-lb release of cyclohexane from a decanter system at a polymer production facility. The occurrence did not harm any people and did not noticeably damage the environment although reporting of the release to regulators was required. The occurrence and the actions taken after the release caused the process to be shut down for about 9 hours and caused 3000 lb of product to be rejected. (The values in Table 5-2 are from a qualitative scale, where 10 would be very high impact and 0 would be very low or no impact.)

From the view of both actual and potential impact, the cyclohexane release affects all business aspects. The incident is a near miss for safety, and a minor-major incident for other aspects of the business. Performing six or more investigations would be fruitless. Performing one investigation that meets the needs of all business aspects is ideal and simpler. The near miss definition and related training will need to explain the potential impact of an

TABLE 5-2
Examples of the Impacts of a 1000-lb Cyclohexane Release

Business Aspect	Actual Impact of the Incident	Potential Impact of the Incident
Safety (harm to people)	0	10
Environment (harm to nature)	1	3
Quality (harm to product)	3	3
Reliability (harm to process efficiency)	5	10
Capital (harm to property, facilities, equipment)	1	10
Customer Service (harm to relationship with clients)	2	10

occurrence in relation to each business aspect, so that the users of the system can identify a near miss. Therefore, the solution includes:

- Emphasize two things during training:
  (1) How to report near misses and
  (2) Where to go for an answer if you are not sure an occurrence is a near miss.
- Consider having *one* incident reporting system with *one* approach for teaching employees the definition of a near miss and with *one* approach for doing incident investigations including one approach for root cause analysis.

5.3. Legal Aspects

There is legitimate concern that near miss reports can be used against a company. This may occur when

- an outsider claims that near misses show a history of unsafe conditions that is apparently tolerated by the company,
- a near miss report is used to show that a company knew that a certain incident was possible at one site, but failed to take effective action to prevent its occurrence at all sites, or
- a near miss report is used to incriminate a company due to inaccurate wording.

Liability is an issue without clear boundaries. An incident that occurs in one country may be raised in litigation in another country to show a pattern of unsafe conditions or lack of management follow-through on the experience. Even when liability is clearly absent, opponents of a company can use reports to sway public opinion. Possible solutions include:

- Take appropriate action on findings to limit future liability.
- Ensure, through thorough investigator training and auditing of reports, that investigators refrain from broad conclusions and that the language used in the final report is appropriate.
- Involve legal counsel on major near misses and incidents where liability could be high to ensure the results are protected as much as possible under attorney client privileges.

Chapter 12 provides additional guidance on legal issues.

Despite the possible liabilities, many companies decide that it is better to get near misses reported and to learn how to prevent incidents than to discourage near miss reporting or recordkeeping. A solution for some companies is to ensure that technical and business managers understand:

- Legal liability concerns should never discourage reporting and investigation.
- Proper investigation and documentation of near misses demonstrates that the company is behaving responsibly to learn lessons and continually improve risk management.
- Follow up for near miss recommendations deserve the same priority as accident-related recommendations.

Investigation and follow-through of accidents and near misses shows the intent by the company to discover and correct problems in the system. However, investigations without follow-up may be perceived as a demonstration of neglect by company management.

Endnotes

1. Bridges, W. G. "Get Near Misses Reported," *Proceedings of the International Conference and Workshop on Process Industry Incidents, Orlando, FL.* New York: Center for Chemical Process Safety (CCPS), AIChE, October 2000.
2. Urian, R. K. "Organizational Unlearning: Detrimental Behaviors Present in Chemical Process Incident Investigation Teams." *Proceedings of the International Conference and Workshop on Process Industry Incidents, Orlando, FL.* New York: Center for Chemical Process Safety (CCPS), AIChE, October 2000.
3. Oth, H.-J. "General Collecting and Evaluating of Major Accidents and Near Misses in the Federal Republic of Germany." *Proceedings of the International Conference and Workshop on Process Industry Incidents, Orlando, FL.* New York: Center for Chemical Process Safety (CCPS), AIChE, October 2000.
4. Hammer, W. *Occupational Safety Management and Engineering*. Englewood Cliffs, NJ: Prentice-Hall, 1985.

6

The Impact of Human Factors

Some people have forgotten the limitations of management systems. All that a system can do is harness the knowledge and experience of people. . . . Knowledge and experience without a system will achieve less than their full potential. Without knowledge and experience, even the best system will achieve nothing.

—*Trevor Kletz*

Studies indicate that the vast majority of incidents involve human performance issues. Incident investigations should not simply identify human performance as a cause; the management system failure that did not prevent or compensate for the potential human performance must be identified and corrected in order to prevent recurrence. An incident investigation management system should provide the tools to identify and address applicable human factors issues. It is important to recognize that human performance problems can occur in all management system areas—not simply to the human operator or maintenance worker closest to the scene.

Reduction in incidents and improved management systems are obvious results of a high quality incident investigation management system. However, when an incident investigation program addresses human factors issues, the results can provide even greater returns. Attention to human factors can improve employee morale, increase productivity, and complement cultural change. This chapter addresses the following human factors topics:

- Introduction to human factors and references for detailed human factors information

- Human performance problems and resulting incidents
- Incorporating human factors into the incident investigation process

6.1. Defining Human Factors

Human factors is the scientific discipline concerned with the understanding of interactions between humans and other elements of a system, and the profession that applies theory, principles, data, and methods to design in order to optimize human well-being and overall system performance.[1]

Implementing an effective human factors program or incorporating human factors into an existing incident investigation system requires *management commitment*. Figure 6-1 illustrates one company's integration of human factors into the company culture.

Operating Environment and Culture

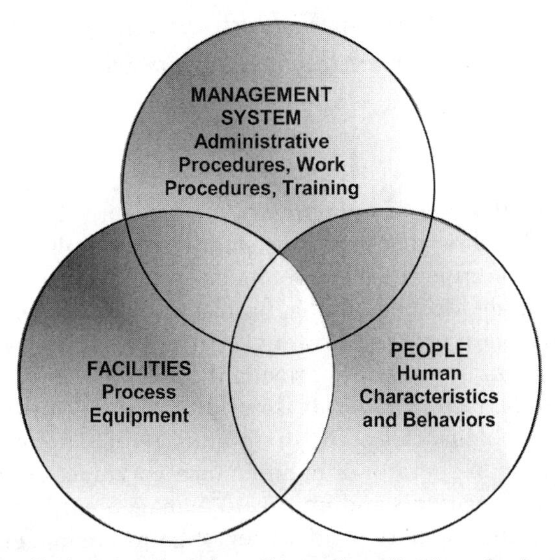

Human Factors are the integration and application of scientific knowledge about:

—People
—Facilities
—Management Systems

to improve their interaction in the workplace.

FIGURE 6-1. **Example human factors definition and concept[2].**

Workers interact with technology, the environment, and organizational factors every day. Sometimes human performance problems involve the individual, but most often, human performance problems are the result of the way technology, the environment, or organizational factors affect a person's performance.

The following examples, describe this concept:[3]

- ***Technology***—Designs that have a detrimental affect on performance include a piece of equipment meant to be used outdoors designed with data entry keys that are too small and close together to be operated by a gloved hand, or a shutdown valve positioned out of easy reach.
- ***Environment***—The human body performs best within a narrow temperature range. Performance will be degraded at temperatures outside that range and fail altogether in extreme temperatures.
- ***Organization***—Shift size and training decisions affect workload and a person's ability to perform work safely and efficiently. Reward and punishments used by the organization influence what individuals do.

Figure 6-2 illustrates how people relate and interact with technology, the environment, and organizational factors.

6.2. Human Factors Concepts

Human deficiencies may occur when technologies, environments, and organizations are not appropriate for desired human performance. These discrepancies allow human performance deficiencies.

In the past, the solutions to these deficiencies were limited. Management may have tried to threaten or entice workers into not making errors, as though proper motivation could somehow overcome poorly designed equipment, inadequate management systems, and inborn human limitations. In other words, the human has been expected to adapt to the system. This usually does not work. Instead, what needs to be done is to *adapt the system to the human.* [3] Thus, when an incident involves human performance, the recommendations must clearly address the management systems that allowed the deficiencies in the technology, environment, or organization. One company communicates this with the philosophy: *The worker should be set up to succeed, not set up to fail.*

Some examples of how the environment, technology, and the organization affect human performance are shown below.

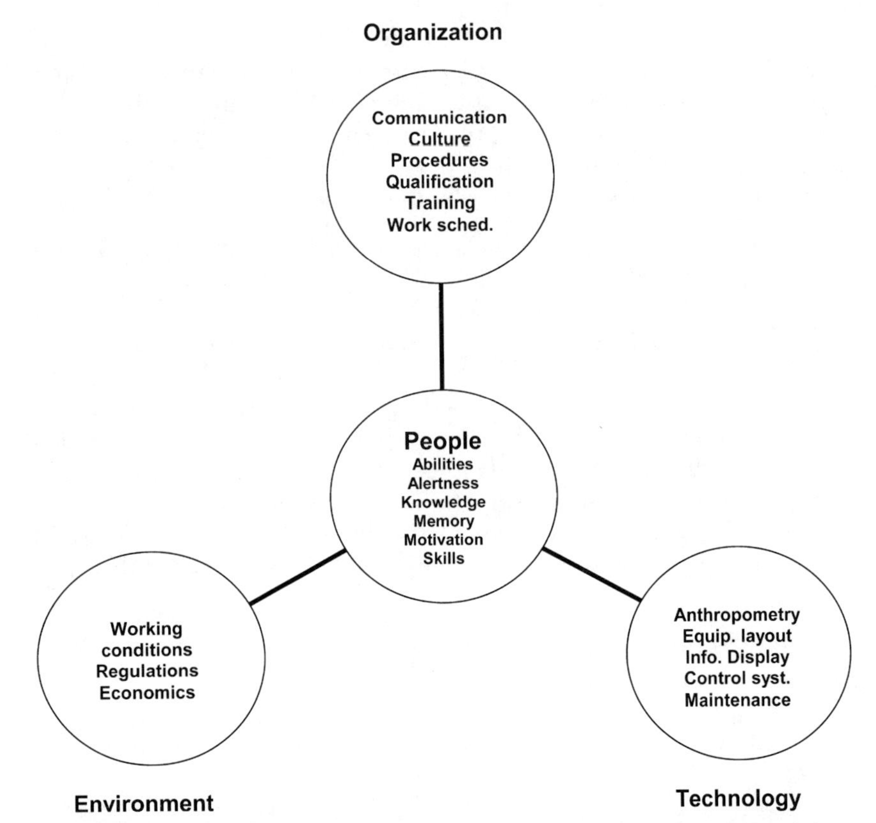

FIGURE 6-2. **How people interact with the environment, technology, and the organization.**

The environment affects:

- *Physical performance* (extreme temperatures, noisy working conditions, or poor lighting could contribute to human performance problems)
- *Mental Performance* (long work shifts, extreme temperatures contribute to fatigue and subsequent human performance problems)
- *Decision-making* (employees may take risks due to conditions in their work environment)

Technology affects:

- *Decision-making and troubleshooting ability* (too many alarms going off in a control room, gauges that are not consistent)
- *Perceptions and comprehension* (designs that go against cultural norms such as red means go and green means stop)

- *Reach, strength, agility* (anthropometry, designing for average size, valves hard to reach)
- *Safety and performance*

The organization affects:

- *Work practices* (procedures, training, work schedules, directly impact work practices)
- *Knowledge and skills* (lack of or inadequate training, cross-training goals, excessive training, all impact a person's knowledge and skill level)
- *Teamwork* (priorities within an organization directly impact the way employees work with one another)
- *Decision-making and troubleshooting ability* (employees may take risks due to conditions in the organization or culture)

Some examples of management systems that influence human performance are shown below.

Workplace Design

- Facility layout
- Workstation configuration
- Accessibility

Equipment Design

- Displays and control panels
- Controls (valves, hand wheels, switches, keyboards)
- Hand tools

Work Environment

- Noise
- Vibration
- Lighting
- Temperature
- Chemical exposure

Physical Activities

- Force
- Repetition
- Posture

Management Factors

- Reward/punishments
- Individual and organizational goals
- Work priorities

Job Design

- Work schedules
- Workload
- Job requirements vs. peoples' capabilities
- Task design

Information Transfer

- Labels/signs
- Instructions
- Procedures
- Communications
- Training
- Decision-making

Personal Factors

- Stress
- Age/culture
- Fitness
- Fatigue
- Boredom
- Motivation
- Body size/strength

Teamwork

- Definition of roles
- Organizational support of teams

Human performance problems occur several ways. To find out the reasons, it is important to understand the job task and the actions involved. Often, incidents occur when humans perform intentional acts without an understanding of the consequences. Human performance problems generally fall into four categories.[4]

1. ***Involuntary or nonintentional action***—For example, an operator leans against a switch, accidentally tripping it.
2. ***Spontaneous or subsidiary action***—For example, the purging process has just been changed, and although the operators have all been trained to purge the reactor for 3 minutes before charging the additives, in the middle of the third shift, the previous process is so ingrained that the operator automatically opens the reactor to charge materials before purging.
3. ***Unintentional action (slip or lapse)***—For example, an operator intends to start pump 1, and as he or she reaches for the pump

switch, the operator unintentionally hits the switch for pump 2, starting it instead.

4. *Intentional but mistaken action*—For example, the operator is required to close valve A but local knowledge suggests that closing valve B provides the same benefit as closing valve A, and is faster to do, so the operator closes valve B.

How these categories relate to a job task is illustrated in Figure 6-3.

In addition to the four categories, there are four common classes of human performance problems:

1. Errors of omission
2. Errors of commission
3. Sequence errors
4. Timing errors

Errors of omission are defined by failure to complete a required, necessary, or appropriate action. There may be a number of reasons for an

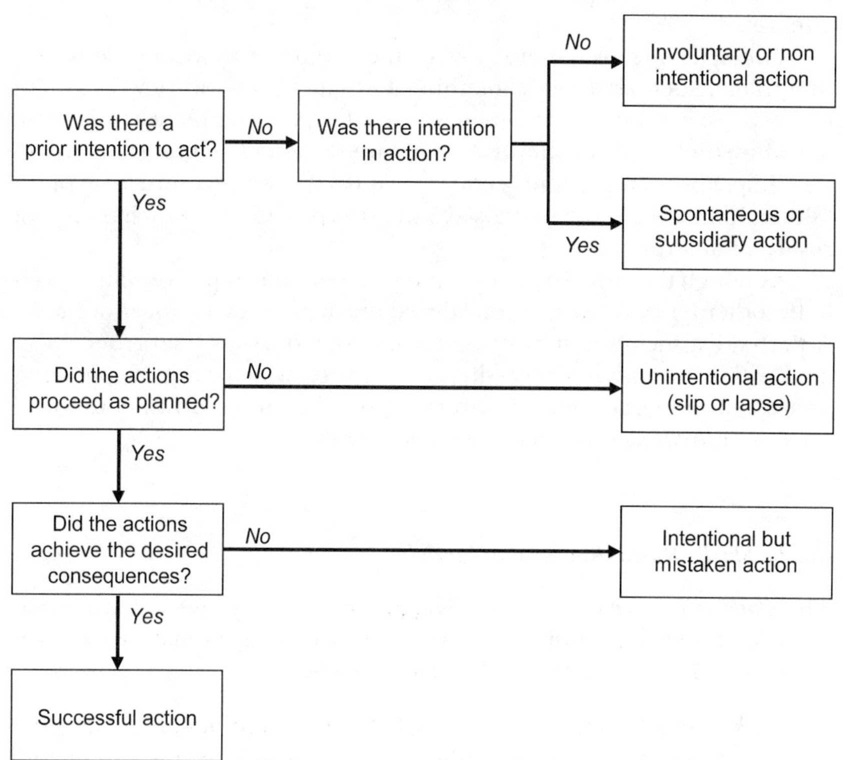

FIGURE 6-3. **Method for determining the type of human error.**[4]

error of omission beyond simply forgetting—lack of knowledge or training, lack of an expected prompt, physical inability to respond, other more pressing duties, etc. For example, an operator is required to close valves A, B and C, but only closes valves A and B.

Errors of commission include failing to act correctly, doing the wrong thing, doing the right thing on the wrong system, or using the wrong procedure in an attempt to do the right thing. Again, there are many reasons for an error of commission beyond simply "making a mistake" or "not paying attention." These include lack of knowledge or improper training, receiving an incorrect prompt, improper instruction, following poor past practices or attempting to make an improvement without properly evaluating the consequences. For example, an operator opens the wrong reactor or dumps in twice as much catalyst as needed.

Sequence errors are tasks done out of order, such as disconnecting a chemical transfer hose before depressuring it. Although someone may make a sequence error simply because of not thinking, other reasons might include improper training, out of sequence prompt, failure to understand the sequence importance, or misguided improvement attempts.

Timing errors are failures to execute certain tasks within the proper timeframe, such as doing something too quickly or too slowly. Timing errors can occur whenever someone is in a hurry to complete a task, is late in completing a task, or simply does not understand the importance of the schedule. For example, a procedure requires the temperature to be raised "slowly," but temperature is raised 5 degrees per minute, which is too fast, resulting in a runaway.

Although these are the major categories of human performance problems, other types should be considered during the search for root causes. A partial list includes high-stress errors; low-stress errors; errors associated with information processing; and errors in detecting, recognizing, and diagnosing deviations. Chapter 9 provides an opportunity to examine how human factors affect root cause analysis.

6.2.1. Skills–Rules–Knowledge Model

The *Skills–Rules–Knowledge* (S–R–K) model has been proposed as a framework for classifying human performance. It divides mental processes in plant operation into three performance levels:

- ***Skill-based:*** At the "lowest" level, these are routine skills of observation, hand–eye coordination, and control skills. Skills also include pattern recognition and actions that are manual, well known, and practiced frequently enough that they can be performed without considerable decision-making or diagnosis required. For example, these are the skills involved when an operator controls the pressure drop across a distillation column during a plant disturbance. Skills-based performance problems are typically associated with slips and lapses. As an example, the operator may be distracted while controlling the pressure drop or the operator's level of experience (skill) can have an impact on performance.

- ***Rule-based:*** At the "middle" level, these are performance tasks that involve actions governed by training and procedures that are executed consistently by different operators across a facility. These typically correspond to written procedures learned in operations and maintenance classrooms and on-the-job training. An example of a mistake associated with rule-based performance is the omission of a step in a formal written procedure.
- ***Knowledge-based:*** At the "highest" level are knowledge-based processes such as diagnosis, decision-making, and planning. These require knowledge about the plant and the processes, deductive and inductive skills, as well as the ability to decide and monitor results. Mistakes during attempts to find a solution to a plant disturbance can be associated with knowledge-based performance. An example is the misinterpretation of alarms and indicators.

These levels of mental processing continue in parallel and often support each other. Each mental process can result in different types of performance with different impacts on process equipment. One of the main successes of the S–R–K model was its prediction of previously unrecognized mental process performance levels and the problems associated with using the wrong level. The original S–R–K model has been modified and extended by several researchers, but it remains a useful concept for relating human performance to specific (potential) root causes.

Having defined a conceptual framework such as the S–R–K model for human performance, it is proper to turn to the underlying reasons behind a deficiency. As is implied, deficiencies associated with rule- and knowledge-based performance are directly linked to root causes associated with organizational deficiencies and human factor deficiencies. A special factor contributing to human deficiency can be the design error. A previously unrevealed design flaw can make it impossible for an operator to diagnose plant status correctly during a process upset. Design error can also be a direct result of an organizational error. Figure 6-4 shows how the skills–rules–knowledge model relates to the spectrum of human error.

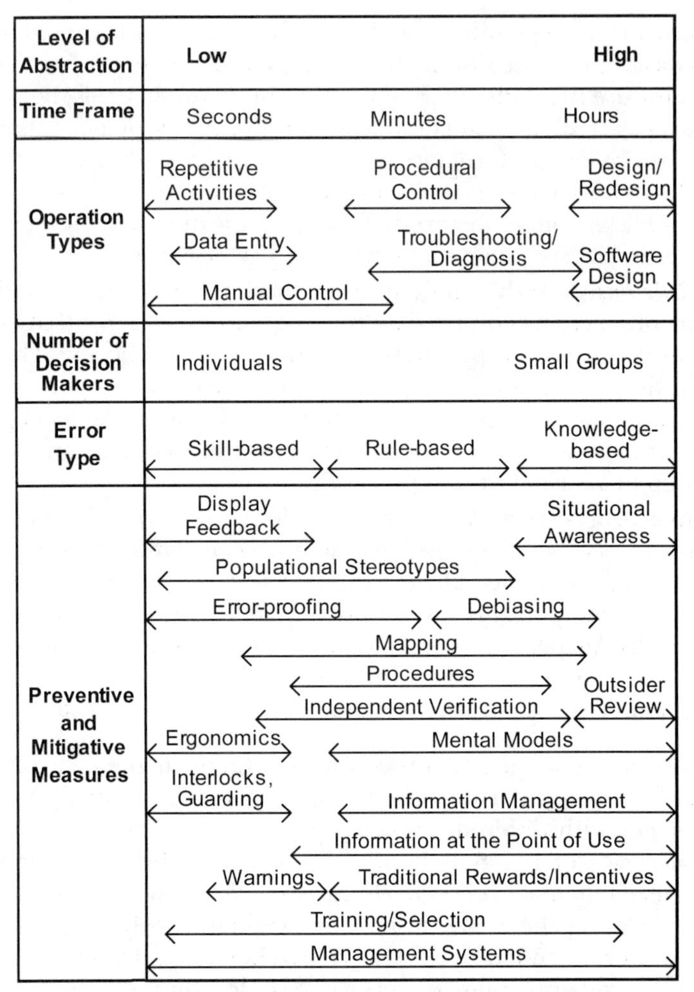

FIGURE 6-4. **The spectrum of human error.**

6.2.2. Human Behavior

Historically, companies and agencies that investigate incidents have overlooked human factor causes almost entirely. Material deficiencies in incidents (for example, equipment malfunction or a deficiency in the structural integrity of the vessel) can often be identified easily (for example, a shaft is broken). However, the real difficulty in incident investigation is to answer *why* these deficiencies occurred, and the answer is often related to human behavior. For instance, the shaft may have broken because of company management decisions, such as cutting back on maintenance, purchasing a less costly (and less well-made) piece of equipment, or selecting less-experienced engineers. The shaft may have broken due to poor supervision of operations or maintenance, an error made during maintenance, or use of equipment outside its safe operating range. Each of these underlying factors also needs to be probed for *why* it happened. (Did the company cut back on maintenance to save money? Did the operator operate the equipment outside its range due to inexperience or willful violation?) Meaningful solutions can be developed only after the investigator understands the true underlying causes. In typical investigations, however, the *why* as it relates to human factors is sometimes ignored.

Incident investigation teams should attempt to determine what management systems improvements could be made to remedy the particular human performance problem associated with the incident under investigation. Oversimplifying a human error is an easy mistake to make but can be avoided if the proper technique is used. In almost every case, there are underlying reasons for the human error beyond the simple assumption that the operator failed to follow procedure. A system failure, design flaw, or training deficiency may be the foundation of the performance problem. A good root cause identification process should identify the underlying reasons.

Designers and manufacturers of aircraft often use a *performance envelope* concept to summarize the expected and intended performance specifications of their product. If the limits of the envelope are exceeded, reliable performance cannot be expected. In the same way, workers function best in a well-identified performance envelope of working conditions. Their performance envelope includes:

- temperature
- lighting
- reach limits
- exertion limits
- data input burden limits
- decision-making load limits
- what they have been taught
- what they have seen work in the past
- their own personal habits, experiences, and beliefs

Exceeding these known limits will result in unpredictable actions usually resulting in significantly reduced performance, and a resulting increase in human error. Investigators and designers should be constantly on the alert for opportunities to make a system more reliable by considering the worker–machine interface and search for every opportunity to make the system less error-prone and more error-tolerant.

Some examples of commonly encountered human performance deficiencies include:

- ***Errors that resulted in the component being unavailable when needed***—For example, equipment not restored to full operability following maintenance, testing, or inspection. Valves left closed, actuation or protective systems not reconnected, slip-blinds left in the line, or failing to return pump selector switches to the standby auto-start position are common examples.
- ***Maintenance errors that led to the component being left in a condition in which the likelihood of failure was increased***—For example, loose bolting, misaligned shafts, and bearings/gaskets not sealed, internal clearances incorrect, foreign materials left in a system component, or bearings not lubricated.
- ***Operator errors that initiate process upsets due to valve line-up errors***—For example, closing the wrong valve, opening a bypass valve at the wrong time, or lining up the valves to allow pumping from or to the wrong tank.

6.3. Incorporating Human Factors into the Incident Investigation Process

Nurturing a blame-free, open culture within an organization is essential for the success of the incident investigation process. The investigation must focus on understanding:

- What happened?
- How did it happen?
- Why did it happen?
- What can be done to prevent it from happening again?
- How can the risk be reduced?

Investigators are not out to assign blame. Actions taken to "blame and shame" generally do little to prevent similar incidents from occurring. Therefore, it is necessary to foster an open and trusting environment where people feel free to discuss the evolution of an incident without fear of reprisal. Without such a supportive environment, involved individuals may be reluctant to cooperate in a full disclosure of occurrences leading to an incident[(3)] and the incident investigation may be concluded prematurely with the root causes left uncovered.

For example, failure to follow procedures is not a root cause. "Failure to follow established procedure" is a common premature stopping point for incident investigation related to human factors. In many cases, the investigation team identifies the fact that a person failed to follow established procedures, then does not attempt to investigate further and determine the underlying reason for the behavior. In most, but not all instances, there is an underlying correctable root cause that remains undetected and unfixed. The *failure to follow established procedure* behavior on the part of the employee is **not** a root cause, but instead **is a symptom** of an underlying root cause. Chapter 9 addresses root cause analysis in detail.

In almost every case, there is an underlying reason why a person did not follow the established procedure. The investigation team has an obligation to try to find and fix the underlying cause for the *failure to follow established procedure* behavior. Typical symptoms and corresponding underlying system defects that can result in an employee failing to follow procedure include:

- Out-of-date written procedure that no longer reflects current practices or current configuration of the physical system, ***due to*** defects in the process safety information, or operating procedures management systems
- Employee perceives that his or her way is better (safer or more effective), ***due to*** deficiencies in the system for establishing and maintaining a specific competency and qualification level
- Employee previously rewarded for deviating from the procedure, ***due to*** a culture of rewarding speed over quality, resulting from and reflecting a defective quality-assurance management system
- Employee following personal example set by his supervisor, ***due to*** a defective system for establishing and maintaining supervisory performance standards
- There are multiple accepted practices (daytime versus weekends for example), ***due to*** the presence of dual standards, ***due to*** defects in the supervisory or auditing management systems
- Employee is experiencing temporary task overload, ***due to*** defects in the scheduling and task allocation system, and/or ***due to*** ineffective implementation of downsizing
- Employee has physical/mental/emotional reason(s) that causes him or her to deviate from the established procedure, ***due to*** defects in the fitness-for-duty management system
- Employee believed he was using the correct version of the procedure, but ***due to*** defects in the document management system, he was using an out-of-date edition
- Employee was improperly trained ***due to*** defects in the training system

In some instances, the *failure to follow established procedure* may be due to inadequate knowledge. The classic recommendation that accompanies this symptom is to provide training (or refresher training) to ensure the person understands how to follow the established procedure. An example of a typical recommendation associated with this mistake is "review the procedure with the employee to ensure that he understands the proper action expected." The training activity may be beneficial to the person(s) who receive it, but in most cases, the training fails to identify and address the underlying cause(s) of the deficiency in the knowledge/competency system that resulted in the person failing to follow the procedure. In many cases, the other employees remain in a state of inadequate training. The training recommendation assumes that the person did not know/understand the proper procedure.

6.3.1. Finding the Causes

Incident investigators should be alert for human performance problems caused by a mismatch between the system design and reasonable expectations of human cognitive performance. Sometimes the designers of chemical processing systems fail to consider reasonable human capability limits and patterns of habit. The result can often be a system that promotes human errors rather than discouraging them. Donald Norman addresses these mismatches comprehensively in the book *The Design of Everyday Things.*[5]

Time is the enemy when considering human actions. As a plant ages, minor modifications and changes can individually or collectively cause human performance problems. For instance, in almost every plant over 5 years old, a search would probably reveal a series of control circuits arranged in illogical sequence. The most common reason is due to a minor modification, and the most convenient solution was to rearrange the control switches slightly.

Example:

Four pumps in the field may be arranged from north-to-south, A-B-C-D. Yet in the control room the switches may now be configured in D-A-B-C sequence because there was no room on the control board for the new switch to be added after the "C" switch, but there was room to add it before the "A" switch. In an emergency, the operator is likely to mistakenly flip the "D" switch when really trying to flip the "A" switch. This ergonomic trap proliferates as time goes on and changes are made without consideration for operator habits, tendencies, and normally expected actions.

In order to comply with the expected conventions for a specific situation, the designer must know what the conventions are. Some of these vary by country. For example, light switches are pushed up to turn on in the US, but down to turn on in Europe. Color-coding schemes may vary from plant to plant. The best policy is to ask the end users what their expectations are for equipment operation.

For most normal operating conditions, the human operator can cope with the incremental additional mental load of inconsistencies, but during emergencies or other high-stress periods, each additional mental task is an opportunity for error.

Conforming to certain expected conventions and meeting normal patterns of actions and habits can enhance human performance. The incident investigation team should be alert for built-in design deviations from normal conventions. These deviations are often an underlying cause for human error.

Example:

People expect the hot water tap to be on the left side and the cold water on the right side. When this is not the case, they are confused and make mistakes. Rising-stem gate valves are expected to close if the handle is turned in a clockwise rotation and open if turned counter clockwise. When this does not happen, we must mentally compensate. We then have a tendency to make an error, particularly when under stress or during an emergency. Deviating from normal convention, expected actions, and established habits can be an underlying cause of human error.

6.4. How an Incident Evolves

There are usually multiple causes of an incident, with multiple people and occurrences contributing to its evolution. The Hazard–Barrier–Target (HBT) concept described in Chapter 3 provides an interesting view of the multiple-cause or system theory.

James Reason offered another useful model, often referred to as the "Swiss cheese model,"[(4)] that explains how the many factors can converge, resulting in an incident (Figure 6-5). A company tries to promote safety and prevent catastrophic incidents by putting into place layers of system defenses, depicted in Figure 6-5 as slices of Swiss cheese. Essentially, the term "system defenses" refers to the safety-related decisions and actions of the entire company: top management, the line supervisors, and the workers. This model recognizes that each defense layer has weaknesses or holes.

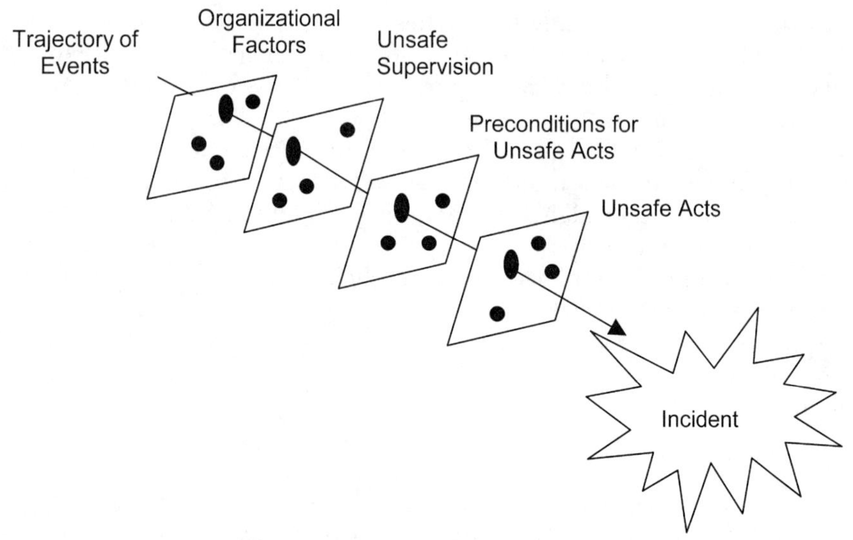

FIGURE 6-5. **An incident in the making (after Reason[4] as adapted by Weigmann and Shappell[6]).**

The following sections describe human factor issues (causal factors or weaknesses) that may be associated with each layer in Figure 6-5.

6.4.1. Organizational Factors

The Organizational Factors layer (slice) represents the defenses put into place by top management. This level of system defenses might include a company culture that puts safety first, and management decisions that reinforce safety by providing well-trained employees and well-designed equipment to do the job. [6]

Resource Management	Organizational Climate	Organizational Process
• Inadequate management of human resources • Inadequate management of monetary resources • Inadequate design and maintenance of facilities	• Inadequate organizational structure • Inadequate organizational policies • Inadequate safety culture • Inappropriate reward/punishments	• Inadequate established conditions of work • Inadequate established procedures • Inadequate oversight

6.4.2. *Unsafe Supervision*

The second layer of defenses is the "supervision" slice. This refers to the first-line supervisor and his or her safety-consciousness as well as organizational factors as displayed by the operational decisions he or she makes. [(6)]

Inadequate Supervision	Planned Inappropriate Operations	Failed to Correct a Known Problem	Supervisory Violations
• Failed to provide guidance • Failed to provide operational doctrine • Failed to provide oversight • Failed to provide training • Failed to track qualifications • Failed to track performance	• Failed to provide correct data • Failed to provide adequate briefing time • Failed to provide proper staffing • Failed to provide adequate operational procedure or plan • Failed to provide adequate opportunity for worker rest	• Failed to correct document in error • Failed to identify an at-risk behavior • Failed to initiate corrective action • Failed to report unsafe tendencies	• Authorized unnecessary hazard • Failed to enforce rules and regulations • Authorized unqualified worker

6.4.3. *Preconditions for Unsafe Acts*

Certain substandard preconditions may foster a climate where incidents can occur. Those preconditions may be related to human factors, practices, or interface with work conditions or the environment. [(6)]

Adverse or Inadequate Conditions of Operators		
Adverse Mental States	**Adverse Physiological States**	**Physical/Mental Limitations**
• Focused attention • Complacency • Distraction • Mental fatigue • Haste • Loss of situational awareness • Misplaced motivation • Task saturation	• Impaired physiological state • Medical illness • Physiological incapacitation • Physical fatigue	• Insufficient reaction time • Poor vision/hearing • Lack of knowledge • Incompatible physical capability

Adverse or Inadequate Practices of Operators	
Shift Resource Management	**Personal Readiness**
• Impaired communications due to language difference • Interpersonal conflict among crew • Failed to use all available resources • Misinterpretation of traffic calls • Failed to conduct adequate briefing • Impaired communication/conflict due to cultural difference	• Impaired due to medication • Inadequate rest

Adverse or Inadequate Work Interface	
Design Issues	**Maintenance Issues**
• Ambiguous instrumentation • Inadequate layout or space • Substandard illumination • Substandard communications equipment • Equipment substandard for job	• Poorly maintained equipment • Poorly maintained workspace • Poorly maintained communications equipment

Adverse Environmental Conditions
• Low visibility • Storm • Extreme temperature (heat or cold) • Wind

6.4.4. *Unsafe Acts*

The following list divides types of unsafe acts into those that are skill-based, perceptual, and decision or knowledge-based. (Note that this division is different from the skills-rules-model knowledge-based model described earlier.) Examples are given of each type of error that may result in an unsafe act.[6]

Errors		
Skill-based	**Perceptual**	**Decision (Knowledge-based)**
• Failed to prioritize attention • Inadvertent use of controls • Left out step in procedure or executed step out-of-sequence • Left out checklist item or completed checklist item out-of sequence	• Misjudged distance/rate/time • Misread dial or indicator • Failed to see/hear/otherwise sense	• Improper procedure or maneuver • Misdiagnosed emergency • Wrong response to emergency • Poor decision

6.5. Checklists and Flowcharts

To be successful in finding all causes, an incident investigation must include human performance considerations and human factor mismatches. An appropriate and easily available technique to do this is the development of extensive checklists and flowcharts to help the investigators address human performance problems or concerns. For example, checklists can be built using the information in the tables shown in Section 6.4. The investigator does not have to be a psychologist or an expert on human reliability analysis to do a reasonably competent job. However, checklist systems are typically built from experience, so it is important to have very experienced and knowledgeable people participate in the checklist development process. Numerous interface devices have been developed that translate theoretical models of human error causation into easy-to-understand engineering terms. Some of these devices are in the form of logic trees or checklists such as the one in Figure 6-6 on the next page. Examples of checklists and flowcharts (see Root Cause Map™ and Comprehensive List of Causes) are available on the accompanying CD ROM.

Chapter 9 describes the use of human factors checklists in root cause analysis.

Endnotes

1. International Ergonomic Association (IEA). http://www.iea.cc/ergonomics/, 2000.
2. Theriot, T. "Exxon-Mobil's Approach to Human Factors." 2nd International Workshop on Human Factors in Offshore Operations, 2002.
3. Rothblum, A., et al. "Human Factors in Incident Investigation and Analysis." 2nd International Workshop on Human Factors in Offshore Operations, 2002.
4. Reason, J. *Human Error.* New York: Cambridge University Press, 1990.
5. Norman, D. A. *The Design of Everyday Things*. London: MIT Papers, 1988. (Also published under the title *The Psychology of Everyday Things*.)
6. Shappell S. A., and Wiegmann D. A. "The Human Factors Analysis and Classification System-HFACS." FAA Office of Aviation Medicine. Washington, DC: Department of Transportation, Federal Aviation Administration, 2000. (Report # DOT/FAA/AM-00/7)

Did the incident involve a human error related to...?

- ☐ **Inadvertent Operation**
  - ☐ *Of electric switch or control device*
    - ☐ Access too difficult
    - ☐ Access too easy allowing for accidental activation
    - ☐ Not clearly labeled—ambiguous
    - ☐ Arranged such that errors are more likely

- ☐ **Error in Executing Procedure**
  - ☐ *Written procedure was not appropriate*
    - ☐ Procedure was not available
    - ☐ System physical changes not incorporated into procedure
    - ☐ New method or sequence not used
    - ☐ Procedure overlooked a hazard
    - ☐ Procedure overlooked a precaution
    - ☐ Procedure was incomplete, overlooked a step
    - ☐ As written, the desired action is not clearly described
    - ☐ Consequences of safety deviations not included
    - ☐ Responsibilities of tasks/assignments are unclear
  - ☐ *Written procedure not understood*
    - ☐ Original training inadequate and/or no verification of understanding
    - ☐ Refresher training not done
    - ☐ Written procedure no longer matches the actual actions in the field
    - ☐ Written procedure does not match physical equipment or control philosophy
    - ☐ Written procedure does not match verbal instructions and/or company current philosophy
  - ☐ *Procedure correct, but not followed this time*
    - ☐ Inconsistent enforcement
    - ☐ Worker knowingly deviated from SOP
      - ☐ In a hurry
      - ☐ Perceived that the action he took was actually more appropriate than what written SOP states
      - ☐ Temporary emotional state
      - ☐ Temporary physical state
      - ☐ Response to peer pressure
      - ☐ Response to perceived supervisor/management expectations
      - ☐ Following supervisor's instructions
    - ☐ Communications breakdown

FIGURE 6-6. **Sample human factors checklist.**

Additional References

Bea, Holdsworth, Smith. "Summary of Proceedings and Submitted Papers." Workshop on Human Factors in Offshore Operations, 1996.

Center for Chemical Process Safety (CCPS). *Inherently Safer Chemical Processes: A Life-Cycle Approach.* New York: American Institute of Chemical Engineers, 1996.

Smith, K. and Franklyn, C. "Conquering Cultural Change in Incident Investigation." International Conference and Workshop on Process Industry Incidents, Center for Chemical Process Safety (CCPS)/AIChE, Orlando, FL, October 2000.

7

Building and Leading an Incident Investigation Team

A thorough and accurate incident investigation depends upon the capabilities of the assigned team. Each member's technical skills, expertise, and communication skills are valuable considerations when building an investigation team. This chapter describes ways to select skilled personnel to participate on incident investigation teams and recommends methods to develop their capabilities and manage the teams' resources.

7.1. Team Approach

Whether investigating a process safety related incident with a large team or a personnel safety incident using a small team, all incident investigations benefit from a team capable of applying the chosen investigative methods effectively and consistently. Organizations with capable incident investigation teams can realize benefits in other aspects of their business besides safety. Performance can be enhanced in product quality, productivity, and environmental responsibility. An organization can add to its accumulated knowledge to prevent future incidents from each well-performed incident investigation.

The composition of an incident investigation team should relate to the type and severity of the incident. It may not be practical or essential to draft a team of senior experts in every situation. For example, it may not be necessary to assign the most experienced technical personnel to conduct the investigation of a minor incident. However, an organization may consider assigning these senior experts to a team investigating a *series* of similar minor incidents that have contributed to a significant negative safety trend or business loss. In addition, a catastrophic occurrence with a very simple system may not require as many investigators as an occurrence involving a highly complex system.

7.2. Advantages of the Team Approach

There are several advantages to using a team approach when performing incident investigations.

1. ***Multiple technical perspectives assist in analyzing the findings***—A formal analysis process must be used to reach conclusions. Individuals with diverse skills and perspectives best support this approach.
2. ***Diverse personal viewpoints enhance objectivity***—As opposed to a single investigator, a team is less likely to be subjective or biased in its conclusions. A team's conclusions are more likely to be accepted by the organization.
3. ***Internal peer reviews can enhance quality***—Team members with relevant knowledge of the analysis process are better prepared to review each other's work and provide constructive critique.
4. ***Additional resources are available***—A formal investigation can involve a great deal of work that may exceed the capabilities of one person. Quality may be compromised if one person is expected to do most of the work.
5. ***Scheduling requirements are easier to meet***—Deadlines set by management, outside parties, or the team leader may require several activities to be performed in parallel. This demands a team approach.
6. ***Regulatory authority may require it***—Specific regulations such as Process Safety Management call for a team approach. Management must be aware of whether a facility falls under such regulations.

The composition and mandate of a team will vary depending on the specific incident. Within a large organization or a company with several very different processes, it may not be practical or desirable to preselect one team to investigate all incidents. Personnel should be selected to participate in investigations based on their specific skills, experience, availability, and the team roles that need to be filled for a particular investigation. Over time, this approach will produce a pool of trained and experienced employees familiar with the investigation process.

7.3. Leading a Process Safety Incident Investigation Team

A successful incident investigation management system, as described in Chapter 2, depends on several factors. These include management involvement and support, management credibility, local organization, corporate culture, employee willingness to participate, and the history of correcting problems. Ultimately, the most important factor is the successful performance of the investigation team. A management system that provides strong team leadership and organizational support will help the investigation team succeed in determining root causes and developing plans to prevent recurrence.

An investigation team leader is generally selected based on three factors related to the incident:

1. its apparent magnitude and complexity;
2. its health, safety, environmental, or business interruption implications;
3. The depth to which the investigation is expected to probe.

A team leader should be objective, independent, and competent in both administrative and managerial skills as well as technical incident investigation skills. Many of the leader's activities involve coordination and communication with both the team and others, including site management.

Often the investigation team leader's first task is to systematically identify the resource requirements and recommend individuals and organizations that should participate. As with management's selection of the team leader, the leader's selection of the members will be based on the magnitude and nature of the incident. The team leader may choose to involve experts on an as-needed basis. These experts may be internal to the organization or contracted from an outside source. Arranging for part-time team member involvement on an investigation team is one way to reduce the investigation's impact on normal operations and limit costs. Part-time participation should be managed to ensure competing priorities are considered properly.

The leader may become aware of confidential information while simultaneously facing outside demands for premature release of the team's findings. The leader should have interpersonal skills suited to interacting with strong personalities while avoiding clashes and maintaining confidentiality. The leader must be able to influence changes within an established operation.

First-time team members providing technical support may also present a challenge to the leader. The new members may need time to become accustomed to their role as part of a close-knit team and may be sensitive to having their technical judgments questioned. The team's independence can create challenges for coordination and teamwork. The leader must strive to guide team members in avoiding duplications, omissions, and in maintaining task priorities.

The organization's incident investigation management system should specify the team leader's responsibilities and authority. Upon selection of the team leader, the leader and senior management should review and agree on all assigned responsibilities. The agreed responsibilities should be written in formal terms of reference. Typical leader responsibilities may include:

- Directing and managing the team in its investigation and assuring the objectives and schedules are met
- Identifying and controlling the restricted access zone
- Serving as principal spokesperson for the team and point of contact with other organizations
- Preparing status reports and other interim reports documenting significant team activities, findings, and concerns
- Organizing team work including schedule, plans, and meetings
- Assigning tasks to team members in accordance with their individual skills, knowledge, capabilities, and experience
- Setting priorities
- Procuring and administering resources needed for the investigation
- Ensuring incident scene safety
- Ensuring that the investigation team activities result in minimum disruption to the rest of the facility
- Keeping upper management advised of status, progress, and plans
- Initiating formal requests for information, witness interviews, laboratory tests, and technical or administrative support
- Controlling proprietary information and other sensitive information

7.4. Potential Team Composition

Team composition can vary considerably depending on the nature, type, and size of the incident. A typical team composition may consist of the following:

- Team leader
- Process operators (at least one worker from the unit experiencing the occurrence)
- Process engineers
- Process safety specialist
- Instrument technicians, inspection technicians, and maintenance technicians (as needed)
- Contractor representatives (if the incident involved contractors)

At least one team member must be a competent facilitator for the particular investigative method the team will use. This person does not necessarily need to be the team leader.

Other participants can be involved in a full time or in a limited consulting role depending on the nature of the incident. It is important to include people who know what actually happens in the field—not just what is supposed to happen. The team selection should be able to withstand third party scrutiny such as that of employees, departments, union representatives, community groups, and legal. Some companies include trade union safety representatives in the incident investigation. Positions to consider include:

- Process control (electrical/instrumentation) engineer
- Industrial hygienist
- Maintenance engineer
- Materials engineer/metallurgist
- Human factors specialist
- Recently retired employee with pertinent knowledge, skill, or experience
- Environmental scientist or specialist
- Chemist
- Quality assurance specialist
- Purchasing or stores
- Construction department
- Contractor participant
- Collective bargaining unit participant
- Civil or structural engineer
- Human Resources representative
- Arson investigator—for expertise to help find probable point of origin in cases involving fire
- Original Equipment Manufacturer (OEM) representative—a factory or team services engineer
- Explosion investigator—for expertise in understanding the ignition source and physics involving explosion
- Research technical personnel
- Emergency response personnel such as fire chief
- Technical consultant or equipment specialist

On some occasions, the following may be appropriate as full time or consultant team members:

- Insurance carrier (underwriter) representatives
- Community representative

Team members who come from another part of the organization, experienced contractors, and part-time staff may bring an unbiased, fresh, objective perspective to the investigation, and they understand general company policies, procedures, and approaches. Some companies choose to avoid selecting managers as team members since they may inhibit open dialogue among other team members and might bias the conclusions and recommendations. Supervisory personnel from the area involved may also exhibit these qualities. As with any team member, managers or supervisors on the team should exhibit the qualities on the list below.

The team organization, size, experience, and skills need to match the size and complexity of the incident. The team leader should become familiar with each member's particular competencies and strong points. Likewise, team members need to feel free to admit they require help or that they do not have the particular competence needed for a task. Team members may not be forensic investigative professionals and should not be expected to contribute beyond their level of competence or experience. The team leader must be flexible in making and modifying job assignments. Team size is also a consideration. Some companies recommend the core team consists of a minimum of two and a maximum of eight people for a workable size group, but large or complex occurrences may have numerous part-time participants and additional help. Large investigation teams are generally more difficult to manage. Additional time may be required to reach consensus and closure on findings and recommendations.

Some personal and technical characteristics to consider when selecting team members are provided below.

Seek out personnel with:

- Open, logical minds
- A desire to be thorough
- The ability to maintain an independent perspective
- The ability to work well with others
- Special expertise or knowledge regarding the incident or the facility
- Experience in technical troubleshooting
- Data analysis skills
- Writing skills
- Interviewing skills

Avoid selecting personnel:

- With preformed opinions on important issues
- Who identify causes of the incident before the investigation starts
- Who are too close to the incident, the facility, or the injured and may be emotionally involved or biased
- Who are offered to the team for the sole reason that they happen to be available
- With conflicting work assignments or other job priorities

- With travel or schedule restraints that are not compatible with the investigation timing and location

The services of outside experts may be needed to provide higher-level expertise in either the site technology or incident investigation methodologies. They often may participate on a part-time basis. Examples include specialists from within the corporate organization, employees, or ex-employees with relevant experience, and outside technical assistance (training, consulting, and evidence analysis). Specific areas of expertise, which are sometimes needed, include rotating equipment specialists, corrosion and materials engineers, instrumentation designers, computer software programmers, metallurgists, chemists, attorneys, or special testing lab services.

Under certain circumstances, the team may also include manufacturers' representatives, technical service personnel, or representatives of collective bargaining groups. They may add valuable perspectives and credibility to the team.

Although they may not be part of the core team, senior management must play a part in the workings of the investigation team. A good practice is to request that a senior manager reviews and comments informally on the work product and progress periodically during the investigation. The practice of keeping senior management informed emphasizes the significance of both the incident and the investigation's results. It can also assist in minimizing red tape enabling faster responses to team requests. The team leader may work with senior management reviewers to determine the format for reviews.

7.5. Training Potential Team Members and Support Personnel

High quality training for potential team members and supporting personnel helps ensure success. Three different audiences will benefit from training: site management personnel, investigation support personnel, and designated investigation team members including team leaders. A written incident investigation management system is one place to describe the topics, methods, and training frequency for each group. (See Section 2.2.5, page 22, for additional information.) A description of each group and suggested topics follows.

Site management personnel—This group can benefit from training on the following topics:

- The company incident investigation management system
- Basic incident investigation concepts
- Company policies related to incident investigation

- Upper-level management's commitment
- Specific job related responsibilities associated with incident investigation
- Media relations, if needed

Investigation Support Personnel—Support personnel may be operations and maintenance employees, first line supervision, and auxiliary staff groups such as technicians, engineers, and middle-level management. They can benefit from training on the following topics:

- An overview of the company incident investigation management system
- The site's incident reporting procedure
- An overview of basic incident investigation concepts
- Formal investigation protocols

This training will help the group cooperate with an active investigation team and provide support activities vital to the success of an investigation. For example, training will help personnel understand why and how evidence needs to be preserved.

Designated Investigation Team Members—This pool of personnel benefits from training on topics such as:

- An overview of the company incident investigation management system
- Incident investigation concepts
- Specific investigation techniques used by the organization
- Interviewing techniques
- Gathering evidence
- Writing effective recommendations
- Documentation and report requirements
- The roles of the team members

Team member training may also include "role playing" for such activities as witness interviews, conflict resolution, and confidentiality issues. Team members should understand they are not expected to perform at the level of full-time professional investigators. They must feel the freedom to request help or training as soon as they recognize a need.

After initial training, brief periodic refresher-training sessions or tabletop role-playing drills involving members of all three groups are a good way to reinforce the training objectives. Refresher training topics may include:

- Site-specific incident investigation plan
- General roles and responsibilities

- Specific assignments for team members such as interviewing, photography, and other roles
- Evidence preservation and handling protocols
- Locations for evidence storage
- Expected communications from team members

This trained pool of site management personnel, investigation support personnel, and potential investigation team members will now be ready to respond when needed.

Potential team leaders learn how to determine the appropriate investigation methodology, how to gather data, how to analyze data for causal factors, how to determine root causes of causal factors, and how to develop effective recommendations and reports.

7.6. Building a Team for a Specific Incident

An incident has just been reported. How is a team activated? The team activation segment of a company's incident investigation management system should guide the user to assemble a team based upon the following two areas of concern:

1. The needs, size, and structure of the organization
2. The type, size, consequences, and nature of the incident

Many companies recognize this and have a system in place to match their resources to the type of incident. Classifying the nature of the incident is one of the first steps. An example of one approach to incident classification follows.

- ***Minor incidents***—Infrequent occurrence of these accidents or near misses have acceptable consequences but recurring occurrences of this magnitude may warrant an investigation.
- ***Limited impact incidents***—Incidents deemed to be controllable with local resources and which have no lasting effects.
- ***Significant incidents***—Incidents that have, or would have in the case of near misses, consequences requiring considerable resources to mitigate and usually involve human injuries and/or major interruptions to operations.
- ***High potential incidents***—An occurrence that, under different circumstances, might easily have resulted in a catastrophic loss.
- ***Catastrophic incidents***—Incidents that have major consequences with unacceptable lasting effects, usually involving loss of human life, severe off-site impacts, and/or loss of community trust with possible loss of franchise to operate.

The upper and lower extremes are the easiest for which to gauge investigation team requirements. One person, with the aid of a reviewer or assistant, or a very small team sometimes investigates minor incidents. Catastrophic incidents that involve a fatality or destruction of a unit or a major portion of the facility will lead to activating a full investigation team. Remember that near misses are incidents and can be classified as a high potential incident for determining the type of team to use when investigating them.

External resources may be needed if the incident investigation work exceeds site capabilities. These resources could include corporate personnel or experts from outside the company. (The team leader may also be external if the incident is major since the leader's independence sets the tone for the investigation.) Company business unit leaders should confer with the team leader to determine whether external assistance is recommended. Factors to consider include significant offsite consequences such as environmental impact or product quality concerns. A team of trained specialists should formally investigate any process incident that could significantly affect the business. At the lower end of the scale, if a near miss or minor incident occurs that has no potential for significant consequences, local supervision or front-line personnel normally may perform the investigation without outside assistance.

7.6.1. Minor Incidents

While a very small team sometimes investigates minor incidents, standard forms and checklists can help ensure that the findings are consistent with reporting protocols. These tools provide the ability to record trends when compiling incident data over the life of a facility. To help with consistency and trending, many companies use one multipurpose reporting form for every initial incident report regardless of the incident's severity or nature: whether it involved fire, property damage, injury, or environmental release. Careful reporting and recording of minor incidents, and near misses, can assist investigators in their search for the root causes of major incidents. An example of a minor incident is a valve-packing leak on a cooling water line. Improper maintenance may have been one cause that could lead to a cooling system failure, and then the cooling failure could, if not repaired, result in damage to a pump. Improper maintenance on the same valve type in another system could cause a catastrophic failure and would be classified as a higher level occurrence.

7.6.2. Limited Impact Incidents

Limited impact incidents may have more significant effects or potential impact than first envisioned in certain circumstances. The site should determine the necessary investigation resources based on site capabilities and *potential* for more severe outcomes. An example of a limited impact incident might be a pump failure in a remote area of the plant.

Regardless of the realized outcome, there is a need and an opportunity to assign the right people to do the investigation. A small team of technicians who know the fundamentals of investigation techniques may be satisfactory for this type of incident.

7.6.3. Significant Incidents

Significant occurrences have greater impact and often have potential for extensive loss if a repeat incident were to occur. Incidents at this level may be complex and could involve a sequence of failures. This team may include both site personnel and expert outside resources to conduct significant occurrence investigations. Due to the complexity and severity of such incidents, determining the appropriate team member qualifications becomes more demanding. Examples of significant occurrences might be:

- a process fire that causes downtime or business interruption,
- a gas release or spill with off-site impact, or
- a serious personnel injury.

Prompt notification is essential in these cases. Management must quickly decide how and when to mobilize the team. Notification procedures are unique to each facility and organization structure. It is the on-site facility management's charge to make the initial decision and it is always preferable to err on the conservative side. It is easier to scale the team back when needed than to add to it later. If the services of a full investigation team are determined to be unwarranted, the leader can document the change of strategy in a brief report based on initial findings.

7.6.4. High Potential Incidents

A high potential incident is an occurrence that under different circumstances might easily have resulted in a catastrophic loss. An example could involve the early discovery of an instrument malfunction (by a diligent operator) that might otherwise have resulted in a process upset and violent explosion.

High potential incidents must be investigated by a complete team with the full rigor of a catastrophic incident as shown in the next section.

7.6.5. Catastrophic Incidents

Examples of catastrophic incidents might include:

- a compressor explosion with shrapnel causing serious injuries,
- a toxic chemical release that exposes employees or the community, or
- a spill contaminating the local water supplies.

Investigating a catastrophic incident normally requires a team of experienced, highly skilled specialists who represent not only process safety but also specialized technologies such as:

- Materials engineering
- Environmental emissions monitoring
- Medical and industrial hygiene (chemical exposures)
- Fire protection

Some organizations establish a highly trained and specialized corporate incident investigation staff as a preparatory measure for rare major catastrophic incidents. The ability to deploy the trained team quickly is essential. The faster the team is in place, the higher the quality of the data they collect.

7.7. Developing a Specific Investigation Plan

The team leader is responsible to plan and direct the team activities. The specific plan for the team should include a designated mechanism for documenting the team activities, deliberations, decisions, communications, and a record of documents requested, received, or issued. The primary objectives of a process safety incident investigation plan are to:

- Identify the physical causes—process and chemistry
- Identify the PSM-related multiple root causes,
- Identify recommendations to prevent recurrence
- Assist in interpreting the recommendations or auditing their implementation as needed

Figure 7-1 offers a typical checklist to use during the planning stage of an investigation of a major complex incident. Low complexity incident investigations do not always call for a formal plan. Some simple investigations may require only 1 to 2 hours to complete.

The team leader usually makes a brief orientation visit and considers several factors including the magnitude of potential outside interest in the occurrence. Outside interest includes three aspects:

1. the potential for media coverage,
2. legal issues, and
3. regulatory impact.

- ❑ Clarify and confirm priorities
    - ❑ Rescue and medical treatment
    - ❑ Secure incident to mitigate further consequences
    - ❑ Environmental concerns
    - ❑ Evidence preservation/Secure the site
    - ❑ Evidence collection (including interviewing witnesses)
- ❑ Plan for witness interviews
- ❑ Remediation and site clean-up
- ❑ Rebuild/restart
- ❑ Team leader selection
- ❑ Team member selection, training, and organization
- ❑ Initial orientation tour/visit
- ❑ Initial photography
- ❑ Plan for evidence identification, preservation, and collection including special handling of time sensitive material such as query control system logs
- ❑ Plan for documentation
- ❑ Plan for coordination and communication with other functions
- ❑ Identify and plan for procurement of team supplies and equipment
- ❑ Plan for any special or refresher training needed by team
- ❑ Establish checkpoints, timetables, and schedule of progress

FIGURE 7-1. **Checklist for developing an incident investigation plan.**

The initial site visit is the first opportunity to establish the physical boundaries of the investigation. The team leader should:

- ensure that access to the area is minimized as much as possible and
- verify that the personnel who enter the incident area are aware of evidence preservation considerations.

One of the most critical issues is clearly establishing which groups have responsibility for which activities and areas. These responsibilities may change during the investigation. The incident investigation team leader must ensure that these responsibilities are clear to all groups to avoid duplication of effort or omission of critical activities.

Management's charter to the team must include management's expectations for accurately reporting investigation outcomes. However, assigning blame or recommending disciplinary actions should not be part of a team's charter.

A high performance team will be independent and autonomous. The leader should encourage this awareness. It helps establish an unambiguous signal to all contributors that the investigation process will be implemented impartially. If witnesses perceive, either rightly or wrongly, that the team is in any way inhibited or intimidated by outside influences, participants and reviewers may question the quality, quantity, and credibility of the information collected.

It is particularly helpful to have an hourly employee from the same or an adjacent plant on the team to build confidence with a wider workforce. There has been a tendency in the past to select staff engineers and ignore operators and technicians. Operators and technicians generally know what really happens better than anyone does, and their involvement on the team can produce facts that would otherwise not become known.

7.8. Team Operations

A complex or significant incident will involve a great deal of work by many people. It is not reasonable to expect the team leader to interact directly with everyone who performs work on an investigation. To assist in establishing internal lines of communication, it is appropriate to organize the investigation team by function. The appropriate human and material resources are required within each function to succeed. Examples of functional group activities would be forensic analysis and personnel interviews. One representative within each function needs to report directly to the lead investigator. In this way, the lead investigator can manage the information on six fronts instead of ten or twelve. The U.S. National Transportation Safety Board (NTSB) commonly uses this approach. A designated point of contact between the team and all other outside groups can help minimize communication breakdowns, delays, and confusion. Outside groups need to be informed of the point of contact.

The investigation discovery phase should follow a problem solving process sequence. Investigation management activities include interface and coordination with other groups, evidence preservation, dismantling, evidence analysis, and resolution of conflicts and gaps in initial information. Identification and analysis of root causes then proceed. This methodology is discussed in detail in Chapter 9.

It is common for team members to disagree initially as to causes, remedies, probable sequence of occurrences, scope of investigation activities, and sometimes on process technical concepts. Active deliberation and open exchange of ideas, opinions, and experience are critical to the function of the synergistic team approach. As the investigation develops, there will normally be a series of regular team meetings to:

- resolve questions,
- update members on new information,
- report on subtasks,
- conduct preliminary analysis for causes and possible remedies,
- establish new items and questions for resolution, and
- generate short-term action plans.

The length and rigor of these meetings will increase as the team conducts an analysis of underlying root causes using a structured method and reaches conclusions. Recommendations are then developed to address prevalent root causes. Ultimately, the team generates a written report.

As the investigation progresses from information gathering to analysis of the results, the physical location of the team's activities usually changes. In the early stages, most activity takes place in the field using specialized techniques and resources. Many technical people prefer this type of work since it taps their area of expertise. However, the collaboration process is pivotal to reaching closure on findings and completing an effective investigation. All key personnel assigned to the investigation team must be present when key points are discussed or debated and when the formal analysis of logic is conducted. A dedicated conference room or "war" room is often used to draw people together at predetermined times. The team will shift its focus toward activities in this room as the investigation reaches closure. Discipline is required to ensure that outside activities are appropriately delegated to others while the team meets as a whole. Figure 7-2 illustrates the physical function of the team. The large central rectangle depicts those activities that must engage all team members. By contrast, the outside boxes depict activities that are carried out by individuals in support of the team.

The final phase of the team's activities is preparation and presentation of the results and recommendations, usually in the form of a written formal report. Actual implementation and follow-up on resolution of all recommendations is one of the essential components of the process safety incident investigation management system. In some cases, the incident investigation team (or selected members) may retain some responsibility and authority for the final resolution of the recommendations; however, responsibility typically shifts to management personnel not on the team. If necessary, the team can be reconvened at a future date to audit, evaluate, and report on the actual implementation of the recommendations. This would further capitalize on the insights gained by the team during the investigation.

After each incident investigation, the team should ask the following questions to review important aspects of their performance:

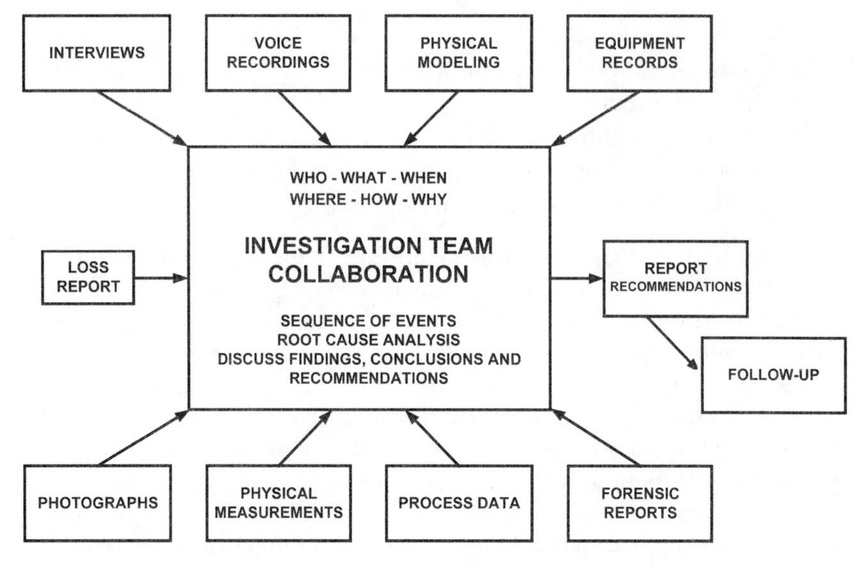

FIGURE 7-2. **Investigation team collaboration.**

- Were we thorough?
- Were our methods effective?
- Was our equipment appropriate?
- Were our technical resources (labs, experitise, sampling) satisfactory?
- Were our supplies adequate?
- Was our preparedness for activation satisfactory?
- What would we do differently next time?
- Was the team size and composition satisfactory?

The team can then recommend or implement improvements to the incident investigation work process.

Team readiness tests should include actual practice drill exercises in activation, communications test (with critique), and periodic check of team supplies and equipment.

7.9. Setting Criteria for Resuming Normal Operations

The decision to restart the facility after an incident is normally up to site or corporate management. However, another important responsibility of the team is to identify recommended criteria and conditions for resuming operations. This activity, of course, must be coordinated with the pre-startup safety review. For major incidents, restart is typically not scheduled until the incident investigation team specifies minimum criteria. In some cases, restart must be coordinated with other groups or some third party such as a regulatory agency. Communications and notifications need to be precise, clear, and verified before resuming operations and restart should not significantly risk recurrence. (See Chapter 2 for additional information on restart issues.)

8

Gathering and Analyzing Evidence

An incident has just been reported and a team has been activated. They must now begin gathering evidence and analyzing the incident's root causes. What are some proven methods that will help them complete this essential phase of the investigation thoroughly? This chapter describes practical guidelines for gathering evidence. The team's field activities will be guided by the questions and issues they identify while analyzing the known data with the techniques described in Chapter 9.

The term *evidence*, as used in this book, is the data on which the investigation team will rely for subsequent analysis, testing, reconstruction, corroboration, and ultimately conclusions. A significant portion of these data is gathered at or around the accident site. Data are also generated during analysis and testing, but such data are not subject to the same time critical activities, preservation, documentation, and other concerns. Thus, this chapter focuses on evidence gathering.

Information derived from data-gathering activities serves as the basis for valid conclusions and recommendations. Without effective data gathering, the incident cannot be defined or analyzed effectively. In some cases, gathering data can consume most of the time and resources spent by the investigation team. Some teams report that it can take up to 70 percent of the investigation effort depending on the nature of the occurrence.

This chapter addresses typical data gathering needs of major investigations. A team may need to augment the activities in this chapter for the unique circumstances of the incident. Performing the activities outlined in this chapter plus special activities provides the incident investigation team with the data needed to complete the next step—systematic determination of the multiple root causes of the incident. However, data gathering and analysis typically involve much iteration as shown in Figure 8-1.

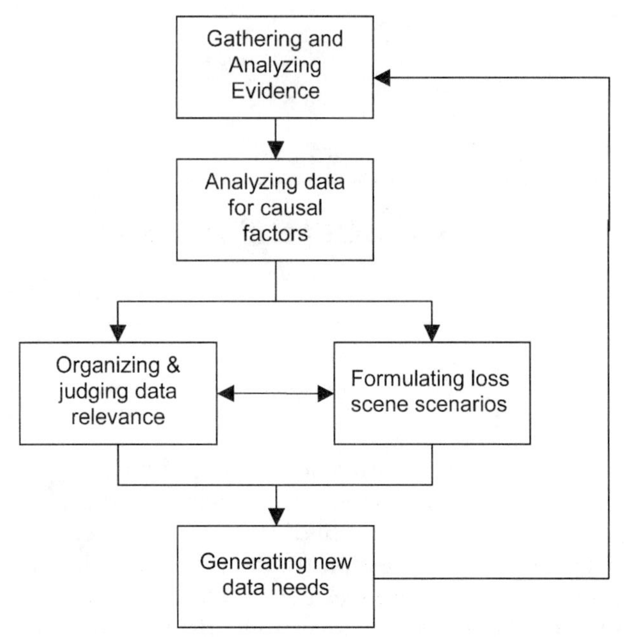

FIGURE 8-1. **Iteration between data analysis and data gathering.**

8.1. Overview

The process shown in Figure 8-1 presents the team's evidence gathering activities in the overall context of a management system designed to achieve certain objectives. In this case, the specific objectives are to gather information for determining root causes, and to support the development and implementation of recommendations. There is no sharp demarcation between the ending of data gathering activities and the onset of root cause determination as implied in Figure 8-1. The U.S. Department of Energy Investigation Manual recognizes this significant overlapping of the data gathering phase and the cause determination phase, as shown in Figure 8-2 .[1] It is common to conduct follow-up interviews and additional special inspections to help confirm, refute, or clarify certain inconsistencies.

8.1.1. Developing a Specific Plan

Each incident investigation is unique and should be accompanied by a customized initial plan. The initial plan, specific to each incident, is continuously revised and updated as new priorities and concerns are identi-

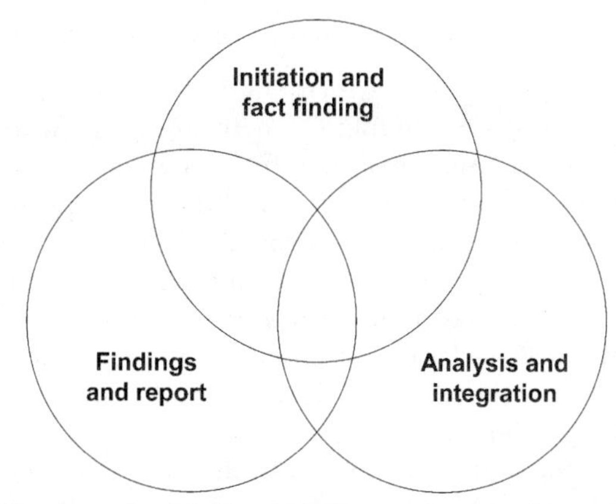

FIGURE 8-2. **Overlap of evidence gathering and cause determination phases.**

fied during the normal course of the investigation. This specific plan is separate from but builds on the general preplanning that has been previously established as part of the incident investigation management system. The team leader can use a general checklist such as the one offered in Chapter 7 to establish initial needs, action plans, and assignments.

The team leader usually develops this initial plan after he or she has made a brief orientation visit. Dynamic risk assessment is required to ensure the safety of investigators. Another factor to consider is the magnitude of potential outside interest in the occurrence. Outside interest includes three aspects: the potential for media coverage, legal issues, and regulatory impact. Although the full investigation team may not have been selected at this early stage, the initial site visit is the first opportunity to establish the physical boundaries of the investigation. The team leader should ensure that access to the area is minimized as much as possible. In addition, he or she should verify that the personnel who enter the incident area are aware of evidence preservation considerations.

One of the most critical issues is clearly establishing which groups have responsibility for which activities and areas. These responsibilities may change during the investigation. The incident investigation team leader must ensure that these responsibilities are clear to all groups to avoid duplication of effort or omission of critical activities.

For most small to medium investigations, the team may only consist of a primary investigator teamed with an assistant investigator. For this type of investigation, all of these field tasks are typically the responsibility of the primary investigator.

8.1.2. Investigation Environment Following a Major Occurrence

Following a major process safety incident, such as an explosion or major fire, the investigation environment can present significant challenges. The starting point for some process safety occurrence investigations can be a crater. Information and evidence useful for determining the occurrence scenario and causes may be destroyed. Although the team may want to rapidly identify and preserve any remaining evidence to prevent the rapid degradation that can occur with time or exposure to the elements, there may be many obstacles to overcome. The site may be under the control of a regulatory agency (US OSHA, US EPA, Fire Marshal, or other Governmental Agency). In many cases, the site is also classified as a HAZWOPER (Hazardous Waste Operations and Emergency Response) site and subject to additional requirements of 29 CFR 1910.120 that severely restrict access and activity. Virtually all serious process incidents will involve litigation. Frequently plaintiff attorneys will initiate legal action that restrict activity at the scene of the incident and affect collection of evidence. Insurance companies often have a significant and legitimate financial interest, and therefore may influence the investigation team's activities. All of these outside groups may limit the investigation team's ability to quickly identify and preserve vital information and evidence. In addition, the infrastructure of the plant may be severely impaired and normal services (utilities, telephones, access roadways, and administrative support services) can be significantly affected. The remaining undamaged portions of the facility may be pressing for permission to resume operations.

Many critical witnesses will be initially unavailable; some in the hospital, some who were on duty at the time of the occurrence may be at home recovering from traumatic and long hours spent during the emergency response. Typically, investigation of occurrences of this magnitude last a number of months. During this period, emotions of personnel may evolve from shock, to disbelief, to sadness, to anger or resentment, especially if the incident involved fatal or permanent injuries. People may also be concerned about their job security. Sometimes there is a legitimate question as to whether or not the plant will be rebuilt. For these reasons, the investigation environment can be challenging, and therefore a systematic approach is necessary for successful investigation of major process incidents.

In the most severe case of a condensed phase material detonation, a substantial portion of a process unit may be completely destroyed with only a crater remaining in the location of the equipment. Where does an investigator start when faced with almost nothing but a crater? This situation presents significant challenges. The investigation team is faced with the challenge of determining what equipment was the source of the explosion and what was collateral damage. Additionally, fragments and debris can be thrown considerable distances, sometimes outside facility boundaries. Loss of plant utilities, chemical spills, and significant blast damage to adjacent process units and buildings may greatly hamper the investigation and may prohibit access to the site for days or longer.

Identifying and capturing time-sensitive evidence is the top priority at the outset of an investigation of this magnitude due to the heightened likelihood of evidence deterioration due to exposure and loss of plant utilities. Electronic process data, chemical samples, witness interviews, fragments outside of facility boundaries, and evidence that may be altered by emergency responders and HAZMAT teams are typically high priority. The loss of electric power to control systems places urgency on the collection of electronic data since battery backups have a limited life span, possibly only 1-3 days. Chemical feed and product samples should be obtained from the area if possible since the material actually in process may have been consumed or ejected during the explosion. Fragments thrown beyond facility boundaries may be picked up by untrained individuals, some of who may not return the fragments to the plant. Offsite damage is also beyond company control, and documentation of the extent of damage may be necessary on an expedient basis before repairs are made.

Evidence that is less time sensitive and within facility boundaries is second priority. Plant personnel can better control such evidence. Nonetheless, evidence may be spread over a large area, and all personnel within the plant must be instructed on the proper manner to communicate the location of evidence for collection by a trained team.

8.1.3. Priorities for Managing an Incident Investigation Team

The incident investigation team has responsibilities for determining the root causes of the occurrence and therefore needs access to the incident scene and other sources of information as quickly as possible. However, there are other priorities especially in the early stages of the investigation.[2] It is extremely important to note that the investigation team's responsibilities are significantly different from those of an emergency response team or search and rescue team. The priorities listed below are for total scene management, not just responsibilities of the investigation team. In fact, the investigation team should not be on site until some of the issues listed below are resolved.

In theory and in practice the data begin to degrade and change as soon as the incident sequence is over. Overall, priorities can be expressed as follows:

Activity	Typical Responsibility
Rescue and give medical treatment to any victims	Emergency response team
Decide if further onsite or offsite evacuation is needed	Emergency response team
Complete head count	As assigned
Address environmental concerns (runoff, verification sampling for contamination from toxic and hazardous materials such as asbestos, PCBs and other possible hazards)	A variety of teams including emergency response, environmental, industrial hygiene, and possibly investigation team
Secure the incident scene to mitigate any further consequences	Emergency response, manufacturing, and incident investigation team
Notify agencies as required	Site function
Preserve physical data and prevent destruction/alteration	Investigation team with assistance
Photograph data and the scene	Investigation team with assistance
Collect data	Investigation team
Have witnesses complete initial witness statements or interview witnesses while incident details are still fresh	Investigation team
Remediate and clean-up the site	Site function
Repair/restart/rebuild	Company function

An overriding responsibility for the incident investigation team leader is to prevent injuries to team members and any other individuals during team data gathering activities. Team members and auxiliary helpers could be exposed to some unusual hazards, such as unstable working and walking surfaces, sharp edges, partially collapsed structures of unverified integrity, unidentified chemicals, residual hazardous materials, blood borne pathogens, and trapped latent energy. Sometimes investigators find stray electricity in a supposedly de-energized circuit even after all known sources are isolated. This is especially notable after an incident where short circuits or fire may have fused conductors or contacts. Double-checking the actual circuit is always worth the additional effort; electrical lockout, locks, and tags are appropriate. It is common for the team to work extended hours in whatever weather conditions prevail. The team leader should watch for signs of fatigue and burnout, as either condition can affect the safety of the team members and the quality of the investigation.

Team members should set a rigorous standard for consistent and proper use of personal protective equipment. They should approach each task with a questioning, skeptical attitude to help prevent additional injuries and minimize unnecessary hazard exposure.

For major investigations, regulatory authorities, such as US OSHA or the US EPA, may deny access to the site until certain precautions are established or until certain tests are completed. Restart may also require agency approval. Often, special measures are required following an incident because of the unusual conditions of the process. One example of such special measures is extensive reliance on fall protection measures as a substitute for normal scaffolding. Insurance or legal representatives may also impose specific restrictions on the team's activities.

When an incident interrupts operation, the incident investigation team will have to deal with the desire to resume production. For small to medium investigations, production may have resumed before the start of the investigation or continued throughout the occurrence if process integrity was not threatened. In these cases, the investigation team may have to rely upon the operations and maintenance personnel to help with initial acquisition and preservation of data. These personnel should be trained in the basics of data gathering and preservation so that the data will be well preserved for the investigators.

For major investigations, production may be interrupted for some period following the incident. Pressures to resume production will begin to increase right from the start of the investigation. Typically, once one or two *causal factors* are identified, facility staff will pressure the team to release the system for production. They perceive that "the cause" of the occurrence has been identified, and therefore the investigation must be nearly completed. The team usually has a great deal of work to perform to identify the remaining causal factors and the root causes of the occurrence. The team leader may need to oppose repairs or resumption of operations until the required data is compiled. In some cases, the process, or portions of the process, will be released back to the manufacturing management for repair and resumption of operations before the collection of data is complete. The decision to release these portions, begin clean up, and start rebuilding should be based on two questions:

1. Is it safe to reenter the area?
2. Have sufficient data been collected to analyze the occurrence?

The decision to restart a process should be based on whether or not enough has been learned about the incident to prevent reoccurrence as discussed in Chapter 10.

8.2. Sources of Evidence

8.2.1. Types of Sources

Potential sources of useful information can extend far beyond the area of the process in which the situation occurred. The search for information may lead to obscure and nontraditional places. The data analysis performed using the techniques discussed in Chapter 9, along with the information in this chapter, should lead the team to identify these data sources.

There are five basic types of data that are useful for the investigation team. The five types of data are:

1. ***People***—Testimony or written statements from witnesses, participants, or victims are examples
2. ***Physical***—Examples are mechanical parts, equipment, stains, chemicals, raw materials, finished products, results of analysis of parts, and chemical samples.
3. ***Electronic***—All electronic format data are included in this category. Examples are operating data recorded by a control system, both current and historical, controller set points, and email. Email may provide a record of what and how people were thinking when decisions relating to the incident were made. This can be an important and powerful source of information.
4. ***Position***—This is the depiction of locations of people and physical data such as valve positions, tank levels, and explosion fragments and debris. Position data is related to both people data and physical data.
5. ***Paper***—Operating logs, policies, procedures, alarm logs, test records, and training records are examples.

The priorities for gathering the data should be guided by how fragile the data are. The more fragile or changeable the data, the more rapidly the team should focus on its collection. Forms of fragility for each source of data are shown in Figure 8-3.

The fragility of the five data types will depend on the specific circumstances of the incident. It is not possible to offer a prescribed priority. In general, historical paper data such as procedures, maintenance records, and drawings are less fragile than people and physical data. The team should identify time-sensitive data as one of its first tasks, prioritize the data, and implement measures to collect or preserve the data.

If the team contains enough members, it can separate the data collection tasks. For example, two team members can perform personnel interviews, two can identify and preserve physical data and its associated position data, and two can gather electronic and paper data. The individual

	Form of Fragility		
Data Source	**Loss**	**Distortion**	**Breakage**
People / Position	Forgotten Overlooked Unrecorded	Remembered wrong Rationalized Misrepresented Misunderstood	Transferred Influenced Personal conflicts
Physical / Position	Taken Misplaced Cleaned up Destroyed	Moved Altered Disfigured Supplemented	Dispersed Taken apart
Paper	Overlooked Misplaced Taken	Altered Disfigured Misinterpreted	Incomplete Scattered
Electronic	Overwritten RAM lost in power outage Destroyed	Data averaged and individual samples overwritten	Incomplete

FIGURE 8-3. **Forms of data fragility.**

investigator's skills and experience should be taken into consideration when assigning tasks.This allows the team to progress more rapidly.

Some examples of time-sensitive data are outlined below.

- Data stored in software files may be very fragile. Process computer system records are sometimes structured such that the level of detail diminishes over time. Therefore, the team may need to assign a high priority to preserving these data. Computers may have a battery backup that will preserve data for a finite time when power is lost.
- Paper data in the form of paper charts on control room and other instruments. These items should be controlled immediately to ensure these items are not lost, damaged, or destroyed by environmental conditions.
- Decomposing materials can change state rapidly and the physical properties are altered over time. Again, the team may have to place a high priority on obtaining samples of these materials.
- Metallurgically significant items can change rapidly, such as oxidation of fracture surfaces.

One approach that can speed up the collection of data as well as ensure a more complete collection of data is to develop a generic list of data to be collected for investigations. This list can then be customized for each investigation. See *The CCPS Investigator's Companion* CD for a sample checklist. For example, the list may include the following:

People data:

- On-shift operators
- Off-shift operators
- Maintenance personnel (company and contract) assigned to the area
- Process engineers
- Operations management
- Maintenance management
- Chemistry and other laboratory personnel
- Warehouse personnel
- Procurement personnel
- First responders/emergency response personnel
- Quality control personnel
- Research scientists
- Personnel involved in initial startup of the system
- Manufacturer's representatives
- Personnel previously involved in operation/maintenance of the system
- Personnel involved in previous incidents associated with the process
- Janitorial, delivery, and other service personnel
- Relevant off-site personnel and visitors
- Original design/installation contractors or engineering group
- Security force (roaming guards or sentries)

Physical data:

- Tanks
- Valves
- Pressure boundary equipment such as gaskets and flanges
- Samples from all relevant vessels and piping
- Raw material samples
- Quality control samples
- Residuals or wastes generated (liquids, solids, gases)
- "New" chemicals present
- Portable and temporary equipment
- Undamaged areas and equipment
- Relief system devices including rupture disks
- Metallurgical samples
- Conductivity measurements

- Missing physical data—items, stains, and other items, that should be present but are missing
- Explosion fragments
- Data recorders
- Sensors
- Process controls
- Electrical switch gear
- Blast damage
- Pieces of process equipment

Electronic data:

- Volatile control instrumentation records (distributive control systems—DCS—data) perform complete system backup of data
- PLC set points
- Security camera tapes

Position data:

- As found position of every valve related to the occurrence
- As found position of controls and switches
- Position of relief devices
- Tank levels
- Pointer needle positions from locally mounted temperature, pressure, and flow devices.
- Location of flame and scorch marks
- Position and sequence of layers of materials and debris
- Direction of glass pieces
- Missile mapping
- Locations of parts removed from the process as part of maintenance
- Locations of personnel involved in the maintenance and operation of the process
- Locations of witnesses
- Location of equipment that should be present that is missing
- Smoke traces
- Location or position of chemicals in the process
- Melting patterns
- Impact marks

Paper data:

- Process data records—strip and wheel charts
- Operating procedures, checklists, and manuals
- Shift logs
- Work permits

- Lockout–tagout procedures and records
- Maintenance and inspection records
- Repair records
- Run histories
- Batch sheets
- Raw material quality control records
- Retained sample documentation
- Quality control (qc) lab logs
- Emergency responder logs
- Process and instrumentation drawings and detailed instrument and electrical drawings
- Equipment drawings and specification sheets
- Design calculations and design basis assumptions and stipulations
- Alarms and set points for trips
- Scenarios for the sizing of relief, venting, and emergency equipment
- Safe operating limits
- Prior proactive analyses of the systems involved (such as a process hazards analysis)
- Material safety data sheets (MSDS)
- Descriptions of normal and abnormal chemical reactions including incompatibilities
- Material balances
- Corrosion data
- Site maps and plot plans
- Electrical classification drawings
- Instrument loop diagrams
- Interlock drawings
- Ladder logic drawings
- Control system software logic
- Management of change records
- Prior incident investigation reports
- Training manuals and records
- Meteorological records
- Dispersion calculations
- Consequence analysis study results
- Phone logs
- Emergency response logs
- Process or product development data and reports
- Process hazard analyses

Starting with a generic list and customizing it for the occurrence under analysis can allow the team to quickly collect a more complete set of data. The more quickly the data are collected, the less likely it may have been compromised.

The investigation team must realize that some of the data collected will not reflect the condition of the equipment immediately after the occurrence. Emergency response activities and post-event stabilization of the system may have altered much of the data. For example, the as-found position of every valve should be recorded. However, some valves will have been operated during emergency response or mitigation activities; thus, their position at the time of the incident may not be able to be determined with complete certainty.

Other information that may be useful does not necessarily relate to the process equipment. For example, there are standard damage assessment references [3,4] for correlating the damage to conventional construction to the corresponding overpressure wave experienced. However, to use these requires collection of data related to the structures that are damaged, not just the source of the pressure wave.

A process safety information (PSI) package is now mandated by most process safety management programs, such as the:

- US EPA Risk Management Program regulations (40 CFR 68)
- US OSHA Process Safety Management regulations (29 CFR 1910.119)
- American Petroleum Institute (API Recommended Practice 750)
- American Chemistry Council (ACC Responsible Care® Code)
- CCPS *Guidelines for Technical Management of Chemical Process Safety* [5]

A high-quality PSI package is uniquely valuable to the investigation team for conducting a successful investigation. Unfortunately, in some cases the PSI package may be partially damaged or even destroyed in the course of the incident itself. One helpful general practice for incident investigation is to maintain a back-up duplicate package in a less vulnerable location. Typical contents of a PSI package will often include many of the items in the list above. Incidents will also occur in systems not covered by these regulations and programs. In these cases, the available information may be more limited. The team will have to do the best they can with the data available.

In addition to the data sources typically available within the facility or organization, other sources of information for the investigation team may include:

- news media videotape footage
- contacts with other manufacturers with similar processes
- university research organizations

- access to proprietary data bases such as those maintained by insurance carriers
- freedom of information document access to government records
- former employees of contract maintenance companies who have personal experience (but not necessarily any vested interest) in the unit of interest
- transcripts of police and other emergency service communications
- Center for Chemical Process Safety incident database (PSID) (See the list of databases in Chapter 11, page 284.)

In most cases, it is best for the team to make and work with reproductions of documents (such as recorder charts and alarm printouts) rather than the actual documents. This avoids damage, alteration, or loss of the original documents.

8.2.2. Information from People

8.2.2.1. Identifying Witnesses

Any person who has or may possibly have information relating to an incident should be considered a potential witness. This concept extends beyond the traditionally identified people who were direct participants or eyewitnesses to the occurrence. Indirect witnesses, such as regularly assigned service personnel who are not operations personnel, often contribute valuable information. Examples of this group include workers from maintenance, contract maintenance, laboratory, janitorial, and shipping. Occasionally regular delivery personnel who routinely visit the process unit are familiar with some aspects of what was normal routine and may have noticed some unusual condition, remark, or actions. These people should not be ignored, because they have the potential to contribute valuable information that can help resolve mysteries that would otherwise remain unresolved.

Emergency response personnel should also be interviewed. During emergency response activities, emergency response personnel may disturb, alter, and destroy data. This is a side effect of the activities that they need to perform. Therefore, emergency response personnel should be interviewed in an attempt to determine the original positions and status of equipment and items. Firefighters may be able to comment on many important observations such as flame patterns, areas of fire, location of victims before rescue, whether fires are pool or jet fires, what equipment was already damaged upon their arrival, and what equipment suffered damage in a secondary fire or explosion.

Indirect witnesses may contribute as much information as eyewitnesses may. Off-shift personnel, previous shift personnel, or whoever last ran the process should be contacted. Recently retired or transferred employees are a potential source of valuable information about the dynamics, equipment, and systems involved. They often have unique knowledge based on many years of experience with the particular systems and equipment involved in the incident. Examples of potential contributions include unique knowledge of:

- Actual operating practices or changes not included in the formal written operating procedures
- Insights on little known failure modes and anomalies in system behavior
- Process control system response to various upset conditions
- Subtle changes in process variables
- Unexpected relationships between certain parameters
- Which instrumentation was considered undependable and which instrumentation was considered consistently reliable
- Unexpected problems and associated changes in the process made during the initial startup of the system
- History of previous problems and actions taken to avoid/rectify problems

If a similar incident occurred in the past, it might be appropriate to reinterview those witnesses involved in the previous incident.

8.2.2.2. Human Characteristics Related to Interviews

Humans are unable to record and playback occurrences in perfect detail. Eyewitness accounts should be considered as incomplete. Most of us have received little formal training in observation techniques. The common optical illusion amusements in Figure 8-4 remind us that our minds will often complete the expected or anticipated picture or image, even if it is not necessarily present. Another example of our usually adequate, but still imperfect, observation performance is the optical illusion. Consider the text in Figure 8-4. Most people will miss the repeated extra word. Common optical illusions as shown in Figure 8-4 remind us of our sometimes-imperfect observation skills. The assumptions made in filling in or removing the extra items in these figures is the same process that witnesses often use to fill in or ignore data that are missing from their recollection of the occurrence. In most cases, the witnesses are not trying to provide false data; they are usually trying to tell their story as best as they can recall it. In some cases, witnesses may be emotionally upset after incidents that were particularly dramatic or involved fatalities. These human

A BIRD IN THE
THE HAND IS
WORTHLESS

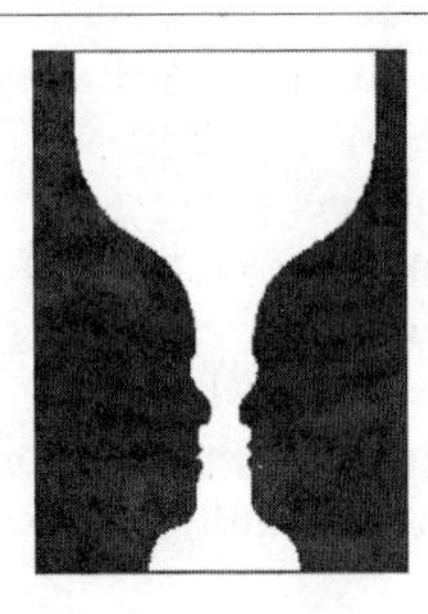

Can you find the typographical error in the sentence? Most of us naturally skip over one.

Look again at the triangle figure. It is an example of an illogical figure. It can be seen and interpreted in different ways.

The figure on the right has become a classic in optical illusions. Is it silhouettes of two faces, or is it a vase?

FIGURE 8-4. **Illustration of human observation limitations.**

performance characteristics are often at the root of apparent inconsistencies and conflicts generated from comparing witness testimony.

Although we have a remarkable capacity to observe, interpret, recall details, and then articulate this information, we are not videotape recorders or computers. No single witness has a complete view or comprehension of the entire occurrence. Each person experiences a unique perspective of the incident. Discrepancies in descriptions of the incident may be due to different perspectives or even different experiences of the individual witnesses. In one way, this concept could be compared to each witness seeing an instantaneous vertical slice or "snapshot" view of a large moving panoramic occurrence. All our incoming information is processed and filtered by our brain as part of the cognitive comprehension process. The information is again processed and "filtered" as we articulate and transmit the information to others.

A classic example of this "filtering" concept is the fable of the four blind men who encounter an elephant as they walk down the road together. Each blind man encounters a different part of the elephant and tries to communicate to his associates what he has found. The first man touches the trunk and believes they have met a boa constrictor. The second man grabs the tail and thinks it is a rope. A third who has encountered a leg begins to argue saying that both of his friends are wrong and that the thing is a tree trunk. The last man, who has hit the side, insists they have hit a wall of some sort. Each blind man was basing his conclusion on the information available combined with his previous experience. The entire picture is not accurately interpreted until the composite information is assimilated. The task of the incident investigation team is to put these four stories together and realize that the men have encountered an elephant and not a rope, snake, tree, or wall.

Another natural human characteristic in interviews is that our memory recall is not always in chronological order. Our replay mechanism does not function in order like a videotape player. This characteristic is one reason why retelling a story several times may help us remember additional details. There is value in repeating portions of the interview. A witness is often stimulated by reviewing his or her own initial testimony. It is often helpful, particularly when reactive chemistry is involved, to have the witness relate his actions, sequence, addition rate, volumes, etc. without referring to the batch or log sheet. This is not an effort to "trip him up" but rather an effort to discover if his field actions did not match what he wrote they were. Sometimes tasks become so routine that they are done without much thought, or an interruption might occur, and it is easy to write one thing and do another.

A witness may have several motives for purposely modifying statements. Witnesses have information that the incident investigation team needs in order to understand the incident and determine the causes. They may choose not to tell the incident investigation team all of the relevant information they have. Sometimes witnesses will purposely modify their testimony or withhold information during interviews. What are some of their motives for doing this? Usually they are the same as those for not reporting near misses. The most significant of these influences is fear of punishment.

Individuals will often modify their testimony or withhold a portion of their testimony when they fear that they or their friends may be punished. They are afraid that if they report the inappropriate or incorrect actions they or others performed, or failed to perform, it will result in punishment for themselves or their friends. They feel that there is no reason to report something during an interview if it can be omitted.

The strategies for dealing with this and other barriers are the same as those for overcoming the barriers to getting near misses reported. One of the prime strategies is to focus on finding the causes, not finding who is to blame. While you cannot promise to eliminate punishment for actions discovered during incident investigation interviews, you can make sure the person being interviewed knows that blame and punishhment are not prime purposes of gathering information. It is common sense that any criminal actions, malicious actions, or sabotage can and should result in punishment. However, if the incident investigation team wants witnesses to tell the whole story, they must make a strong effort to convey the message that they are there to find the causes to prevent recurrence, not to find the guilty for punishment.

It is important to note that violations of safety and other rules are usually not the result of sabotage or a malicious act. In most cases, personnel believe that violation of the safety rules is in the best interest of the individual and the organization. For example, an operator skips a preoperational check of a system because the operator believes the check will not discover any problems and takes valuable time that could be used to produce product. In other words, the operator believes the check is a waste of time, and therefore not a good use of his time or the company's resources. Perhaps the operator had skipped the preoperational check many times, and it had never caused any problems. His supervisor may have known he normally did not perform the preoperational checks and had said nothing because it resulted in increased production. Skipping the checks was not a malicious act or act of sabotage. The operator was trying to do her best for the company. However, the last time the operator skipped the check, the system failed and an environmental release occurred. Will the operator tell the incident investigation team that she skipped the preoperational check? Why should she tell this to the team? What motivation is there? What potential punishments are there? Unless the operator believes that it is in her own best interest to divulge the information, she probably will not. Therefore, we must not make it our mission to punish individuals for the information revealed during incident investigation interviews if we want personnel to tell all of the information they have. The investigation team's responsibility is to gather facts and draw conclusions. Punishment should not be part of the investigation process.

When there is contradictory evidence, such as when interview evidence contradicts paper or electronic evidence, the investigation team will use the various concepts provided in this text to prioritize information. A fact hypothesis matrix is described in Chapter 9.

8.2.3. Physical Evidence and Data

Physical data can provide a source of valuable information for investigators. Investigators should not only consider the process system itself, but also control, safety, support, auxiliary, and adjacent systems as part of the analysis. When examining physical data, typical items of interest include:

- Fractures, distortions, surface defects/marks, and other types of damage
- Items suspected of internal failure or yielding

- Seized parts
- Misaligned or miss-assembled parts
- Control or indicating devices in the wrong position
- Incorrect components
- Raw materials
- Pools of residues of chemicals or materials
- Completed products and chemicals
- Foreign objects
- Part, product, and chemical samples
- Portable equipment (including tools, containers and vehicles)

Not everything in the incident zone will be significant. It is often important to make a distinction regarding equipment, structures, and pipe work that are not damaged. The trick is to quickly sort the wheat from the chaff while causing minimum disturbance to the wheat. This judgment is based on team members' experience and expertise. Key physical items need to be photographed and tagged before any movement, if possible. A guide rule for the decision on what to keep is—*too much is better than too little*. Any known or anticipated dismantling, disassembly, or opening of equipment should be planned and coordinated with the appropriate groups using a test plan as outlined in Section 8.4.5.

Analysis of the data should lead the investigators to the information that they need. However, if the data are not initially preserved, it often cannot be recovered or recreated later. So, keep too much rather than too little.

In addition to the incident investigation team, there will be others with stakes in the data. Additional samples and copies of photographs will often be required.

8.2.4. Paper Evidence and Data

While paper data is typically not very fragile, there is a high priority for securing the data, rapidly collecting it, and preserving it. Often, the most difficult issue with paper data is finding the required documents and finding information within documents. Consequently, it can be very time consuming to obtain information from paper data.

Paper data from instrument charts such as strip chart recorders and disk recorders should be controlled immediately after the occurrence. Strip charts and disk recorders will not all turn at exactly the same rate, so checking the turn rate can be critical in comparing the charts. The measurement range and units for each pen must also be ascertained. For crucial charts, it may be necessary to perform a check of the calibration. If chart recorders are still operating, before removing the charts, mark and document each one with a time, then wait 30 minutes or an hour and mark again. Each item should be marked with the instrument number or name, the date, time of removal, and the last position of data recording. Make sure the charts are hooked back up; key data can be lost on follow-up occurrences if the charts are removed too early or not hooked back up after removal. In some circumstances, it may be advisable to recover and preserve chart recorders as evidence.

Paper data in the form of operator logs, batch sheets, or additions sheets or logs may be particularly important if reactive chemistry is suspected. These may highlight the accidental mixing of incompatible materials, improper sequencing of additions, or improper addition rates or volumes.

The size and scope of the investigation will mandate a special document control procedure, in which each document is given a unique identification number. Records are maintained of which documents have been obtained and from whom. A complete retained document set should be maintained in order to minimize confusion. A special log can be useful in maintaining some degree of control over the flow of paper documents and in finding the answers to questions in the documents when they arise. This is especially true when legal issues or regulatory agencies are involved.

Paperwork may be recovered from locations exposed to an explosion, fire, chemical release, fire fighting materials, and the weather. Documents may have to be dried and/or decontaminated. Some of these documents may be partially destroyed and very fragile. Collecting crucial documents that are deteriorating is a high priority task.

For major investigations, outside agencies can have a voracious appetite for documents. It may be necessary to dedicate one full time person to execute and manage this single area of the investigation, in order to free up the team members for actual investigation activities. In any case, it is a good idea to assign the task of document control to a single person that has the sole authority, responsibility, and control of all documents that enter or leave the incident investigation location. Accurate records of the documents distributed to outside agencies are necessary when legal or regulatory issues are involved.

Like physical data, paper data may require a chain of custody procedure to ensure that the actual documents used during the incident are the documents being examined by the team.

Examples of specific paper data resources that may be useful during an investigation include:

Useful Paper Resources	
Management policy and programs	**Operations and maintenance**
• performance expected	• operating logs
• use of resources	• data recorders
• directives given to supervisors	• maintenance records
• control exerted	• testing records
Engineering files	• inspection records
• facility construction	• operating/maintenance procedures and manuals
• equipment installations	• contractor records/procedures/policy manuals
• modifications	**Personnel development**
– during initial commissioning and development	• laws and regulations
– following commissioning	• professional standards
• areas involved	• job instruction development
Hazard analysis results	• supervisor selection
• identification of hazards	• supervisor training
• evaluation of hazards	• aptitude exams
• risk management decisions	• physical exams
• allocation of priority	• orientation
• completion of actions	• job training
• follow-up	• skill training
• Purchasing	• certification
• standards specified	• counseling
• safety coordination	• enrichment/follow-on/retraining
• supplier conformance	• supervisor appraisal
• exceptions taken	• employment application

8.2.5. Electronic Evidence and Data

It is important to perform a data capture operation for electronic data as soon as possible to maximize the quantity and quality of such information. For example, one investigator had to download the data within the preset file purge time of eight hours. The programmer was on vacation and there was no documentation available on the program. Even when computer specialists are available, this may be a difficult task because computer specialists do not frequently encounter the questions asked by investigators. If there is a of loss of electric power, information may be accessible for a limited period due to battery backup for volatile memory. (Email should also be considered as electronic evidence.)

8.2.6. Position Evidence and Data

Position data is the last of the five data types. Position data is associated with people data and physical data. Position data may help answer the following typical questions.

- What failed first?
- Where did the fire start?
- Where was the pressure the highest?
- How far did an object travel?
- Where was the witness at each point during the incident?
- How far apart are the two items?
- Which gaskets failed and which did not?
- Is the distance between the scratches the same as the distance between the protruding bolts?

Examples of position data were listed previously. Examples include:

- As found position of every valve related to the occurrence
- As found position of controls and switches
- Position of relief devices
- Tank levels
- Pointer needle positions from locally mounted temperature, pressure, and flow devices
- Position of flame and scorch marks
- Position and sequence of layers of materials and debris
- Locations of parts removed from the process as part of maintenance
- Locations of people involved in the operation and maintenance of the system

Position data is one of the most fragile types of data. It can be lost through many activities including:

- Emergency response activities
  - Removal of the injured
  - Stabilization of the system including repositioning of valves, switches, and controls, draining of tanks, fire fighting activities
- Witness movement
- Restoration/stabilization/demolition work
- Degradation from weather
- Investigator actions

Typically, position data are recorded by documenting visual observations, photography, drawings, maps, and measurements. Photography can be used to record such items as:

Site orientations

- Positions of valves, switches, and indications
- Witness views
- Manner in which equipment was used
- Assembly of equipment
- Disassembly stages
- Locations of training aids and procedure helps
- Location and position of fragments and debris

Taking photographs as soon as possible after the occurrence helps to document the "original" condition of the equipment and site right after the incident, before post-event response activities such as site cleanup and demolition activities are performed.

The position of all witnesses (immediately before, at the time of, and immediately after the occurrence) should be documented. Special attention should be given to determining (if possible) which direction the witness was facing at the time he first became aware of the occurrence. The investigators should determine by what means the witness first became aware of the occurrence. The investigators should attempt to determine and/or confirm what each witness could or could not see from the positions they were in throughout the occurrence. The investigators should

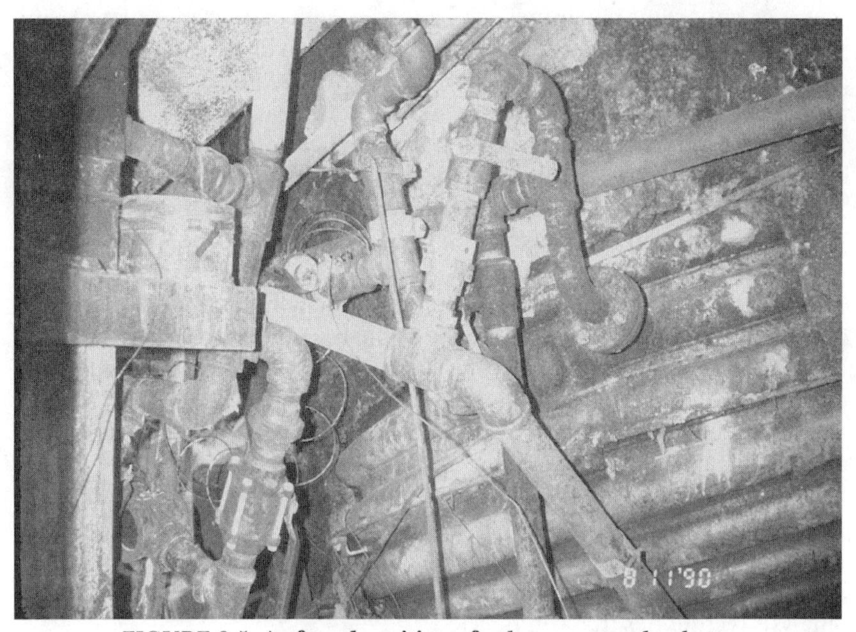

FIGURE 8-5. **As-found position of valves—example photo.**

also determine the position and location of all injured personnel (immediately before, at the time of, and immediately after the occurrence).

The location of marks such as scratches, dents, paint smears, and skid marks that could possibly be associated with the incident should be identified and documented. It is important to determine if such marks were made before the incident, at the time of the occurrence, or afterwards as part of the emergency response or clean up.

Stains or discoloration can be the result of numerous causes including heat exposure, overflow, and release of material from adjacent equipment or some internal occurrence. It is important to determine if the stain or discoloration existed before the incident, was created during the incident, or was created because of emergency response/recovery/clean-up activity.

The team should record the accumulation of soot or airborne fallout debris and the overall deposit pattern. The investigators should also note irregularities, skips, or absence of soot or fallout, especially when there is an anomaly in the pattern. If there are differences in depth, color, pattern, or appearance, these differences should be noted, examined, analyzed, and photographed.

Maps and diagrams can be used to rapidly document the location of items such as people, equipment, materials, and structures. Measurements to reference points can be written right on the drawings. The movement of important personnel can be traced on a map or plot plan. Using color-coding and recording the times the individuals were at each location can help to understand the testimony of witnesses. The position and location of all injured personnel (immediately before, at the time of the occurrence, and immediately after the occurrence) can also be plotted on a site plan. Isopleths (lines of constant value) for important parameters (for example, concentrations, overpressure, heat flux, sound level, light level) can also be plotted on maps and diagrams of an area.

Certain incidents, such as explosions, may require special mapping of fragments and selected debris. By careful documentation, established at the onset of the investigation, it is possible to create an accurate diagram of the position the various pieces of a vessel ended up after an explosion. Using this missile mapping data as a base, knowing the weight of each fragment, and having an indication of the trajectory of fragments, it may be possible to estimate of the energy release of the explosion. The energy release value can sometimes be used to confirm or rule out certain proposed scenarios as to cause or sequence of occurrences. The Peterson (TNO) report on the Mexico City LPG terminal disaster is an excellent example of such a study. [6] Baker provides a comprehensive treatment of analysis techniques that can be used. [7]

8.3. Evidence Gathering

The following sections describe the initial site visit, evidence management, team tools and supplies, tips on photography, and witness interview techniques. Some activities may proceed simultaneously; therefore, the investigation team may have to split up assignments. The team leader should make sure everyone understands their roles.

8.3.1. Initial Site Visit

After the preliminary, specific plan has been established, the next usual occurrence is the initial visit by the investigation team. This is not intended to be primarily a data-gathering activity. It is, instead, an orientation walkthrough to establish perspective, relative distances, dimensions, orientations of equipment, scale or magnitude of damage, anticipated logistic challenges, and to enable planning of specific initial photography/videotaping or sampling activities. In Figure 8-6, the picture may be helpful in determining which way the pressure wave traveled.

The initial visit to the incident location presents unique opportunities for investigation. The clean-up should not be started and the dataare in its

FIGURE 8-6. **Initial site visit—example photo.**

optimum position and condition. While the full team may not have been selected at this early stage, the leader and members of the team who have been selected should make a slow and deliberate circuit from the outside perimeter rather than rushing immediately to the suspected point of origin.

This initial tour will assist in identifying any potential hazards the investigation team may need to address during the later fieldwork. The tour also gives the team the opportunity to note what was not damaged. Most investigators will benefit from intentional pauses during the circuit. During a pause, the brain can catch up with the available input information. Investigators quickly spot the obvious and then desire to move on to the next obvious observation. However, questioning the obvious and looking at all of the equipment is often the key to discovering important data. The longer the investigators stay in one place, the more likely they will become aware of other data. Most investigators need to force a slow pace during the observation circuit, in order to allow the brain to register what the eyes are seeing. Another advantage of this initial perimeter circuit is that it gives an opportunity to see the big picture before focusing on the small significant details.

For fire and explosion occurrences, the team should make a careful and detailed observation visit to the suspected point of origin. One successful technique used by fire and explosion investigators is to face outwards from the suspected point of origin, and then walk forward, away from the point of origin. During this walk, the investigator notes what was exposed to the energy release, taking note of details such as insulation damage on the exposed side. The investigator then turns and walks directly towards the point of origin observing what is not damaged on the side and surface of the items that were shielded from the energy release.

Data gathering is intended not only to provide conclusive proof of what happened, but also may provide conclusive data to reject a speculated hypothetical scenario. If, for example, one potential source of hydrocarbons in a vent header might be the possibility of a leaking rupture disk, and examination of the disks found all to be intact, this data can prove conclusive to reject a particular scenario.

During this orientation tour, the team should use the necessary safety precautions including appropriate personal protective equipment. If it is safe to do so, photography during this stage is normally quite productive; however, serious problems can be created if any physical data is disturbed in the slightest degree. Taking notes and making rough sketches can be useful at this point as long as no physical data is disturbed.

The investigation team should make a conscious effort to determine *what is absent that should be expected to be present* during the operations that were being conducted. This determination requires a relatively thorough understanding of the operation, activities, and physical systems on the part of the investigation team members. In most cases, this determination is not at all obvious. This determination is part of the iterative investigation process. During the initial visit to the incident scene, the team members may not be familiar enough with the operation to be able to identify missing things. Absences in damage patterns can also help to disprove speculated scenarios.

At the time of the initial team visit, the incident scene may still be under the control of the emergency response organization. Any restrictions established by the emergency response organization must be followed. It is common for the team to require an escort for this initial visit. An example of US government rules is contained in 29 CFR 1910.120 US OSHA regulations. Subpart q of these HAZWOPER (Hazardous Waste Operations and Emergency Response) regulations establishes specific requirements that apply to some incident investigation sites. Portions of the investigation area may remain under the administrative control of the emergency management organization for extended periods following the incident. If necessary, the investigation team can ask emergency responders to answer questions about the site, to take photos or to collect data.

Immediately following this initial field tour, the team will begin developing the detailed investigation plan, specifying action items, and assigning responsibilities. Some experts find that it is helpful to repeat the field tour the next morning (before any clean up is permitted). They find that it is amazing how much additional data is observed that was missed on the initial tour. This is a point in the investigation where the need for specialists is identified and evaluated, and when plans are initiated to secure their services. One lesson learned from experienced investigation team leaders is not to assume that the team possesses a particular skill or expertise. Delayed discovery of a missing or incomplete team competency can lead to frustrating delays. Correspondingly, the individual members should be expected to decline an assignment beyond their expertise.

Plans for sharing documents and information among groups are a critical element that must be accomplished very early in the investigation. This plan should have a specific protocol for document control as outlined previously, thus establishing a clear record of which documents came and went from where and to whom. This plan is especially important if regulatory agencies are involved or if litigation is anticipated.

8.3.2. Evidence Management

Chain of custody is an important issue for investigators to address for physical data. This is a concern not only from a legal and regulatory perspective, but is a good practice designed to ensure that testing is performed on the correct items. The incident investigation team should establish a list of parts, samples, and other physical data that are collected during the investigation. Each part should be tagged, numbered and/or spray painted to prevent mishandling or disposal of the items. Access to the data should be controlled to prevent curious personnel from disturbing data.

Early in the investigation and preferably before any field investigation activity is actually started, the incident investigation team must establish a protocol for systematically identifying all types of anticipated data. Some data will be highly mobile (small parts of valves and instruments, personal protective equipment and tools of injured workers). Other items will be perishable (residual liquid and residue inventories for example). Some data may be of interest to multiple groups (regulatory agencies, insurers, fire departments, and representatives of potential plaintiffs). This variety of challenges requires a clearly understood, well-communicated, and publicized method for data identification.

A numbering system that can be applied to a variety of types of physical and documentary data is critical to success. An up-to-date log of assigned numbers is necessary.

Color-coding via tags or spray paint can be helpful to those engaged in moving or removing debris. One method is to have the demolition crew move only material that has been clearly marked. The guiding rule is: *if it is inside the investigation zone and it is not marked, then it is to be left alone.* Long intermittent runs of piping should be marked at regular intervals, especially where the piping passes across the boundary of the investigation zone

Tag attachment can be best-achieved using plastic tie-wrap type devices. It is a good practice to photograph the attached tag in place on important items and to log each of the tags.

8.3.3. Tools and Supplies

The following equipment has been found useful for incident investigation. Not all is needed or appropriate for every investigation, but should be potentially available on short notice.

An inventory of all of the equipment should be maintained and periodically reviewed to ensure it is available when needed.

Personal Equipment

The items below can be packed into a single soft pack container that can be carried with shoulder straps or attached around the waste, thus leaving both hands free.

- notepad, clipboard, pens, pencils

- small plastic bags (sandwich size)
- duct tape
- string
- toothbrush (for cleaning soot/debris off selected evidence)
- Swiss-Army knife, scissors, Phillips and regular screwdrivers
- flashlight (explosion proof)
- pocket extension mirror
- magnifying glass
- 25-foot retractable tape measure
- 6-inch or 1-foot ruler
- permanent marker

Protective Gear

- hardhat, goggles, gloves (rubber and latex), safety shoes or boots meeting site requirements
- extra pair of socks and gloves
- respiratory protection (1/2 mask air purifying devices with a supply of organic vapor/acid gas (OVAG) and general purpose cartridges or other as required)
- slicker suit, acid suit, disposable Tyvek, etc.
- tie-off belt/harness

Team Supplies

- quality 35-mm or digital camera, capable of taking sharply focused photos (both close-up and wide angle), flash, film or data cards, extra sets of batteries
- barrier tape
- tags with plastic ties
- 1-gallon (3.785-liter) plastic bags, self closing
- small first aid kit
- plastic jar (1-quart or 1-liter size) with tightly closing cap
- level
- video camera with extra battery pack and a blank cassette
- pocket dictating recorder with extra batteries and blank cassette
- pair of walkie-talkie radios with extra batteries
- instant print camera and film
- 100 feet (30.48 meters) synthetic 3/8-in. (9-mm) rope (for fall protection tie-off of personnel)
- thermometer
- compass
- 100-foot (30.48-meter) steel measuring tape
- spray paint, paint stick markers, grease pencil (waterproof, indelible marking pens, dark and white)

- small tool kit, nonsparking type tools (channel lock pliers, needle nose pliers, screwdrivers adjustable wrenches, clamps, tie-wire, valve wrenches)
- large supply of duct tape
- plastic drop cloth (100 ft^2/9.2 m^2) for data preservation/protection
- masking tape
- Post-It® notes—various sizes and colors
- data collection forms
- notebook computer for documentation tasks
- electronic media for file backup (disks, CDs)

8.3.4. Photography and Video

Photography can be used to capture a great deal of information about the condition of equipment and the relative positions of items following the incident. Photography should be used throughout the investigation process to capture information. The term *photography* is used in this section in the broadest sense and includes videotape. Since the earliest days of image reproduction, investigators and documenters have applied this powerful tool in continuously more creative ways. Recent technological advances, such as reductions in size and price of videotape devices, and high quality digital still cameras, have created new horizons of capturing, preserving, and even presenting data.

Although photography of the scene as soon as possible after the incident should be a high priority for the team, emergency response activities including treatment of injured personnel, containment of chemical spills, securing unstable equipment, and deenergizing systems should always be the top priority. Some hazard reduction activities could take days or weeks; nonetheless, photography may be possible in selected locations as designated by the incident commander.

Incident investigation involves varying levels of photographic expertise. For most minor incidents, the team or a company employee can adequately meet the photographic needs. Incidents that are more serious require an experienced individual, such as a forensic specialist, who systematically documents the scene, equipment involved, damage, evidence collection, and position data. For specialized photographic needs, the services of a professional commercial photographer or other specialists are necessary and are justified. Examples of such needs include:

- Microscopic analytical views
- Magnaflux
- X-rays

- Infrared
- Complex sequences
- Incidents involving legal, insurance, or regulatory impacts
- Extremely close-up views of machinery or equipment
- Nighttime shots

It is obviously desirable to photograph objects of interest before they are disturbed in any way. This includes moving, turning over, or even lifting to tag or affix an identification number. A thorough and up-to-date log of all photographs is invaluable. Whenever possible, identification of the data should be incorporated as part of the photograph itself. Data preservation concepts can and should be included in the initial and periodic refresher training given to personnel involved in incident investigation. Photographic equipment containing electrical components should not be used in any location that has flammable concentrations of vapors. Plant safety procedures will frequently dictate the atmospheric monitoring requirements and types of equipment that can be used.

The 35-mm and digital cameras are standard tools for investigations. Both are relatively simple to use, inexpensive, reliable, and can perform most tasks needed by the incident investigation team. SLR (single lens reflex) cameras with good close-up capabilities may be needed for specialized documentation, such as fracture surfaces. Simpler 35-mm cameras are now available which are compact, rugged, weatherproof, have built-in flash, and automatic focus and settings. These smaller cameras are more easily carried and suitable for general documentation and many macro photography needs.

Digital cameras have many advantages. Once purchased, there is no need to buy film. The digital recording component can be used repeatedly. The only ongoing cost associated with using the camera is the cost of the batteries. For incident investigation documentation, a camera with a resolution of at least 1 mega pixel is recommended and resolution of 3 to 5 mega pixels is suggested to allow for enlargements without significant loss of clarity. The camera should also have the capability for extreme close-ups and a zoom capability to take pictures of distant objects. Another advantage of digital photography is that the pictures can be easily incorporated into reports and presentations. Although digital photography has many advantages for most investigations, digital photographs may not be admissible in court proceedings.

Professional photographers will often use a $2\frac{1}{4} \times 2\frac{1}{4}$-size format. With the larger size, there is a considerable increase in image size (3266 mm^2 compared to the 864 mm^2 for the standard 35-mm format). The larger image size results in clearer detail and less grain on enlarged prints. These extra benefits are desirable, but usually not necessary for most incident investigation work.

Additional lenses are sometimes useful. The wide-angle lens can show relationships between equipment, and, for close up work (less than 3 feet or 1 meter), a macro lens has great advantages.

Detailed discussions of photographic technology such as depth of field, shutter speeds, filters, and F-stops, film speeds, and types can be found in other publications. Suffice it to state that the higher the ASA film speed rating, the less natural light is needed. Some investigators have found ASA 200 and 400 speeds to be good all around choices. The investigator should have some prior experience with the particular camera used in the investigation. An avoidable mistake is to use the camera for the first time during the actual investigation. Shooting 20 to 30 different types of photos in advance, for example, macro, zoom, and area shots under various conditions such as outdoors, indoor, and poor light is a good investment of time.

The normal practice is to designate a single person on the incident investigation team to coordinate photography. This person works closely with the team member responsible for documentation and record keeping and coordinates with other groups outside the team.

An accurate, complete, and up-to-date log of photographs is a necessity. For most process-safety-related incidents, each photograph should be identified with the following information:

- Time and date taken
- Key item of interest (content)
- Orientation of the photo (for example, "looking east from reactor r-123")
- Identity of the person taking the photograph
- Sketches, drawings, and plot plans to document the perspective of each photograph rapidly if needed to augment or as an alternative to the orientation entry

A camera's automatic date and time feature of a camera are useful, especially when conditions are changing. The investigation team should understand, however, that such camera imprinted date/time markings are generally *not* accepted in most courts unless auxiliary documentation is provided (for example, logbook). When using the date and time feature, the photographer should be aware of possible interference with the composition and background. Sometimes the date stamp obscures or confuses the image of the object of interest.

Another highly useful application is the instant print camera. This is especially helpful during the first several days following a major incident when there is a high demand for initial information by outside parties. These prints should also be logged and identified. This method is relatively expensive when compared to 35 mm, digital photography, or videotape, but it does provide a rapid method for producing hard copies of photographs where they are taken. These prints are especially useful when communicating to demolition workers who are assisting the incident investigation team.

Video cameras represent another powerful tool for visually recording data. They are easy to use, small, and relatively inexpensive. The original tape should be clearly marked and preserved. Copies of tapes should be made for working purposes. One major advantage of videotape is the ability to have narration commentary, thus reducing the clerical load on the investigator. Another unique benefit is the capability to capture motion as a particular investigative action unfolds, such as the opening or disassembly of a particular piece of equipment. A common error in amateur videotaping is inadequate lead-in time before panning the camera. Allow a full 15 seconds at the beginning of each shot. This lead-in time is needed if the tape is later edited for reports or training. Videotape does not capture sufficient detail to substitute for 35-mm or digital photography. In incidents where media coverage is likely, it is a good practice to record newscasts showing footage of the incident.

A special application of photography is to record the viewpoint perspective of a particular witness. This can sometimes enhance the witness testimony, clarify apparent inconsistencies, and verify key items in question.

Medical and legal personnel may have need for photographic documentation of injuries. This area is usually best left to the medical and legal experts.

Before-the-event photographs will usually be extremely rare and difficult to find. One possible source is construction progress documentation shots. Company annual reports and advertising departments can sometimes produce a useful picture although usually not of the exact view desired. Current and retired employees sometimes possess photographs of the area in which they spent years. Sometimes if the need or request is publicized in a productive and positive manner, illicitly taken "before" photographs will show up anonymously in the mail. If you do not now retain photographs of most portions of plants and major hazardous processes, it is a good practice to get started.

Photography guidelines for investigators are listed in the Appendix C. Additional sample photographs and diagrams are included on the CD ROM.

8.3.5. Witness Interviews

8.3.5.1. General Guidelines for Collecting Information from Witnesses

The accuracy and extent of witness information is highly dependent on the performance of the interviewer. The interviewer's ability to establish rapport and recreate an atmosphere of trust affects the quality and quantity of information disclosed.

Promptness is a critical factor, which cannot be overemphasized. Data from people are among the most fragile. For most people, short-term memory for retaining and recollecting details rapidly degrades in a nonlinear fashion. This drop-off is dramatic and has been measured in various controlled studies. The second reason for promptness is rooted in the fact that contact and communication with others can significantly affect our "independent" recollection of occurrences. It is best to prevent any exchange of information among witnesses, if possible. In most cases, complete isolation is not practical so as a minimum the witnesses should be asked to refrain from discussing the incident with anyone until their initial interview.

The interaction among witnesses causes modulation of details and changes emphasis both consciously and subconsciously, especially at the edges of our memory. Recollection is affected by our emotions, by perceived unfairness, by fear of embarrassment, by fear of becoming a scapegoat, and by preexisting motives, such as grudges and attitudes. Many people are so reluctant to be identified as different from their peer group that they may withhold information if they perceive the peer group would desire them to do so.

When documenting initial witness statements, it is a good practice to request the witnesses to separately and simultaneously write down their observations and recollections in an informal narrative statement. The statements should focus on the sequence of occurrences that occurred and their first-hand observations. This may help the witness clarify and focus their thoughts. However, some people do not like to write and may prefer to talk into a tape recorder for transcription later. Still others find tape recorders intimidating or are self-conscious about speaking into a recorder. The investigator needs to remember that most eyewitnesses are not trained nor accustomed to giving such statements clearly. Unclear and incomplete passages must be expected. The interviewer should insert narration into the audio record when appropriate to clarify what is physically happening during the interview. For example, if the witness points to a chart or a diagram, the interviewer should narrate, "Mr. Witness is pointing to Reactor K-13 on Chart XYZ," so that listeners, or those reading a transcript, will be able to follow the action.

Initial Witness Statements are Important

The initial witness statements address three needs. First, the incident investigation team cannot interview all the witnesses promptly. It takes time to work through the list of all the witnesses. The initial witness statement helps to capture the basic thoughts of each witness before too much times passes. Second, they help the incident investigation team prioritize the witness interviews, putting those with the most fragile and valuable information first and those with the least fragile and least valuable information last. Finally, the statements can be used to trigger the witnesses' memories when the interview is actually performed. Because of the high degree of variability in length, amount of detail, clarity, etc. of the statements, they can provide a misleading perception of value of the information a witness holds. The investigation team should not only look at the statements, but all the data collected in setting interview priorities and in conducting the interviews.

In the following paragraphs, we use the pronoun "she" when referring to the interviewer and the pronoun "he" when referring to the witness. This is for convenience of reference only and no gender preference is implied.

Be Aware of the Influence of the Interviewer on the Witness

There is sometimes a tendency for the witness to relay what he thinks the interviewer is expecting (wanting or waiting) to hear. There is also a corresponding possibility for the interviewer to "lead the witness" by sending various response signals. Sometimes the interviewer is not even aware that she is leading or steering the discussion. Therefore, the interviewer must be careful not to ask leading questions. Leading questions contain some hint of the answer in the question. For example, "After you check the pressure you then adjust the inlet valve, right?" It implies to the witness that the correct action is to adjust the inlet valve although the witness may believe otherwise. The witness may answer yes, just to make the interviewer happy.

Interviewers can also influence responses by repeatedly asking about the same issue or topic. For example, if the interviewer always asks "Is that consistent with the procedure?" "What does the procedure say next?" and "Is that in the procedure?" The witness will start to relate all his answers to the procedure because he realizes that this is important to the interviewer.

Questions asked by the interviewer should be carefully worded to be as neutral, unbiased, and nonleading as possible. A common core group of questions should be asked of all witnesses to provide a control sample and to cross-confirm key information.

The interviewer's comments made in response to statements by the witness can also influence what they say. For example, a mechanic admits taking a short cut in the lockout/tagout process. In response the interviewer says, "Wow, no kidding! You did what?" This is certainly going to influence the other information the witness communicates during the remainder of the interview.

Even the nonverbal reactions of the interviewer can influence the witness. If, for example, the witness admits making an error in performing an operation and the interviewer shakes her head back and forth and lets out an exasperatedsigh, the interviewer has communicated her perception, "what a stupid operator," without words. Thus an interviewer must remain constantly aware of the potential influence she can have on the witness.

Limit the Number of Interviewers

Select the interview style to maximize results from witnesses. During the initial interview, a one-on-one or two-on-one interview style is the best. By limiting the number of interviewers to one or two, the stress level for the witness is lowered and it seems less like an inquisition. If two interviewers are present, one should lead the interview by asking questions and interacting with the witness. The other interviewer should play a background role, primarily acting as a note taker. This division of tasks allows the primary interviewer to concentrate on listening and asking questions and speeds up the interview because less time is spent waiting for notes to be completed. This also prevents the witness from feeling ganged-up on, which normally occurs when multiple interviewers start asking questions.

For follow-up interviews and general information gathering (fact-finding type meetings), the ratio of interviewee/interviewer is less critical. A group interview can come across as more open, honest, and less covert. A team atmosphere can be created. Later in the investigation, it may be acceptable to have multiple witnesses present as details and inconsistencies are resolved. The team will have to make this judgment based on the specifics of the occurrence and the workplace atmosphere.

Maintain Confidentiality of the Interviews

In most cases, it is unreasonable to tell witnesses that the information provided during an interview will remain confidential. The team should make reasonable efforts to protect the identity of each witness and the information provided by each witness. For example, in reports, the names of witnesses should not be used. However, the report will have to show the sequence of occurrences and the identity of witnesses may be apparent to personnel at the facility. The notes from each witness should not be shown or released to anyone outside the incident investigation team.

A list of the individuals interviewed can be included in an appendix of the report to show the thoroughness of the analysis and the level of effort expended during the analysis. However, this can also be accomplished by indicating the titles of the personnel interviewed. This provides some anonymity for the individuals who provided data. Again, there is no reason to distribute this list widely, as most individuals should be able to determine the general level of effort by the examination of the remainder of the report.

8.3.5.2. Conducting the Interview

An overview of the interviewing process is shown in Figure 8-7. The interviewing techniques discussed in the following section are generic to any interviewing activity, but have been modified to incorporate specific issues unique to incident investigation.

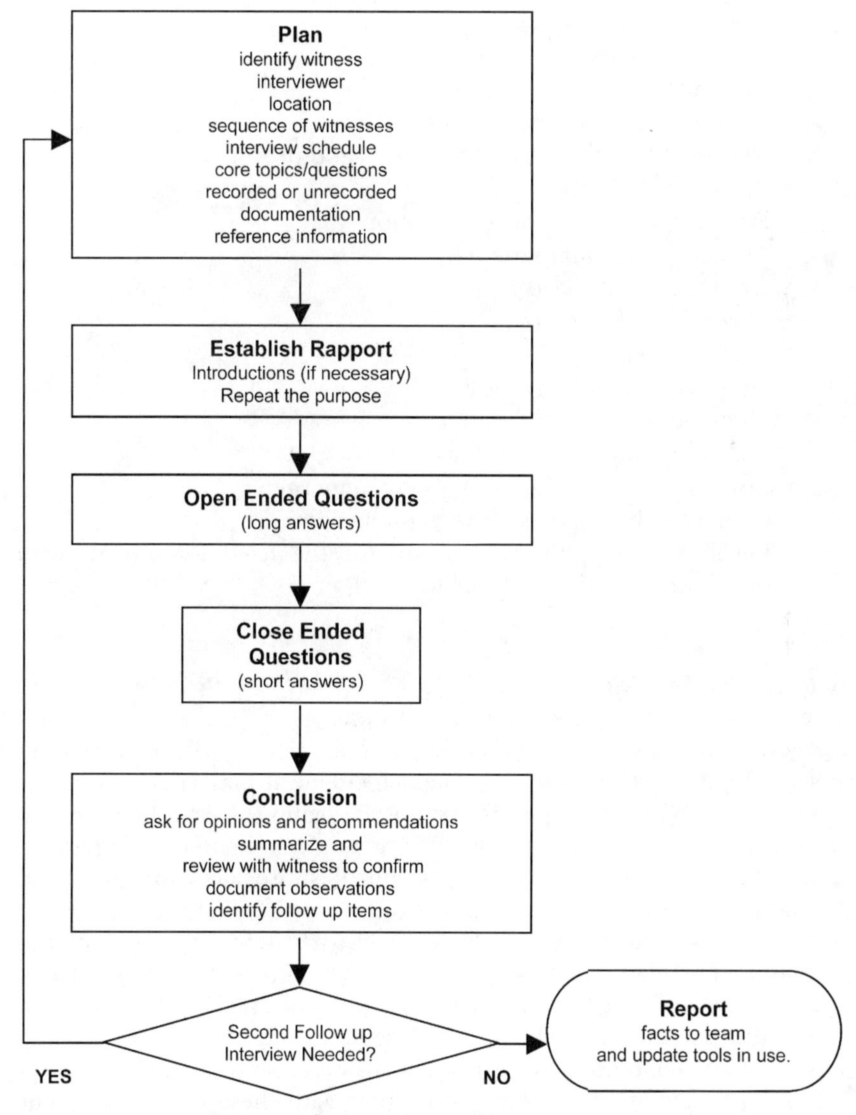

FIGURE 8-7. **Overview of interview process.**

Locating Potential Witnesses

The first step is to locate potential witnesses. The list of potential people and data sources previously listed should be reviewed to identify potential witnesses. In addition, the following sources may also be consulted to help identify witnesses:

- List of people associated with the system
- Operator's and other logs
- Work schedules
- Computer access records
- Employee and visitor sign-in sheets
- Names of personnel on work orders and procedures
- Purchasing records
- Design and drawing documentation
- Training documentation
- Organizational charts
- Lockout/tagout records
- Audit records
- Hospital admission records
- Phone logs or records
- Referrals made by current witnesses
- List of personnel responding to the emergency
- Contact with people outside of the facility
- Responses to public advertising for the need for anyone with related information to come forward

Selecting an Interviewer

The interviewer should be someone with whom the witness will feel comfortable. Sometimes, this is someone at a similar level in the organization, not too high up or too low. In a major incident, it may be necessary for company interviewers to come from another facility, or for a third party to be involved. One important qualification is good interviewing skills. These skills can sometimes overcome hurdles associated with being an "outside" person conducting the interview. Being familiar with the system and the terminology used in the facility is also a benefit. In some cases, the operations and maintenance personnel on the team can be of assistance in the interviews of other operators and maintenance personnel. There is a risk in this, however, since being too close to the incident or person being interviewed often tends to a "leading the witness" syndrome. On the positive side, they may have a built-in rapport with these individuals, thus leading to more information during the interview.

Selecting Interview Locations

Choose a neutral interview location with convenient access for the witness. This will allow the witness to feel more relaxed. In general, any location the witness is familiar with will work.

The incident scene is sometimes a desirable location for the interview. When the interview is conducted at the incident scene, it will seem less like an interview and more like an informal discussion; this will help to relax the witness. The visual cues at the incident scene may help the witness to remember information. The witness will also have something to do during the interview, walking around and pointing out equipment, which will put him more at ease so he will tend to talk more. There is also a great deal of information that the witness will communicate at the incident scene that they may not otherwise think to tell the interviewer, for example, the distance between a valve and indicator. However, there are some reasons not to conduct the interview at the accident scene. They include the potential distractions at the scene such as other personnel, repair activities, demolition activities, the presence of other potential witnesses, unsafe conditions. In addition, poor weather can also be a problem. Finally, the accident scene may be too emotionally painful for the witness, especially if a friend of the witness was injured or killed.

Other potential interview locations include neutral locations such as meeting rooms, classrooms, and the witness's office. Try not to conduct the interview at a location unfamiliar or uncomfortable to the witness. For example, conducting the interview in the office of a high-level manager may unnecessarily increase the tension and stress on a worker.

Minor arrangements can help in achieving a more thorough, accurate, and useful interview. These include such items as arranging transportation home for the witness if necessary, and providing overtime meals or refreshments.

Arranging the Interview Room

If the interview is not at the accident scene, arrange the room to be welcoming to the witness. Have the witness sit on the same side of the table or desk as the interviewer. Having the interviewer sitting across the table or desk from the witness establishes a more confrontational atmosphere. If using a scribe to take notes, have the note taker sit next to the interviewer, so that the witness can see both individuals at the same time, making the witness feel more at ease. Have reference information readily available (for example, flow diagrams, plot plans, procedures, and work orders). This will give the witness something to point to and something to do during the interview, making him more relaxed and willing to talk. Be careful not to inundate the witness with documents, but have them available so that they can be referred to as they come up in conversation or are requested by the witness.

Eliminate other distractions from the room if possible. Do not allow the witness to see any documents, such as causal factor charts, fault trees, showing the incident investigation team analysis of the occurrence. This may be appropriate for later interviews when only specific information is needed or a specific time gap is being filled in.

Hand-drawn sketches (regardless of the artistic quality) are a valuable tool in the interview process and should be encouraged by the interviewer. It is a good practice to have paper, flip charts, and pencils in the interview room for use by the witness.

Scheduling Interviews

Establish the witness interview order. Your first witnesses should be those with the most fragile information, those with the most detailed information, and those most likely to want to provide information. It could take considerable time to gain information from uncooperative witnesses. This is time not spent interviewing other witnesses. If you encounter significant resistance from a witness, interview the other witnesses and come back to the uncooperative witness later.

Schedule interviews at a convenient time for each witness. Make appointments with witnesses through the appropriate channels, such as with union and contract personnel. Select a schedule that minimizes contact between witnesses to minimize the sharing of stories between witnesses. For example, schedule each initial interview for 30 minutes. Allow 30 minutes between interviews to complete the documentation of the last interview and prepare for the next one. This gap minimizes contact between witnesses. Adjust the schedule and interview list based on data as they appear.

Although it is undesirable, witnesses will talk to each other about the occurrence and about the interviews they have had resulting in contamination or blending of information. The incident investigation team should avoid contributing to this problem. Do not have witnesses waiting in a common area for their interviews. This simply allows them to spend the time together talking about the occurrence.

Do not exceed the witness's interview time without the witness's consent. If more time is needed, consider scheduling a follow-up interview if continuance is inconvenient for the person being interviewed.

Telephone interviews may be appropriate as an initial interview to determine if a face-to-face interview is required. For example, a telephone interview may precede an interviewing trip. It may also be appropriate to conduct an interview by telephone if the witness:

- is not promptly available,
- will primarily provide factual information related to the chain of events,
- has little information related to contributing and root causes, or
- is not key to the occurrence.

Develop a List of Core Topics and Issues

Develop a list of specific topics to cover and issues to resolve during the interview. This is not a list of questions to ask, just topics to cover or issues to resolve. It is hoped these topics and issues will be addressed and resolved by the open-ended questions asked at the beginning of the interview. The list of specific topics and issues can be developed from the questions and data needs identified using the analysis techniques from Chapter 9. Typical questions for an interview are listed below.

Example Questions for Witnesses

- In your own words, please, tell me everything that you saw and everything that you did.
- What were the initial conditions?
- What were you doing just before the occurrence?
- What were you doing during the occurrence?
- What was the timing of occurrences?
- What indications did you have of the occurrence?
- How did you know what to do when you saw_______?
- What communications did you have with others in the area?
- What other individuals were in the area?
- Where were they?
- What were they doing?
- What were the environmental conditions?
- What was different this time?
- Did you notice any equipment that did not operate properly?
- Any training or preparation issues?
- What are your opinions, beliefs, and conclusions related to the causes of the occurrence and the recommendations that should be implemented? (*Note:* The investigation team should remember and note that this information is opinion, not fact.)

Example Questions for Emergency Responders

- What were the initial conditions when you arrived?
- Did you or others move or reposition anything?
- What emergency response activities did you perform?
- Have there been similar occurrences in the past?
- Whom else should we talk to?
- Who else might have information?
- What are your opinions, beliefs, and conclusions related to the causes of the occurrence and the recommendations that should be implemented? (*Note:* The investigation team should remember and note that this information is opinion, not fact.)

Documenting the Interview

During the interview, interviewers should document the interview as unobtrusively as possible. The primary interviewer (or secondary interviewer if present) should take notes during the interview. Other options include a video/audio recorder or the use of a dedicated court reporter or stenographer. However, these last two options are less desirable, as they tend to make the witness uncomfortable. This also makes the process seem more like an interrogation and a legal proceeding. Often the presence of a microphone creates an extra degree of stress for the witness. The extra gain in accuracy of recording the information is usually offset by the decrease in participation by the witness.

Documentation of the interview should not be a covert, hidden process. The witness should not believe that hidden, secret notes are being taken during the interview. One way to address this issue is to review the interview notes with the witness near the end of the interview, although this may not be practical in a long interview. Documentation of the interview should at least include the witness's name, date, time, statement, and recorder's name.

During telephone interviews, take as many notes as possible. The witness will not be able to see what you are doing, so if you need a moment or two to complete note-taking efforts, tell the witness. If practical, read the notes back to the witness at the end to confirm your understanding of the witness's comments.

Establish and Maintain Rapport

The interview can be a source of considerable stress, even if the witness is sincere, cooperative, and was in no way responsible for the incident. Each witness brings his own unique collection of emotions (fears, anxieties), motives, attitudes, and expectations into the interview. On some occasions, these emotions can include reactions to the death or serious injury of a friend or co-worker. To help reduce the stress level, start by introducing everyone present. Having a "mystery" person taking notes or sitting in the background listening may add considerable stress to the situation. Next, explain the investigation process to the witness and describe his role in the effort. Explain the purpose and objectives of the interview. Characterize the witness's important contribution to the investigation.

Warm up with nonbusiness issues and routine matters such as the witness's name, position, and years at the company. This allows the witness to answer some easy, simple questions and overcome initial jitters before getting into the body of the interview.

Before moving on to the body of the interview, ask the witness if he has any questions. Typical questions include confidentiality of the information provided during the interview, how long the interview will take, and what the investigators know about the occurrence. Answer all questions about the interviewing and investigation process as completely and honestly as possible because deceiving the witness will generally cause problems with later interviews and investigations. However, responses to questions about the status of the investigation should be answered by stressing the work in progress, not the information currently known. Providing the witness with information from other sources tends to contaminate and influence the witness. Finally, ensure that the witness understands the recording process being used whether it is simply an investigator's notes or a court reporter.

This first phase may appear on the surface to be shallow chitchat, yet it can determine the entire outcome of the interview. It provides an opportunity for the interviewer to explain the purpose, format, expectations, confidentiality, and to deal with any special concerns of the witness. Yet, the most beneficial aspect is the opportunity to establish a constructive atmosphere in which communication can begin.

Throughout the interview, the investigator *should:*

- be friendly and respectful
- listen attentively and reflectively
- show compassion
- avoid attitudes that destroy rapport
- remain as neutral as possible

During an interview the investigator *should not:*

- act surprised when the witness tells you new information
- act happy or pleased when the witness confirms other witnesses' testimony or your current theory of the causes of the occurrence
- be overbearing, commanding, proud, overly confident, overeager, timid, or prejudiced
- judge the information that is being presented by the witness, even if you know it to be incorrect
- rush the witness, even if little new information is appearing
- make promises to the witness

Remember that the point of the interview is to obtain as much information from the witness as possible, not to show the witness how smart the interviewer is.

Promote an Uninterrupted Narrative

Using open-ended questions (questions that require more than one word yes or no answers) ask the witness for an initial statement. Examples of open-ended questions include:

- Tell me what you saw or did when you first discovered the problem.
- What were you doing just before the occurrence?
- What was different this time?

It is important during this portion of the interview to remain quiet. Allow the witness to talk. As long as the interviewer is talking, the witness will be quiet. The interviewer should avoid the urge to interrupt with follow-up questions after asking an open-ended question. Closed-ended questions, those that only require short answers, should be avoided during the initial portion of the interview. Too many closed-ended questions at the beginning of the interview will tend to condition the witness into giving short answers.

Resolve to remain unbiased and to avoid any actions or questions that may lead the witness. Avoid leading and accusatory questions throughout the interview. The interviewer must not project the direction she thinks or hopes the interview should go.

If the specific issues the investigator needs to resolve are not contained in the answers to the initial open-ended questions, the investigator should pursue these areas of interest with more detailed questions about the following:

- Timing of occurrences
- Location of personnel
- Environmental conditions
- Positions of personnel and victims
- Anything moved/repositioned
- Emergency response activities
- Indicators of conditions
- Actions of other people
- Training and preparation
- Histories of similar incidents
- Information gaps
- Inconsistencies in data
- Management and staff involvement
- Possible causal areas
- Beliefs, opinions, and judgments that led to unadvisable actions

As the interview moves more toward closed-ended questions, the interviewer should periodically restate what she thinks the witness has said. This gives the witness a chance to correct any errors or misinterpretations or add further details.

This interactive dialog portion of the interview is most like the common image of an interview, as popularized by TV personalities. Specific, objectively worded questions are asked in this stage. During this portion of the interview, there is significant potential for the interviewer to influence the witness. This risk is constantly present and demands continuous recognition by the interviewers.

Interviewers will always experience apparent inconsistencies in incoming information. Invariably, conflicts will be generated during the initial information gathering interviews. All incoming information will not be fact (that is, objectively verifiable), but may be perceived as fact by the person supplying the information. In most cases, it pays to delay judgment on the apparent inconsistencies. Just as the interviewer must strive not to come to conclusions on accident causes until the facts are fully developed, she must delay judgment on these apparent inconsistencies. Often a scenario will emerge which reveals both the apparently conflicting items to be true, but at different times during the incident sequence. Even if information is found to be wrong, the reason for the misinterpretation can frequently reveal other important information. At other times, the source of the apparent inconsistencies can be traced back to the interviewer herself, who inadvertently modified the incoming information from the witness based on what the interviewer knew to-date about the incident.

Different people may have different definitions in mind for the same word. Thus, it can be advantageous to ask questions to clarify the ideas expressed by the witness.(8)

When a noun is used, the interviewer may need to clarify by asking, "*What,* exactly?" For example, a motor valve may be electric or air operated; the difference may be important to the investigation.

When a verb is used, the interviewer may ask, "*How,* exactly?" For example, *shutting down the reactor* may mean, *gradually reducing the feeds in normal shutdown mode* or it may mean *hitting the emergency stop button.*

Sometimes, rules or values may be mentioned, such as, *the outside operator should always close the drain valve*. It may be helpful for the interviewer to ask, "What would happen if he didn't?"

A witness may generalize, by using words such as:

- All
- Always
- Everybody
- Never
- They

The interviewer can clarify these generalizations by asking, "All?" "Always?" "Everybody?" "Never?" "Who are they?"

Sometimes a witness may use a comparator without an antecedent; for example, "Pump A is better." The interviewer can gain clarity by asking, "Better than what?"

The best practice is to let the witness lead the exchange, but it is important for the interviewer to explore apparent paths of new information. More than one witness has said after the interview that they knew a certain fact, but since the interviewer did not ask about it, the witness did not mention it. The witness made a judgment that the information was not important or was not relevant.

Conclude the Interview

Ask the witness for his opinions and recommendations. Most witnesses want to tell the investigator their ideas about what caused the occurrence and how to fix the problems. However, this should only be done at the end of the interview to minimize influencing the information provided by the witness. Asking for opinions earlier in the interview adds another filter to the data presented. Ask who else may be able to contribute valuable information and invite additional input if new information is remembered or discovered. Finally, the witness should be asked in as nonthreatening a manner as possible, "Is there anything else you want to add regardless of how unimportant you think it might be?" This question is then followed by an extended pause.

The interviewer should express appreciation for the witness's time, information, and cooperation and gain consent to contact the witness later if necessary for a follow-up interview even if she is confident she will not need it. If the interviwer asks permission for follow-up interviews only with some witnesses, those witnesses may feel they are being singled out. Finally, the investigator should review her notes with the witness. During this review, numerous clarifications and additional details are usually provided.

It is common for a witness to recall additional information after the interview is over. Astute investigators anticipate this human trait and provide a clearly understood and easily accomplished mechanism for the witness to contact the interviewer later. Always close an interview by inviting the witness to return or contact the investigator if he remembers something else, or would like to otherwise modify or add to the interview results. Provide the investigator's contact information to the witness.

Conducting Follow-up Activities

Once the interview is complete, the investigator must perform a few additional tasks. Immediately after the witness leaves the room, the interviewer should take the time to do several things:

- Evaluate the interview
- Organize the information received

- Identify any key points that confirm or conflict with previous information
- Record her findings.

Findings would include such items as observations, specific insights, and items to be followed-up on in later interviews or investigation activity.

In addition, the team should also update the in-progress analysis of the occurrence. For example, update the fact list, logic tree, or fact hypothesis matrix. Finally, the information from the interview must be communicated to the remainder of the investigation team.

Conducting Follow-up Interviews
Conduct follow-up interviews in the same general manner as other interviews, but use a more structured, straight-to-the-point interview style. Initially, the interviewer may use open-ended questions, but follow-up, closed-ended questions are usually asked sooner than they would be asked during the initial interview. Ensure that witnesses do not believe that the follow-up interview indicates the interviewer doubts their credibility. Focus on the gaps in information and apparent inconsistencies.

8.4. Evidence Analysis

Evidence analysis is a distinctly separate function from evidence gathering. The evidence analysis phase is an iterative activity and overlaps with evidence gathering. Evidence analysis is conducted over a typically longer time frame and can last for several months as additional tests and data generation is done. Evidence analysis activities often identify the need for additional specific information, and the evidence gathering cycle begins again. Evidence analysis is conducted under different circumstances from the initial evidence collection. Evidence analysis must follow a systematic and thorough approach.

8.4.1. Basic Steps in Failure Analysis

The basic steps in failure analysis include the seven steps shown in Figure 8-8.

Step 1—Evaluate conditions at the site.
This can be the most important step in the analysis. Understanding the conditions at the site and how the parts or items were used can eliminate some potential failure mechanisms and support other failure mechanisms. Typical questions to ask include:

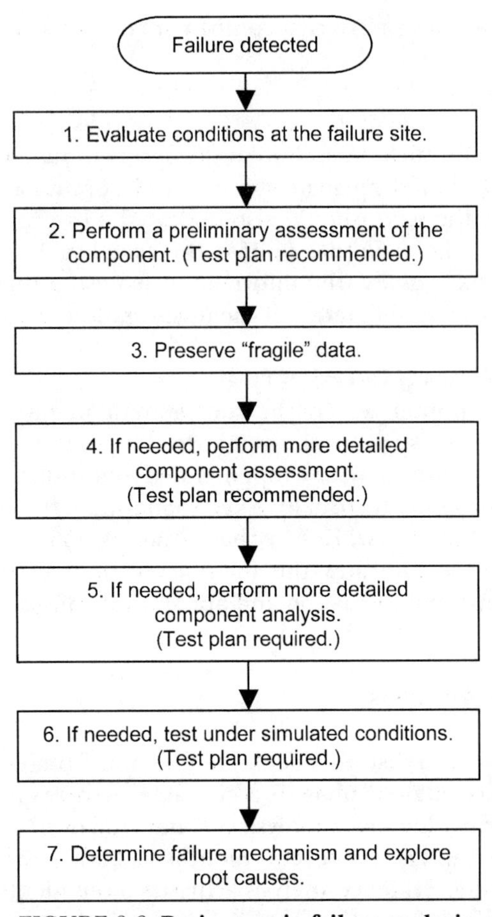

FIGURE 8-8. **Basic steps in failure analysis.**

- How long had the part been in service?
- What were the environmental conditions?
- Did the failure occur during startup, shutdown, abnormal or normal operations?
- Was it a rotating piece of equipment?
- Did it rub against something?
- Was there any fluid or gas flow past the device?
- What are the chemicals to which the part is exposed?
- What is the part made of?
- What activities were taking place in the area?
- Are generally recognized reactive materials present at the site? If so, what are they?

- Is there a potential for reactive interactions (caused by inadvertent mixing of incompatible materials) at the site? If so, what are the materials?
- Does any portion of the process utilize "controlled" reactive chemistry?
- Is any equipment used for more than one service, requiring cleaning before reuse?

The answers to these questions should allow the investigators to focus their subsequent data collection efforts.

Step 2—Perform a Preliminary Assessment of the Component

During this step, a preliminary analysis of the parts is performed. Typically, the focus is a visual examination of the items. The investigators should avoid disturbing data until necessary, conducting their visual examination without alterations. Take pictures of the items and mark positions in the field if immediate removal is necessary. Remove the parts in a controlled, careful, and methodical manner. Evaluate the importance of coatings/residues/deposits/impurities. Samples of the chemicals, soil, deposits, and coatings may be taken at this point.

Macro visual examination is done with the unaided eye or low power (up to 20X) magnification. An item may need to be cleaned prior to examination. Fracture origins and location of samples to be removed for further analysis are determined by macro visual examination. Nondestructive evaluation methods may be necessary for more accurate determinations (see Step 4).

Step 3—Preserve Fragile Data Sources

Provide a safe, secure, and controlled storage location for the physical data. Consider special storage features that might be needed such as temperature control, humidity control, wrapping, and others. Prepare the parts for further evaluation and avoid actions that may destroy or degrade data.

There are two distinct phases of data preservation: the immediate and the long term. Some data will need to be retained for an indefinite period and may require special storage measures. A documented "chain of custody" has become more significant due to legal considerations. Provision must be made for long-term storage of large items in order to minimize effects of weather (especially ultraviolet light and rust). Items can easily become lost unless special measures are taken using a dedicated, fenced, locked, posted, or routinely inspected storage area. Smaller size items can present even greater challenges, because most facility warehouses do not normally have a special section that can be controlled and dedicated exclusively to this purpose. In process incidents, previous batch (raw material and finished goods), quality control results, and retained samples should be secured and protected from degradation. If possible, post-incident samples should be taken and similarly protected.

Short-term preservation (several weeks) can usually be successfully accomplished by generous application of polyethylene plastic wrap and duct-tape. Be careful in applying the tape to avoid direct contact with the items. The adhesive can be left behind on the part and when removed, can take with it important dust, fibers, or other data. Even short-term preservation requires some control of access, chain of custody documentation and conspicuously posted signs. Many workers are genuinely curious to learn and see for themselves what the damaged or essential part actually looks like, and may unintentionally cause changes.

Step 4—Perform a More Detailed Assessment of the Component (as Necessary)

Perform a more detailed analysis of the items. This stage may include field-testing, field disassembly, and shop disassembly. Additional pictures of the component should be taken, especially during testing and disassembly activities. All of these activities should be performed in a careful and controlled manner using a test plan as discussed below.

In some major investigations, reconstruction of the damaged equipment may be required to understand the physical relationship between the various items that are recovered. A dedicated area or warehouse space may be required to effectively reconstruct and analyze the physical data.

Step 5—Perform A More Detailed Analysis of the Component (as Necessary)

Some items and components may require examination that is more detailed. Specific techniques for data analysis are beyond the scope of this guidebook. Entire volumes have been written on specific issues such as the fracture patterns of alloys and the corresponding clues for determining the actual cause and mechanism for the failure. Known specific materials and alloys perform and fail in consistent and predictable ways. This area of expertise is normally supplied to the incident investigation team via the use of specialists, from either within the parent organization or from outside experts or labs engaged specifically for the task.

Detailed fracture and metal failure analysis is usually a very reliable and extensive aspect of investigations of major loss incidents. For most small to medium investigations, macroscopic evaluation is typically sufficient. Macro evidence, such as indications of shear or brittle failure on fracture faces, lines showing detonation direction, and the chevron (herringbone) pattern all provide valuable clues to sequence, type, and cause of the failure.(9) (See Figure 8-9.)

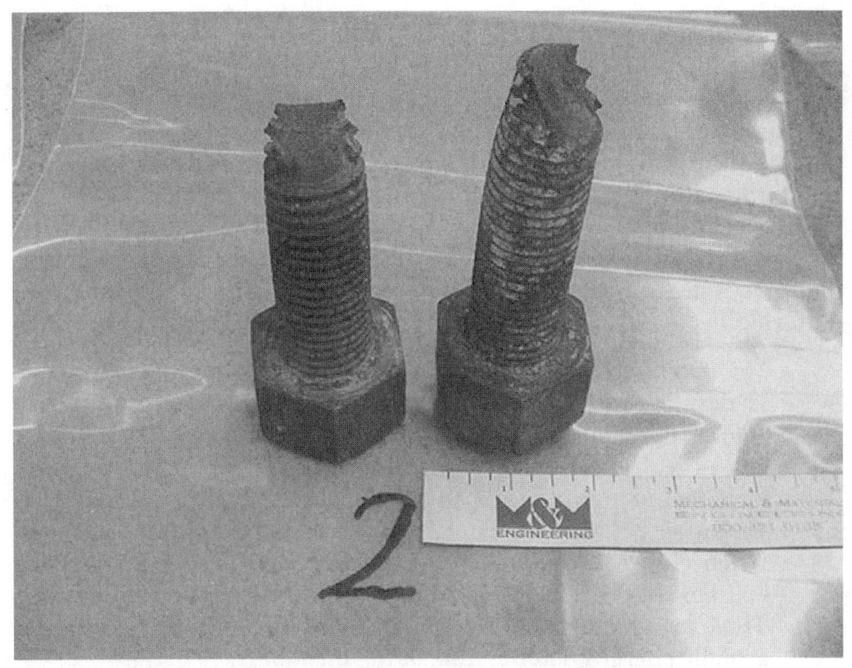

FIGURE 8-9. **An example photograph of fractured bolts.**

Investigators can often distinguish between the ages of various segments of a crack in the metal. This age difference can often allow credible determination of whether the crack was present before the catastrophic incident. Metal fatigue and related failures are well-established phenomena. Three recommended references on metallurgy aspects of investigations are: *Assessment of Fire and Explosion Damage to Chemical Plant Equipment/Analyzing Explosions and Pressure Vessel Ruptures*, by D. McIntyre,[10] *Defects and Failures in Pressure Vessels and Piping*, by H. Thielsch,[11] and *Understanding How Components Fail* by D. Wulpi.[12]

Examination of the fragmentation and tear patterns may be instructive to evaluate an internal explosion, such as providing evidence of the general magnitude of pressure at the time of failure and distinguish between deflagration and detonation occurrences. Macroscopic examination of the deformed shape and fracture pattern is the first step and may be sufficient for many investigations. Detailed metallurgical examination may be needed to interpret tear patterns, identify failure modes, and provide insight into other effects such as localized strain and metal temperature.

In piping systems, the damage is most pronounced at elbows, tees, or other constrictions such as valves. Sometimes an occurrence will initiate as a deflagration and then accelerate into a detonation. There is usually a distinctly different damage pattern at the location where the transition from deflagration to detonation occurs. This change may be determined by examination of the metal fragments. Blast pressures involved in detonations of condensed phase material tend to shatter metal pipe into many small fragments. Elongation, deformation pattern, and failure modes can be useful clues when analyzing failure mechanisms.

Professional arson investigators have developed highly effective methods of deducing facts from a systematic study of burn, char, and melt patterns. Typical examples include:

- Most woods will burn at a steady rate of 1.5 in./hr (3.6 cm/hr).
- Hydraulic fluids usually exhibit a consistent response of smoke color, flame color autoignition temperature, and a whitish residue.
- Glass breakage patterns can be used to estimate the overpressure wave that in turn can be used to estimate the energy released in an explosion.
- When electrical conductors break while *not* carrying current, the break is different from the pattern shown when current is flowing. Thus, investigators can often determine if a particular device was actually energized at the time of the incident.

Not all evidence is simple to diagnose. Steel weakens at approximately 1100°F (575°C).[13] Steel exposed to 1500°F (816°C) for a short period can begin to fail and show the same degree of damage as steel exposed to a lower temperature for a longer period of time. Thus, a sag pattern can be a relatively reliable indicator that the steel was exposed to a temperature of *at least* 1100°F (575°C), but, the maximum temperature above 1100°F (575°C) cannot be accurately determined without additional evidence.

General Examination Methods

Microscopic visual examination is done on mounted, polished, and etched samples prepared by metallographic (metals), petrographic (ceramics, glasses, and minerals), and resinographic (plastics and resins) techniques. The microstructure of the material and the nature of the damage can be determined by microscopic visual examination. Reflected light microscopy can be used with opaque materials. Transmitted light microscopy, often with polarized light, is used to examine transparent or translucent materials at magnifications up to 2000×. The scanning electron microscope (SEM) and transmitting electron microscope (TEM) are used for examination of specimens up to 150,000× magnification.

Fractography is the examination of fracture surfaces with no sample preparation other than cleaning to determine the fracture mechanism.

The SEM at magnifications from 5× to 15,000× is the primary instrument for fractography due to the large depth of focus. Sometimes fractographic examination with the unaided eye or low power optical microscope (up to 100×) is conclusive. The TEM is used to examine replicas from fracture surfaces when magnification greater than 10,000× is necessary to characterize the fracture.

Dimensional Measurements

The extent of corrosion or wear can be determined by measuring the remaining thickness and comparing it to the original thickness. The extent of distortion in deformed components and elongation of fractured components should be determined. Common machine shop measuring tools provide adequate accuracy.

Nondestructive Evaluation (NDE)

Various NDE techniques are used to locate defects and flaws in the failed or similar equipment that may not be apparent during the macro visual inspection. An analysis of cracks and other damage during the initiation or progressive phases often provides more information regarding the failure mechanism(s) than the same analysis would at locations where complete failure occurred. Considerable secondary damage to a worn, fractured, or corroded surface may occur after failure. The most common methods of NDE are given below.

- *Visual Examination:* With the unaided eye or assisted by borescopes, fiberscopes, television cameras, and magnifying systems.
- *Leak Testing:* Location of through wall defects and flaws while under pressure or vacuum. Various fluids and gases are used for pressure testing; several types of leak detectors are used to locate a leak.
- *Liquid Penetrant Inspection:* Useful to find surface discontinuities. Can be used with any material that has a reasonably smooth nonporous surface.
- *Magnetic Particle Inspection:* Useful to find surface discontinuities and some discontinuities very near the surface in ferromagnetic materials. More sensitive than liquid penetrant inspection.
- *Eddy Current Inspection:* Eddy current inspection is the most widely used of several methods of discontinuity detection for electrically conductive ferromagnetic and nonferromagnetic materials. Both surface and subsurface discontinuities can be detected.
- *Ultrasonic Inspection:* Ultrasonic inspection is used to detect surface and subsurface discontinuities in metals and sometimes other materials. Skilled inspectors are needed for ultrasonic testing.

- *Radiography:* Primarily used to detect subsurface discontinuities in metals by use of gamma or X-rays. Gamma or X-ray radiography can detect discontinuities in nonmetallic, inorganic materials and the location of many metal components in an assembly. Neutron radiography is used to detect changes in density of organic materials.
- *Acoustic Emission Inspection:* Quite useful for locating defects in fiberglass and other composite materials. Once the location of a defect is determined, other NDE methods can help determine its severity.
- *Other NDE Methods:* The following methods also have applications: magnetic field testing, microwave inspection, thermal inspection, and holography.

Mechanical Testing

The information presented in this section provides general guidance on the analysis of physical data. Often many factors affect the data. Experts in the analysis of physical data may be needed if an accurate determination of a failure mechanism with a high degree of certainty is required for the success of the investigation. However, for most occurrences, investigators with some training in the analysis of physical data can make reasonably solid conclusions about the failure mechanisms. If other data (people, position, paper, and other physical data) are all consistent with the investigator's conclusions, the use of experts may not be required. Mechanical testing to support failure analysis can help determine:

- if the failed component met original product specifications,
- if changes have occurred in the component over time, or
- if material exposed to simulated service conditions behaves similar to the failed component.

The last reason for testing may take a considerable time to complete and can be complex. Mechanical testing to determine if the material met specifications or if it changed over time should be conducted where possible, measuring against the requirements of the original product specification.

Metals

Tensile testing or a hardness test is a basic requirement of most metal specifications. Some product specifications also require impact-testing, bend and other ductility tests, proof testing, flange or flare tests. The size of the sample may limit which tests can be performed. Macro, superficial, and micro-hardness tests are routinely done in failure analysis even if the original product specification did not require them.

Ceramics, Concrete and Glass

The mechanical tests required by the original product specification should be conducted. Changes in the physical properties occur with many of these materials over time regardless of service.

Plastics, Elastomers, and Resins

The specification requirements for mechanical testing should be performed. Degradation of the physical properties of organic materials is the best indication of chemical, radiation, or thermal degradation. When organic materials are removed from contact with solvents and other chemicals, the physical properties immediately after removal may be quite different than those measured several days later.

Chemical Analysis

Chemical analysis may be applied to a material in bulk usually to determine if it has met product specifications. Chemical analysis can also be conducted on individual phases in a material, deposits on a surface, or wear particles. Most of the chemical analysis techniques are used to identify or quantify elements, ions, or functional groups. It is also very useful in many cases to identify and quantify compounds.

Optical examination of etched polished surfaces or small particles can often identify compounds or different minerals by shape, color, optical properties, and the response to various etching attempts. A semi-quantitative elemental analysis can be used for elements with atomic number greater than four by SEM equipped with X-ray fluorescence and various electron detectors. The electron probe microanalyzer and Auer microprobe also provide elemental analysis of small areas. The secondary ion mass spectroscope, laser microprobe mass analyzer, and Raman microprobe analyzer can identify elements, compounds, and molecules. Electron diffraction patterns can be obtained with the TEM to determine which crystalline compounds are present. Ferrography is used for the identification of wear particles in lubricating oils.

Usually, the level of testing described in this step is only required for one or a few items. Again, the analysis of the data performed in Chapter 9 should help guide these analyses. Some of the analyses at this stage will require experts from outside the organization. These tests can be costly. The method of documenting the results of the testing should be thought out before the tests are conducted.

Step 6—Test the Items Under Simulated Conditions (if Warranted). In this step, experiments may be performed such as operational tests, mixing experiments, combustion experiments and other types of experiments. Simulations can be performed with similar parts or samples in an attempt to recreate the situation at the time of the failure. Pilot runs of the process or system may also be performed.

Simulations can be very simple or very complex. Example simulations could include:

- mixing two liquids in a beaker to see if they stratify without agitation,
- measuring heats of reaction for reactions with various types and amounts of impurities,
- using a computerized simulator to assess the potential impact of a flow variation, or
- measuring the heat-up rate of a tank of liquid exposed to the sun

Information gained from simulations can reveal key insights that explain gaps or contradictions in information. The time line is a useful tool in this development. For incidents of unexpected chemical reactions, it is common to attempt a lab scale simulation of the conditions involved in the exotherm or explosion. Many chemical processes can be modeled and duplicated dynamically by computer algorithms. Accelerated rate calorimeters (ARC) have proven to be highly useful tools for studying exothermic or overpressure runaway reactions.

Two important concepts must be kept in mind when considering use of simulations. First, top priority should be the prevention of a second injury or incident, which may be caused by the simulation. This classic error happens with surprising frequency when investigating partial amputations involving cycling of machinery. Considerations for safer simulations include:

- Volume and concentration of reactants
- Amount and type of reaction initiator (catalysts, ignition source)
- special barriers such as personnel protective equipment, overpressure relief devices (capacity, type, and discharge point)

Second, these simulations only mimic and do not exactly duplicate the occurrence. The information obtained can be useful, but it is narrow in scope and by nature is obtained under ideal and known conditions. Investigators should be mindful of the limitations and should use discretion when applying the data from these sources.

Investigations involving complex human performance problems can benefit from simulations. Process simulators are often used for operator training. In some cases, these process simulators can be excellent tools for learning more about human error causation. The incident investigation team can expose operators to simulated process upsets and gain valuable insights into the operator's response to rapidly and accurately diagnose the problem and execute the proper action.

The talk-through exercise is a technique sometimes used by investigators to gain insight and to verify conclusions drawn from verbal testimony. The technique is similar in format to a talk-through emergency drill. To be effective, such exercises must be planned by the investigator. The actual talk-through itself is seldom very time-consuming, but the burden is on the investigator to take good notes and observe any potential problem areas.

The talk-through technique, often used by human reliability analysts, has particular application for learning more about specific tasks or occurrences. It is a method in which an operator describes the actions required in a task, explains why he or she is doing each action, and explains the associated mental processes. There is a normal protocol for how to organize the talk-through. When the procedures call for the manipulation of a specific control or for the monitoring of a specific set of displays, the operator and the investigator approach them at the control panels, and the operator points out the controls and displays in question. If the performance is simulated, the operator touches the manual controls that would be operated and describes the control action required. The operator:

- points to displays,
- states what reading would be expected,
- describes any time delays and feedback signals, and
- describes implications of an action to the process function.

A talk-through of control room operations can reveal previously undisclosed information. In a control room analysis, an operator and the investigator actually follow the path taken by the operators during the performance of the procedure being analyzed.

Step 7—Make a Failure Mechanism Determination and Explore Root Causes.

This step is always performed. Using analysis tools and methods such as fault trees, causal factor charting, checklists, predeveloped trees, or alternative methodologies will help to identify the root causes of the failures.

8.4.2. Aids for Studying Evidence

8.4.2.1. Sources of Information

The following discussions are intended as an introduction with illustrative examples of some special techniques used by professionals for technical analysis of evidence. Novice investigators and individuals who are not experts in these fields should use caution when applying the information contained in these references to an investigation. For most minor investigations, review and application of the information in these materials is adequate for the investigation team to analyze the data. However, if legal concerns may arise out of the investigation, experts in the forensic analysis of data should always be used to ensure proper preservation and analysis of the data. Failure to do so can result in severe financial consequences to the organization when opposing attorneys accuse the organization of destroying data supportive of the plaintiff's case through sloppy preservation and analysis of the physical data. There are usually many issues to deal with shortly after a major occurrence. Inexperienced team members may become overwhelmed and not be able to think of all the issues. Using experts with experience in this area will help reduce the potential for inadvertent destruction or alteration of the data by team members.

Kuhlman in Chapter 10 of his publication, *Professional Accident Investigation*,[14] presents some excellent information regarding failure modes of metals. These can be helpful in analyzing the physical data for determining sequence, mechanism, and reason for failure.

Wulpi provides an excellent overview of failure mechanisms for mechanical parts in his book *Understanding How Components Fail*.[12] The book contains a number of photographs that compare the different types of failures. It is a good reference for mechanical component failures including shafts, tanks, piping, and boilers.

Fire investigators can reach remarkably quick and accurate determinations as to fire origin. Using carefully observed burn, char, melt, and damage patterns, the incident investigation team can gain a significant amount of information.[15] Even after a high-energy occurrence such as an explosion, examination of damage patterns can reveal sequences of occurrences. The National Fire Protection Association (NFPA, Quincy, MA) has established a series of specialized guides to assist investigators in gathering and analyzing evidence that include the following.

- **NFPA 906M—Guide for Fire Incident Field Notes.**[16] This section contains a series of forms that can assist the investigator in taking notes in an organized manner while collecting data about the incident. These forms can be supplemented by other data that each facility will find useful to collect.
- **NFPA 921M—Fire and Explosion Investigations.**[17] This section contains information to assist in improving the fire investigation process and the quality of information on fires resulting from the investigative process. Examples of the content of this section include:
  - Chapter 4, Fire Patterns, has some very useful information related to melting points, char patterns and the "V" burn pattern marker used by arson investigators.
  - Chapter 7, Source of Information, presents a discussion of possible sources of information where investigators can go prospecting for useful data, although many may not be relevant to internal company evaluations.

- Chapter 8, Recording the Scene, gives advice on how to record the fire scene through the use of photography and drawings.
- Chapter 9, Physical Evidence, includes collection, preservation, and documentation of physical data.
- Chapter 11, Origin Determination, is devoted to determining the origin.
- Chapter 12, Cause Determination, focuses on "cause determination".
- Chapter 16, Management of Major Investigations, provides some general guidance on the organization of the investigation.

Melt and autoignition temperatures for many materials are known, as are normal flame temperatures. Table 8-1 gives selected temperatures of interest to many investigators. Soot will normally *not* affix itself to surfaces at more than approximately 700°F (371°C). Therefore, areas of high fire intensity may have little or no soot deposits. Flame temperatures are

TABLE 8-1
Temperatures of Interest to Process Safety Incident Investigation Teams

Telltale	°F	°C
Paint begins to soften	400	204
Zinc primer paint discolors to tan	450	232
Zinc primer discolors to brown	500	260
Normal paints discolor	600	310
Zinc primer paint scorches to black	700	371
Lube oil auto-ignites	790	421
Stainless steel begins to discolor	800–900	427–482
Plywood autoignites	900	482
Vinyl coating on wire autoignites	900	482
Rubber hoses autoignite	950	510
Aluminum alloys melt	1125–1215	610–660
Glass melts	1400–1600	750–850
Brass melts (instrument gauges)	1650–1880	900–1025
Copper melts	1980	1083
Cast iron melts	2100–2200	1150–1250
Carbon steel melts	2760	1520
Stainless steel melts	2550–2790	1400–1532

Sources: Perry's Chemical Engineer's Handbook [20]; *Marks Mechanical Engineer Handbook* [18]; NFPA 422M Table[13]; *NFPA Fire Protection Handbook* [21]

dependent on the amount and type of fuel and oxygen (air) present and whether these are pressurized or not. For example, normal flame temperatures (match, candle, methane/air) are in the range of 1000–2000°F (550–1100°C). Yet, with pressurization and pure oxygen, a methane flame can approach 3000–5000°F (1650–2700°C).[18,19]

8.4.2.2. Additional Engineering Analysis Tools

In addition to the physical data analysis methods, traditional engineering analysis tools and methods are also useful during incident investigations. Traditional analysis tools can be used to determine the following.

- Concentration of gases in a tank or a release
- Flow rates of gases and liquids through piping and through release points
- Change in levels of tanks over time
- Temperature rise or fall of fluids through a heat exchanger
- Rates of chemical reactions
- Weight of a partially filled tank
- Dispersion of a gas
- Strength of a component or platform
- Number of barrels of raw materials used during a time period

Traditional engineering analysis tools will usually be used at some point during the investigation. The investigators will use these methods to support and refute the various theories that are put forth during the investigation. Often rough calculations may be all that is needed to determine if a scenario is possible. For example, even if the entire contents of a tank are released, it may not be sufficient to cause an overflow in another part of the process. This calculation may be sufficient to eliminate certain theories that have been proposed.

8.4.3. New Challenges in Interpreting Evidence

Technology advances in electronics such as process control instrumentation systems, computer capabilities, programmable logic controllers, and the use of independent PC's (personal computers) at field locations for special dedicated functions present new challenges to incident investigation. Some of the advances are so rapid that the team may not have the internal expertise to determine failure scenarios, sequences, and modes. The suppliers and manufacturers of these high-tech devices are sometimes the only source of credible information on failure modes of these devices.

Reliance on outside expertise may be the most feasible option for some of these issues. The incident investigation team may act as facilitators and advisors in a similar mode to a PHA (process hazard analysis) study. The outside expert would supply the failure mode information on which possible failures are credible.

Another new challenge for incident investigators is capturing, preserving, and retrieving electronically stored data. As previously mentioned, electronic computer control systems have limited storage capacity for detailed process information. Capturing this detailed information is often an overlooked task.

Computer-controlled systems are becoming ever more complex. With this complexity comes a greater difficulty in testing the equipment and a greater probability of errors, especially in unusual conditions. Determining software errors in systems that are tightly coupled and highly complex can be an arduous task for the incident investigation team. Experts in software analysis and troubleshooting may be needed to determine the causes of these failures.

Multi-level computer security measures such as software and hardwired keyed systems can be a blessing or a headache for investigators. If a well-designed and well-functioning management of change system is in place, then following the electronic trail can be greatly aided by rigorous security measures. On the other hand, an incomplete (or inconsistent) security system and management of change system can present impossible obstacles to determining the exact causes of a particular electronics occurrence. The basis for some decisions can be permanently lost and thus become undeterminable if the management of change system is inadequate.

Lasers, radioactive devices, complex chemical reaction kinetics, fiber optics, biological hazards, self-diagnostic equipment, and high-tech laboratory devices represent additional unique technologies that are normally outside the incident investigation team skill level. For such investigations, it is reasonable, appropriate, and cost effective to engage outside resources for selected tasks during the investigation.

8.4.4. Evidence Analysis Methods

Specific techniques for evidence analysis are beyond the scope of this guidebook. Entire volumes have been written on specific issues such as the fracture patterns of alloys and the corresponding clues for determining the actual cause and mechanism for the failure. This section is intended to provide an overview and general understanding of some of the common concepts and issues associated with evidence analysis. Known specific materials and alloys perform and fail in consistent and predictable ways.

This area of expertise is normally supplied to the incident investigation team via the use of specialists, from either within the parent organization or from outside experts or labs engaged specifically for the task.

8.4.5. The Use of Test Plans

Developing a test plan for the analysis of a part is like preparing for an interview. However, unlike the open-ended questions that are part of the interview of a person, parts can only answer closed-ended questions. In addition, equipment has a limited capability to answer past-tense questions (that is, most equipment and physical items do not have a memory). Finally, the order that questions are asked is important. Getting the answer to one question may prevent the investigators from obtaining the answers to other questions. For example, once a cover is opened to see what is inside, the cover cannot be replaced in exactly the same manner it was originally. The oxidation layers and adhesives used to seal the cover cannot be replaced exactly as they were. Once a pump is hand rotated, it cannot be disassembled to see the position it came to rest in following the failure. Consequently, the investigators must be careful to think about the questions that need answering and what data are destroyed when each answer is obtained. Test plans are designed to help the investigators think ahead. Test plans also serve to gain agreement from multiple parties on how, by whom, and when the test should be performed.

Typically, test plans are designed to answer one or more of the following questions:

- How does the part work?
- Did the part function as intended?
- How did the part fail?
- Why did the failure occur?

Test plans should be developed before the analysis of physical data is started. Test plans help to:

- Ensure complete collection of required data
- Ensure complete analysis of the data
- Prevent inadvertent destruction of data by the investigators
- Gain agreement from all parties involved in the investigation concerning the analysis processes and methods
- Ensure the test is worth doing before it is done
- Identify decision points in the analysis

The test plan should include:

- The objective of the test

- The methods for performing the test
- A description of the methods/procedure
- Names of the persons who will perform the test
- Scheduled times and locations of the test
- How the test results will be recorded
- Information on multiple tests of the same item
- Disposition of the test specimens after the test
- The order in which the different steps of the plan will be executed
- Which organizations, both internal or external, will approve the test plan

Test plans are not intended to be long, complex documents. They should be concise documents that lay out the test process in sufficient detail to allow all involved parties to understand what will be done to the physical data.

Endnotes

1. U.S. Department of Energy. *Accident/Incident Investigation Manual*, DOE/SSDC 76-45/27, 2nd edition. Washington, DC: U.S. Department of Energy, 1985.
2. Ferry, Ted S. *Modern Accident Investigation and Analysis*. 2nd ed. New York: Wiley, 1988.
3. Lees, F. P. *Loss Prevention in the Process Industries*. Vol. 1. London: Butterworths, 1980.
4. Stephens, M. M. *Minimizing Damage to Refineries*. Washington DC: U.S. Department of Interior, Office of Oil and Gas, 1970.
5. Center for Chemical Process Safety. *Guidelines for Technical Management of Chemical Process Safety*. New York: American Institute of Chemical Engineers, 1989.
6. Pieterson, *Report TNO—Mexico City LPG Terminal Disaster*.
7. Baker, W.E. et al. *Explosion Hazard and Evaluation*. New York: Elsevier Scientific, 1983.
8. Laborde, G. Z. *Influencing with Integrity Management Skills for Communications and Negotiation*. Palo Alto, CA: Syntony Publishing Co. Pp. 92–106.
9. Bulkley, W. L. *Technical Investigation of Major Process-Industry Accidents*. New York: AIChE Loss Prevention Series #0823.
10. McIntyre, D. *Assessment of Fire and Explosion Damage to Chemical Plant Equipment/Analyzing Explosions and Pressure Vessel Ruptures*. Materials Technology Institute of the Chemical Process Industries, Publication #30, National Association of Corrosion Engineers.
11. Thielsch, Helmut. *Defects and Failures in Pressure Vessels and Piping*. Malabar, Florida: R.E. Klieger Publishing Co., 1975.

12. Wulpi, Donald J. *Understanding How Components Fail,* 2nd ed. Materials Park, OH: ASM International, 1985.
13. NFPA 422M—*Manual for Aircraft Fire and Explosion Investigators*. Chapter 5 Evidence, Chapter 6—Determining Incident Sequence. Quincy, MA: NFPA, 1989.
14. Kuhlman, R. "Professional Accident Investigation." Loganville, GA: Institute Press-International Loss Control Institute.
15. U.S. Department of Commerce. *Fire Investigators Handbook*., Washington, DC: National Bureau of Standards, 1980.
16. NFPA 906M—*Guide for Fire Incident Field Notes*. Quincy, MA: NFPA, 1998.
17. NFPA 921M—*Guide for Fire and Explosion Investigations*. Quincy, MA: NFPA, 2001.
18. Avallone, Eugene A. *Mark's Standard Handbook for Mechanical Engineers,* 10th ed. New York: McGraw-Hill, 1996.
19. Althouse, et. al. *Modern Welding*. South Holland, IL: Goodheart-Willcox Co. Inc., 1980.
20. Perry, Robert H. *Perry's Chemical Engineer's Handbook*. 7th ed. New York: McGraw-Hill, 1999. Available in hard copy and CD-ROM version.
21. Cote A.E., and Linville, J.L . *Fire Protection Handbook,* 17th ed. NFPA principal editors. Quincy MA: NFPA., 1997

Additional Reference

Gano, D. L. *Apollo Root Cause Analysis—A New Way of Thinking*. Apollonian Publications, September 1994.

9

Determining Root Causes—Structured Approaches

Process safety incidents are invariably the result of multiple causes, which can usually be categorized into three types:

1. Immediate causes
2. Contributing or enabling causes
3. Root causes

Correcting only the immediate cause is a simplistic approach that may prevent the identical incident from occurring again at the same location, but will not prevent similar incidents. Correction of contributing or enabling causes goes further in helping eliminate future similar incidents, but does not solve the problem once-and-for-all. Identifying and correcting the root causes should eliminate or substantially reduce the risk of recurrence of the incident and other similar incidents at the location. More importantly, the new knowledge and corrective methods resulting from the investigation may be translated for use throughout a company and possibly apply to an industry as a whole.

A thorough incident investigation identifies and addresses all of the causes of an incident, including the root causes. It provides the mechanism for understanding the interaction and impact of management system failures. This analysis provides the means for fully addressing the incident, similar incidents and even dissimilar incidents, caused by the same root causes, either realized or not yet realized, throughout the facility, company, and industry. Finding and addressing management system failures is the ultimate solution yielding the maximum benefit from an incident investigation.

Root cause—*A fundamental, underlying, system-related reason why an incident occurred that identifies a correctable failure(s) in management systems. There is typically more than one root cause for every process safety incident.*

As an example, consider a scenario where a worker steps into a puddle of oil on the plant floor, slips, and falls. A traditional investigation might identify "oil spilled on the floor" as the cause, with the remedy limited to cleaning up this particular spill and possibly admonishing the worker for not being more careful. By using the tools described in this chapter, it will be clear that the oil on the floor is actually a symptom of the underlying causes, rather than the root cause itself. A structured root cause investigation explores the underlying causes and examines the systems and conditions involved in the incident. This approach would consider issues such as:

- How did the oil come to be on the floor in the first place?
- What is the source of the oil?
- What tasks were underway when the oil was spilled?
- Why did the oil remain on the floor?
- Why was it not cleaned up?
- How long had it been there?
- Was the spill reported?
- What is the usual condition of walking surfaces in that unit?
- What influenced the employee to walk in the oil?
- What type of shoes was the employee wearing?
- Why didn't the employee go around it?
- Was the area barricaded to prevent entry?
- Are there training or consistency of enforcement issues involved?

As these questions are answered, the continuing prompt for a better understanding of why the incident occurred should be, "Why? Why did this particular occurrence occur?" These answers take the investigators deeper into the origin of the incident. For instance, if the oil was determined to have leaked from a defective container, one might ask:

- Why was a defective container used?
- What are the procedures for inspecting, repairing, or replacing the containers?
- Are the procedures clearly understood and enforced?
- Is the system to manage the containers properly designed or are there gaps?

> ***. . . It is from identifying the underlying causes that the most benefit is gained. By addressing only the immediate cause, the identical accident is prevented from occurring again; by addressing the underlying cause(s) (root and system failures), numerous other similar incidents are prevented from occurring. . . .***
>
> —*Guidelines for Technical Management of Chemical Process Safety (AIChE/CCPS 1989)*

If a failure occurs and no changes are made in the operation, then the failure will likely occur again. Often corrective action is taken — yet the failure still recurs. Frequently this is because the corrective actions address symptoms rather than causes.

The objective of incident investigation is to prevent a recurrence. This is accomplished by establishing a management system that:

- ***Identifies and evaluates causes;***
- ***Identifies and evaluates recommended preventive measures that reduce the risk (probability and/or consequence); and***
- ***Ensures effective implementation and follow-up (completion and/or review) of all recommendations.***

Determining the causes is a necessary precursor to detailing findings, formulating recommendations, and implementing actions. This chapter presents two overall approaches to structured root cause investigation and develops a set of tools to identify root causes.

9.1. The Management System's Role

Management systems consist of many components such as technology, process, plant equipment and other physical systems, employees, supervisors, managers, policies, and procedures. The human is an integral part of the system with human factors, reliability, and performance playing important roles in management systems. Each person has an individual responsibility to perform within a preestablished framework. For example, individual managers have distinct roles and responsibilities in the overall management system. Additionally, there is a shared responsibility to function as an interdependent team.

Most root causes are associated with weaknesses, defects, or breakdowns in management systems. The majority of incident recommendations should address modifications to management systems.

. . . Experienced incident investigators know that such specific failures are but the immediate causes of an incident, and that underlying each such immediate cause is a management system failure, such as faulty design or inadequate training. . . .

—Guidelines for Technical Management of Chemical Process Safety (AIChE/CCPS 1989)

To give further insight into the role of management systems and the distinction between multiple root causes and non-root causes, consider the following actual case histories.

FLIXBOROUGH

In 1974, 28 people died in an explosion resulting from a large release of cyclohexane in Flixborough, U.K. The source of the hydrocarbon release was a failed expansion joint in a section of 20-inch (508-mm) diameter pipe. Investigation revealed the pipe had been "designed" with little technical input as a temporary bypass for a reactor that had been removed after it cracked.

The immediate cause was a failed expansion joint. Fixing or replacing the expansion joint was the apparent corrective remedy. However, a more thorough root-cause analysis looked deeper into the reasons why the joint failed. Here are some of the identified underlying root causes:

- The management system for reviewing, approving, and managing changes to process equipment was inadequate and needed substantial improvement. Temporary modifications were not reviewed by the appropriate technical discipline.
- Originally, the reactor failed due to stress corrosion cracking from nitrates. The source of the nitrates was water sprayed from an external hose used for supplemental cooling. The inadequate cooling capacity was resolved with a less than adequate technical solution that caused unexpected and unwanted consequences. Management of change was not properly applied.

CHALLENGER SPACE SHUTTLE

The *Challenger* space shuttle disaster (January 1986) was the culmination of a series of occurrences each with its own root cause.

The immediate cause was failure of the ring joint seal on the solid rocket booster. Yet, a root-cause analysis revealed a much more complex scenario. According to information published after the investigation, post flight evidence from as far back as early 1984 showed that the joint seals were failing to meet design specifications.

Almost 2 years before the incident, engineers knew that holes were being blown in the putty that shielded the *primary* O-rings from hot gases. In addition, evidence from 1983 showed the *secondary* O-rings were experiencing problems due to joint rotation during launch conditions. The reduced flexibility of the O-rings at temperatures below 50°F was also known. In July 1985, the concerns had grown to the point that further launches were postponed until an attempt was made to remedy the situation. But these remedies were only a patch and did not deal with the immediate, contributing, or root causes of the problem joint. Despite all that was known about the O-ring problem, a decision was made to launch the Challenger on a cold January morning. The *Challenger* space shuttle disaster is an excellent example of the principle that apparently simple mechanical problems are related to more complex underlying causes rooted in management systems.

The recommendations submitted by the presidential commission focused on *root* causes. These involved changes in management systems that would not only fix the ring joint problem, but the systems, procedures, and overall approaches to identifying, evaluating, resolving, monitoring, and auditing safety-related concerns as well.

9.2. Structured Root Cause Determination

Best practices in incident investigation have evolved substantially in the last 20 years. The general approaches to incident investigation are discussed in Chapter 4. This chapter will describe more thoroughly the tools and techniques used in incident investigation.

There are two structured root cause approaches presented in this chapter.

1. *Method A* involves a deductive search for all credible ways an occurrence could arise using timeline construction and a simplified fault tree approach. It can be viewed as an integrated method for systematically searching for all underlying root causes. The structured framework helps the investigator to keep on track, reach sufficient depth, and not stop prematurely at the symptoms or apparent causes.
2. *Method B* involves timeline or sequence diagram construction, identification of causal factors, followed by the use of predefined trees or checklists. A predefined tree provides a systematic approach for analyzing and selecting the relevant elements of the incident scenario. It is a deductive approach, looking backward in time to examine preceding occurrences necessary to produce the specified incident.

Structured root cause methods recognize that incidents have multiple underlying causes. Structured root cause investigations attempt to identify and implement system changes that will eliminate recurrence not only

Method A:
Described in Sections 9.5–9.8

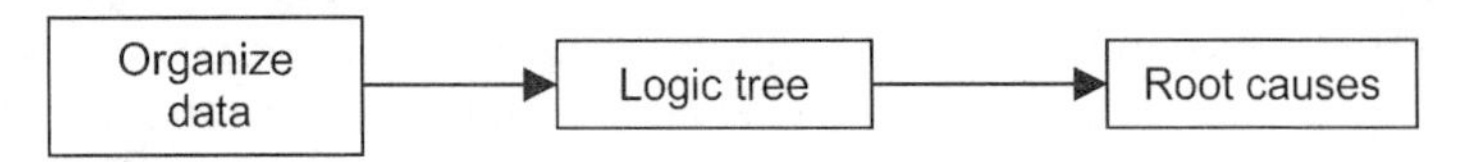

Method B:
Described in Sections 9.9–9.12

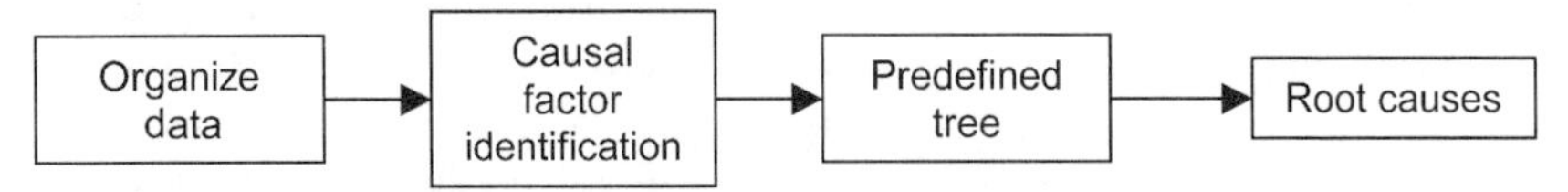

FIGURE 9-1. **Methods described in this chapter.**

of the exact incident, but of similar occurrences as well. These methods improve the quality of investigations by directing the focus past the immediate surface causes to the underlying root causes and mandating a search for multiple causes. One of the strengths of systematic methods is the ability to separate a complex incident into discrete smaller occurrences (segments) and then to examine each piece individually.

Figure 9-1, the two flowcharts describing root cause determinations using Methods A and B, presents general frameworks for root cause determination. Method A focuses on the logic tree method using a simplified fault tree approach. Method B focuses on the predefined tree method.

While some methods use checklists as the logic analysis step, an understanding of the logic tree approach is still helpful because checklists are developed from logic trees. Checklists are especially helpful when related to human factor issues. A sample checklist is included in Chapter 6 as Figure 6-6 (page 94).

The approaches shown here also present tools to test logic, determine if the root causes identified go deep enough, help discern what to do if you get stuck, and other decision aids. These tools work with any logic analysis methodology.

SOME GUIDING QUESTIONS FOR MULTIPLE CAUSE DETERMINATION.

- ***WHY? (Keep asking WHY? WHY?)***
- ***What were the underlying causes? (Why did they exist?)***
- ***Was there a system-related deficiency (or weakness) that caused (or allowed) this condition to exist, or caused or allowed the occurrence to proceed? (Why did such a system failure exist?)***

It is not the intention of the CCPS to endorse one particular method, but to present guidance on the various options and applications available. However, structured methodologies that seek out multiple underlying systems-related causes of an incident and provide the mechanisms for determining and correcting system faults are generally most useful in the broad spectrum of activities in the industry.

9.3. Organizing Data with a Timeline

One of the most effective and easy-to-use tools for getting organized and for cataloging data is the *timeline*. In addition to being an investigative organization tool, it is often very helpful in problem solving by graphically displaying the relationship of various occurrences to each other and to the final occurrence or incident. While the timeline itself may not actually identify the sets of causes of an accident, it helps the investigator focus on the essential areas that may yield such information.

It is also helpful to include conditions in a timeline; however, it is important to distinguish between events and conditions. Conditions tend to be passive items, such as *the pump **was** running*, *the pipe **was** corroded*, or *the operators **were not** trained on the draining procedure*. Condition statements are identified by the words *was* or *were*. *Events* by contrast are active, such as *the pump started up* or *the pipe failed*. Both events and conditions can be facts. They can both remain as suppositions also if indisputable.

The timeline can also include nonevents or omissions, such as *failure to follow a step within standard operating procedure* or *relief valve failed to open at setpoint*.

9.3.1. Developing a Timeline

Developing the timeline is an iterative activity, extending across the entire life of the investigation. The timeline increases in content and accuracy as new information becomes available and inconsistencies are clarified and resolved. Timelines can be developed using various forms and levels of complexity, usually dictated by the particular circumstances of the investigation being conducted. The timeline helps the team to see the events in a chronological order. This can help them understand when—and perhaps why—important events took place. Any pertinent information or evidence should be inserted in the timeline. When gaps are observed, the team can try to find the information to fill in those gaps. Timelines deal with a combination of data to be charted. Some of the data that will go on a timeline is very precise, both in timing and in values. For instance, the print out of operating conditions and alarms from a basic process control system or a safety-instrumented system may show:

- When a particular parameter was exceeded, to the tenth of a second
- The rate of rise for that particular parameter
- The final value before the accident occurred

Figure 9-2 is an example of a simple timeline using precise data from the Fictitious NDF Incident Example discussed in detail in Appendix D.

On the other extreme, a field operator's observations and actions will be less precise. "Sometime around noon," or "right after the 10:00 AM morning break," may express these approximations.

Figure 9-3 is an example of a simple timeline using very imprecise data from the field operator. This timeline uses a portion of the approximate data from the Fictitious NDF Incident Example discussed in detail in Appendix D.

Normally the investigator is presented with a combination of both precise and imprecise data. Mixing these significantly different data often proves to be a challenge—a challenge, however, that can be overcome simply by understanding the source and precision of the data and the use of appropriate graphing techniques. One such technique involves using a line with timing marks as the common boundary between the two different types of data. On one side of the line, the known precise data is logged against the timing marks. On the other side of the line, imprecise or approximate data is listed within the period in which it occurred. This is usually displayed as an event occurring some time between two timing marks.

11:10:21 AM	Heat detector alarms for Kettle area annunciate (DCS, assumed fireball #1)
11:09:30 AM	LEL detectors in Catalyst Prep area alarm (DCS)
11:05:03 AM	Plant wide electrical outage (DCS, time of pump shutdowns)
11:03:45 AM	Kettle #3 high pressure alarm acknowledged (DCS)
11:03:15 AM	Kettle #3 high-pressure alarm alarms (DCS)
11:00:47 AM	Kettle #3 level reaches 90% (DCS) but alarm does not log (later found inhibited)
10:30:33 AM	Control room operator initiates filling of Kettle #3 (DCS)

FIGURE 9-2. **Timeline example based on precise data. DCS = Distributive control system.**

Time	Events
11:15 AM	♦ Plant fire brigade reaches emergency location (plant dispatch log)
11:12 AM	♦ Plant fire brigade called (plant dispatch log)
11:10 AM	♦ Control operator tries to reach outside operator by radio, but there is no response ♦ Heat detector alarms for kettle area annunciate ♦ "Whooshing" noise heard
After outage	♦ Control room operator asked outside operator to visually inspect Kettle #3 due to high LEL alarm ♦ Thunderstorm passed and rain diminishing ♦ Kettle #3 exit piping cracks (concluded from data)
11:05 AM	♦ Plant wide power outage ♦ Kettle #3 high pressure alarm acknowledged ♦ Kettle #3 high pressure alarm sounds
11:00 AM	♦ Severe thunderstorm starts
10:30 AM	♦ Control room operator initiates filling of Kettle #3 ♦ Contractor enters Reactor Area to replace gas detectors

FIGURE 9-3. **Timeline example based on approximate data.**

Figure 9-4 is an example of a timeline that uses a mixture of precise and imprecise data from the Fictitious NDF Incident Example discussed in detail in Appendix D.

One additional benefit of this technique is that the imprecise approximate times can often be narrowed when compared to the precise data. For instance, the operator may realize that when he manually closed valve A, valve B had already been automatically closed. Therefore, the period within which he closed valve A is narrowed.

Timelines do not have to end at the time of the occurrence or incident. Sometimes post occurrence data can be valuable. Often, it is important to understand how the emergency response actions affected the ultimate outcome of the occurrence. This type of data can be used to improve emergency response actions in the future.

When timelines are combined with simulations (see Chapter 8), they become powerful tools in both understanding the sequence of the events

DCS / recorded data	Time	Events
	11:15 AM	♦ Plant fire brigade reaches emergency location (plant dispatch log)
	11:11 AM	♦ Plant fire brigade called (plant dispatch log)
	After heat detectors alarm	♦ Control operator tries to reach outside operator by radio, but there is no response
		♦ Plant fire brigade called (plant dispatch log)
Heat detector alarms for Kettle area annunciate (DCS, assumed fireball #1)	11:10:21 AM	
	11:10 AM	♦ "Whooshing" noise heard
	After LEL detectors alarm	♦ Control room operator asked outside operator to visually inspect Kettle # 3 due to high LEL alarm
		♦ Thunderstorm passed and rain diminishing
LEL detectors in Catalyst Prep area alarm (DCS)	11:09:30 AM	
		♦ Kettle # 3 exit piping cracks (concluded from data)
Plant wide electrical outage (DCS, time of pump shutdowns)	11:05:03 AM	
Kettle #3 high pressure alarm acknowledged (DCS)	11:03:45 AM	
Kettle #3 high-pressure alarm alarms (DCS)	11:03:15 AM	
Kettle #3 level reaches 90% (DCS) but alarm does not log (later found inhibited)	11:00:47 AM	
	11:00 AM	♦ Severe thunderstorm starts
Control room operator initiates filling of Kettle #3 (DCS)	10:30:33 AM	
	10:30 AM	♦ Contractor enters Reactor Area to replace gas detectors

FIGURE 9-4. **Timeline example based on combination of precise and approximate data.**

TIMELINE TIPS

- Use a poster size piece of paper to start your timeline. Use sticky notes for each condition or event item. You can easily rearrange the blocks.
- Identify precise and imprecise data (different colors are one way).
- You may also find it useful to list the source on each note (basic process control system, interview, etc.)

FIGURE 9-5. **Timeline tips.**

leading up to the incident and the development of accurate recreations. This allows a more thorough and comprehensive analysis with associated action plans for correction.

9.3.2. Determining Conditions at the Time of Failure

Factual (or supposed) conditions are included on the timeline. Determining conditions at the time of the failure is an activity bridging the gap between evidence gathering and root cause determination. It is an excellent opportunity for application of the timeline tool. Failures rarely occur without some prior indications or precursor information. However, unless someone specifically is charged with looking for it, the information is frequently overlooked. Therefore, someone should be assigned the project of timeline development and should update it periodically as new information comes available. A goal of the incident investigation team is to search back in time, find this information, and correlate it with the failure occurrence to confirm or refute a postulated failure hypothesis. This circumstantial evidence may be short-term, that is, immediately preceding the failure, or may be long-term and include anecdotal information from earlier failures or from previous operating experience. It should also include post incident occurrences that may have affected emergency response, mitigation actions, or secondary damages.

The information that is gathered as described in Chapter 8 will be most useful in accurately determining conditions at the time of the incident and immediately preceding it. Analyzing evidence and determining preincident conditions begin as parallel efforts, but converge as the investigation progresses. This failure definition is just as valid for hardware or piping systems as it is for a human being.

The incident investigation team must look specifically for evidence that provides the point of initial failure, its progression path, and the preexisting conditions that led to the initiation. Having an understanding of a fundamental failure mode and the sequence of events, the investigator should seek evidence that shows or confirms the actual failure mechanism. The incident investigation team could analyze to confirm material properties and examine the actual failure sites to identify the nature of the failure, such as fatigue, stress corrosion cracking, intergranular stress corrosion, or embrittlement.

The timeline tool pulls all of this information together into a manageable record of events and sequence providing a perspective conducive to proper causal analysis.

9.4. Organizing Data with Sequence Diagrams

Sequence diagrams are a more elaborate graphical depiction of a timeline, and allow the investigator to present related events and conditions in parallel branches. These sequence diagrams are also known as *causal factor charts*.

Like a simpler timeline, a sequence diagram does not identify root causes, and therefore it should be used as part of a combined methodology with other tools. In this respect, a sequence diagram may be used in place of the timeline within the two main incident investigation methodologies presented in this chapter.

As with a timeline, construction of the sequence diagram should start at the earliest opportunity as soon as the initial facts become known about the incident. By starting early, the investigation can spot missing information or inconsistencies in the "facts" and focus upon resolving those gaps. By adopting a format that may be easily updated as new evidence emerges, the sequence diagram can be progressively revised to maintain the focus of the evidence gathering. A popular format is the use of Post-It® notes, on which a single fact (event or condition) is written and arranged sequentially on a large board or sheet of paper.

A diagram depicting the sequence of events leading to an incident has a number of advantages over a simple timeline that may be summarized in three main areas: investigation, identifying actions, and reporting as shown below.(1)

Investigation

- Summarizing the events in the form of a diagram provides an aid to developing evidence, identifying causal factors, and identifying gaps in knowledge.
- The multiple causes leading to an incident are clearly illustrated.
- Diagrams enable all involved in the investigation to visualize the sequence of events in time, and the relationships of conditions and events.

- A good diagram will serve to communicate the incident more clearly than pages of text, and ensure a more accurate interpretation.

Identifying Actions

- The diagram will provide a cause-orientated explanation of the incident.
- Areas of responsibility will be clearly defined.

Reporting

- Summary diagrams can be used in reports to provide a concise, easy-to-follow-representation of the incident for readers.
- Diagrams should help to prevent inaccurate conclusions by revealing any gaps in the logical sequence of events.
- Where gaps are shown, the requirement for further analysis/investigation can be identified.
- Diagrams provide a means of checking the conclusions as the facts are uncovered.
- Recommendations can be evaluated against the events and causal factors identified in the diagrams.

Several investigative tools employing graphic displays of incidents have been developed, but only a few are used in the chemical industry. Multilinear Event Sequencing (MES) and Sequentially Timed Events Plot (STEP) were originally developed for incidents other than process incidents. MORT interpretations of some original MES concepts [(2)] have been adapted through Events & Causal Factor Charting (E&CF) for use within the industry. These tools are discussed below.

9.4.1.1. Multilinear Events Sequencing (MES)

Multilinear Events Sequencing (MES) was developed from work at the National Transportation Safety Board in the early 1970s.[(2)] During the MES process, investigators convert observed data into events and array the events on a matrix with time and actor coordinates. An event is defined as one actor plus one action. Actors can be people or things, and actions are what the actors did. As data defining an actor and what the actor did are acquired, each new event is positioned on its actor row on the matrix, and positioned horizontally under the time it started. This displays what people or things did in the sequence they did it, showing time and precede/follow relationships.

The MES matrix can be a wall, board, or large paper. The investigators use cards or Post-It® notes to record a layout the events. Often investigators transfer the display to a computer for further processing. As new

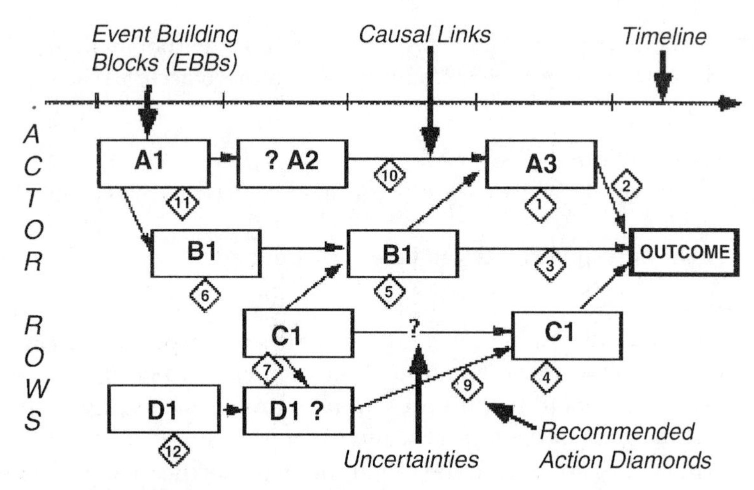

FIGURE 9-6. **Schematic of an MES display.**[3]

information becomes available, the investigators simply update the original card or insert a new card on the matrix.

As events are added, the investigator also adds arrows to link interacting or coupled events. By convention in MES, arrows always flow from the earlier event (causal events) to the later events (effect events) and from left to right. The linking arrows show the flow of the interacting events, or causal flow, from the earliest to the final event on the display. Gaps define data that are still needed. Question marks are used to show uncertainties for data which are needed, or for which no valid data can be developed. A final necessary and sufficient logic test determines the completeness of the display. The tested display is then the best description and explanation of what happened that can be developed by the investigation. Problems and potential changes to improve performance are identified and defined by examining the real relationships within each coupled event pair or set. Each potential change is noted on the matrix with "recommended action diamonds" to indicate where in the accident or incident process improvement opportunities exist (see Figure 9-6).

Johnson's interpretation of MES concepts is known as Events & Causal Factor Charting (E&CF), or Causal Factor Charting for short, and has been adopted as one of the building blocks of several methodologies for process safety incident investigation.

9.4.1.2. Sequentially Timed Events Plot (STEP)

Sequentially Timed Events Plot (STEP)[4] is a name used for the multilinear events sequence-based matrix display. It evolved from the 1975 MES concepts, but only events are displayed because conditions or states are changed by actions. The matrix entries focus on the behaviors or actions, which produced the undesired outcomes and would have to be changed to improve future performance. The STEP procedures are part of the latest MES investigation process.

9.4.1.3. Causal Factor Charting

Events & Causal Factor Charting (E&CF) [5] was adopted by the developers of MORT to identify and document the sequence of events leading to an incident. A number of proprietary process safety incident investigation methodologies, such as SOURCE™ [6] and TapRooT® [7] include E&CF as one of their building blocks.

The causal factor charting method of developing chronological data in graphical format is an excellent tool for organizing process safety incident evidence. The graphical representation of the incident sequence assists the investigator in organizing all the data and understanding the incident. The investigator is then better able to communicate effectively that understanding to others. This is especially important with complex process incidents, although the diagrams themselves can become rather complex too.

The technique for developing causal factor charts share a number of fundamental principles with MES and STEP. Basic principles for constructing sequence diagrams[8] are given below.

Chart Format

- All events are enclosed in rectangles, and conditions in ovals.
- All events are connected by solid arrows.
- All conditions are connected to other conditions and/or events by dashed arrows.
- Each event or condition should be based upon valid evidence or, if presumptive, shown by dotted rectangles or ovals.
- The primary sequence of events is depicted in a straight horizontal line (bold arrows are suggested).
- Secondary event sequences are presented at different levels.
- Relative time sequence is from left to right.

Criteria for Events Description

- Events must describe an action, not a condition.
- Events must be described with one noun or verb.
- Occurrences must be precisely described.
- Events must describe one discrete action.
- Events should be quantified when possible.
- Events should range from beginning to end of the accident sequence.
- Each event should be derived from the one preceding it.

These principles are not mandatory. The most important aspect is that the investigator understands the incident, and these principles are meant to facilitate that objective. Some investigators draw causal factor charts differently; for example, some investigators do not distinguish between events and conditions. It is permissible to violate the above principles provided the method helps the investigator and others understand the incident.

9.4.1.4. Developing a Causal Factor Chart

The first step in developing a causal factor chart is to define the end of the incident sequence. Construction of the chart should start early from the end point and work backward to reconstruct what happened before the incident by identifying the most immediate contributing events.

Starting at the end point, it is then necessary to convert the collected evidence into statements of either fact or supposition. By taking a small step backward in time, the investigator asks, *"what happened just before this event."* It is important to clearly distinguish any assumptions as supposition. Then the investigator writes a statement for what happened, and enters the fact (or supposition) as an event block or condition oval on the causal factor chart at the appropriate location on the timeline. Statements that caused an event to occur should be treated like conditions and added in an oval.

The investigator tests this new event (or condition) for sufficiency by asking, for example,

"Does this block always lead to the next block (in this case, the endpoint)?"

"Are there any layers of protection that should have prevented this sequence?"

The process is repeated slowly working backward in time.

The entire causal factor chart is then reviewed to identify any omissions or gaps in the chronology. Additional effort is required to gather further evidence to close these gaps. If new data are inserted into the timeline, the sequence should be retested for sufficiency. Some gaps may remain even after this additional effort. The causal factor chart review should also identify and eliminate any facts that are not necessary to describe the incident. Detailed rules for causal factor charting are shown in Figure 9-7.

Charting the events and conditions on a causal factor chart assists the investigator in thinking logically through the incident. However, the investigator must exercise care to avoid locking into a preconceived scenario. It is important to keep an open mind and objectively analyze all possible scenarios for the events and conditions leading up to the incident. Initial assumptions can change dramatically during the course of an

Rules for Causal Factor Charting

1. Start at the end of the sequence (at the incident or the near-miss statement) and work backward in time.
2. List all outcomes, if there are multiple impacts/outcomes of the sequence of events.
3. Generate any questions necessary to quantify the outcomes.
4. Use whole sentences and quantify each statement as much as possible (merely stating that the "pump is destroyed" is not sufficient; include a quantification of the damage to the pump).
5. Take a very small step backward in time by asking *"What happened just before this event?"* The answer may be a phenomenological occurrence or condition, or it may have been an action by a human or machine. If there are multiple choices for the size of the step backward, take the smallest step proposed.
6. Write a complete sentence for what happened and add the event to the chart.
7. Test for sufficiency of information by asking one or more of the following questions:
   a. *Will (insert the complete statement of Fact B here) always lead to (insert the complete statement of Fact A here)?*
   b. *Every time Fact B occurs, does Fact A have to follow?*
   c. *Just because Fact B occurs, will Fact A always follow?*
   d. *Are there any layers of protection that should have prevented Fact B from progressing to Fact A?*
   e. *Does anything else have to occur or does any other condition have to be satisfied for Fact B to lead to Fact A?*
   f. *Are there any other potential causes of A other than B?*

   FACT B → FACT A
8. If the answer to either question 7a, 7b, 7c, or 7d is NO, or if the answer to question 7e is YES, then brainstorm what else would have to occur or what other conditions would have to be satisfied for Fact B to lead to Fact A. Generate the questions or list the data needed to answer the hypothetical questions/concerns.
9. Gather data to answer the questions or address the data needs identified in step 8.
10. If any of these facts are relevant, convert them into building block format and insert them into the Causal Factor Chart at the appropriate location on the time line. If any facts are inserted between Fact B and A, then retest each pair of facts for sufficiency as stated in steps 7 and 8, and repeat steps 9 and 10 as necessary.
11. Continue to gather data to answers the questions/needs developed above. Remove the questions from the chart as they are answered.
12. From Fact B, repeat steps 5 through 11 until all data are exhausted (changing the reference to Fact C, Fact D, etc. as necessary, of course).
13. Keep all remaining questions on the chart to demonstrate what is known and not known.
14. Review the entire Causal Factor Chart and eliminate any facts that are not necessary to describe the incident.
15. Find the facts in the main sequence on the Causal Factor Chart that describe a component failure or a human error. Ensure the fact is not describing a management system failure (i.e., ensure the fact is not a root cause, near root cause, or root cause category). The identified negative events/conditions are candidate causal factors. Any candidate causal factor that is not dependent on another candidate causal factor is a valid causal factor.

FIGURE 9-7. **Rules for causal factor charting.**

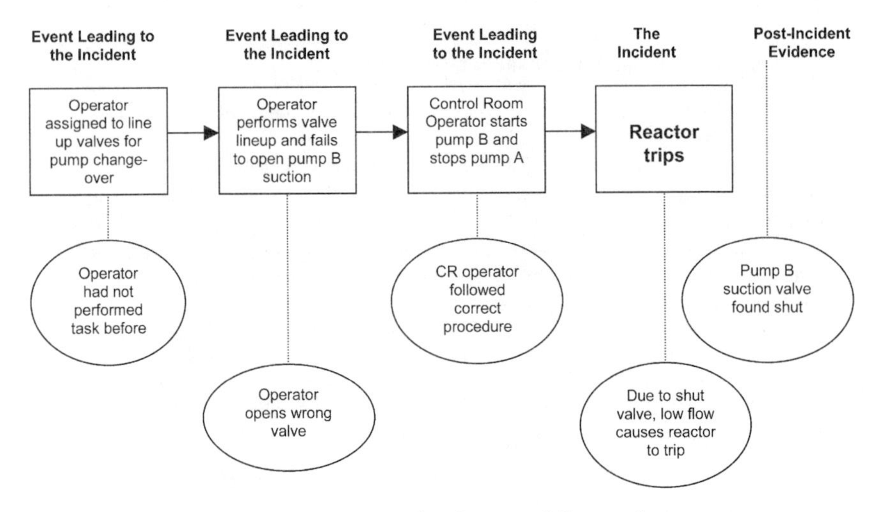

FIGURE 9-8. **Example of a causal factor chart.**

investigation. Note that it is sometimes helpful to reconstruct what immediate actions occurred after the incident.

An example of a causal factor chart for a relatively simple incident is shown in Figure 9-8. In this example, there are two redundant pumps, one of which is required to supply feed to a reactor downstream. The operator is requested to change-over operation from Pump A, which is running, to Pump B, which was previously shutdown. Instead of opening Pump B suction valve, the operator opens the wrong valve, causing the Reactor to trip on low flow detection.

9.4.1.5. Summary

Sequence diagrams provide a versatile and systematic means for an investigator to:

- present collected evidence in a logical and chronological order and
- identify all the issues to be investigated.

The diagram can assist the overall methodology to identify multiple causes, when used in conjunction with logic trees or predefined trees.

Once the sequence diagram has been developed, the investigator may proceed to identify the causal factors and/or root causes. However, the investigator should resist the temptation to begin this step until he is certain that he fully understand *what happened*.

On the basis that a picture is worth a thousand words, a sequence diagram can be used as an effective tool for communication about the incident sequence, the initiating events, and their causal factors. Sequence diagrams for process incidents are often complex, so sometimes it may be appropriate to simplify the diagram or use multiple smaller diagrams for presentation purposes. The diagram can also be an essential part of the incident investigation report and documentation.

9.5. Root Cause Determination Using Logic Trees—Method A

The following section presents a systematic discussion of the concepts and actions depicted in Figure 9-9. The starting point for the flowchart is the accumulation of facts, information, observations, insights, questions, and preliminary speculations gained from the evidence collection activities described previously.

9.5.1. Gather Evidence and List Facts

The first task is to develop a list of all known facts. This list includes not only facts relating to the incident sequence, but also all pertinent background data, specifications, and recent past or external events that could or did have an influence on the overall system.

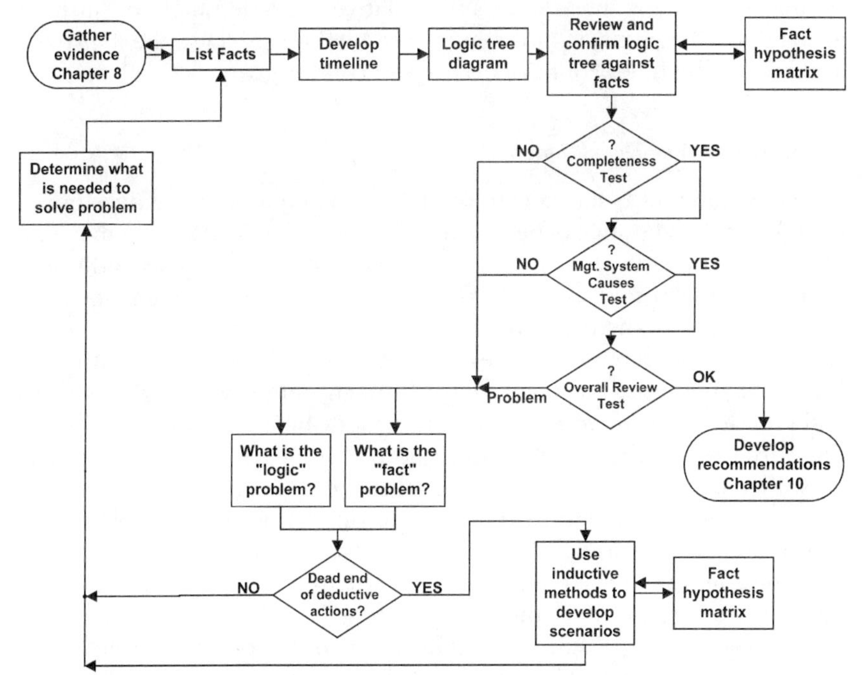

FIGURE 9-9. **Flowchart for root cause determination using logic trees.**

A fact should be just that—fact. Avoid drawing conclusions or making judgments at this stage. An example fact is—*Hearing protection boundaries in an area are not marked*. A conclusion may be that the boundaries are not clear, but do not make that jump yet. Sticking to the facts will help prevent people involved in the incident from becoming defensive and will prevent the team from jumping to conclusions. Sticking to the facts will also assist legal counsel in understanding the total incident report.

The team must take care to avoid being trapped by hidden or erroneous assumptions. All facts should be tested. The facts are essential to selecting the correct incident scenario later. Any apparently conflicting facts should be resolved through additional data gathering. Listing the source of each fact will facilitate conflict identification and resolution.

9.5.2. Timeline Development

Next the team develops a chronology of events based on the available known times and sequences. This document is usually referred to as a *timeline*. (See the detailed description of timeline development in Section 9.3.) Unconfirmed assumptions regarding chronology should be clearly identified as unconfirmed, and action should be initiated to verify assumptions. Many investigators use relatively simple timelines (instead of sequence diagrams) with the logic tree methods because the logic tree itself shows the interactions of events and conditions.

9.5.3. Logic Tree Development

After the initial facts have been listed and the initial timeline developed, the logic tree diagram can be constructed. The tree diagram is a dynamic document; it continues to expand and may even rearrange as additional information becomes available or when new information changes the understanding of the original facts.

Now all the minimum pieces are theoretically in place to confirm or refute a hypothesis. For many simple and straightforward failures, general knowledge of the component failure mode behavior, used in conjunction with the specific information gathered for a particular incident, may be sufficient to diagnose the causes. However, most process safety incidents are complex in nature and have multiple underlying system causes. Therefore, a systematic deductive approach is usually appropriate.

The common format at this stage of the investigation is to conduct a multiple root-cause determination meeting. The participants normally include the investigation team and members of the operating unit where the incident occurred. (There will probably be some of members from the operating unit already on the team.) The meeting should be small enough to be efficient and inclusive enough to have all the information necessary to develop the logic tree. Where possible, try to limit the number of participants to about eight or fewer. Participants need to include those who have the facts and those who understand important elements of the process, operation, chemistry, equipment, and controls. Some specialists may be brought in when needed. In some situations, it may be necessary to include representatives from unit management, employee unions, and legal counsel. The meeting should be as open and factual as possible. When deciding to include people from outside the investigation team in the meeting, consider these questions:

- Are they up to speed on the process?
- Will they hamper the independence of the team?

In the opening segment the facilitator should discuss the importance of and methods for choosing the top event and any preestablished and existing boundaries of the investigation. If multiple events are involved, it is best to start with the last event in the time sequence. It may be appropriate, depending on the nature of the occurrence, to formally review the rules and symbols used in logic tree or fault tree development or whichever other formal method will be used.

At the end of this meeting, a formal critique should be considered to consolidate lessons learned for future meetings. The critique should consider what went well and what changes could be made to improve future meetings. It would also be appropriate at the conclusion of the session to thank the participants for their contributions, to restate the purpose of the meeting, and to recap how and whether it was achieved.

At this point, the logic tree structure is examined to ensure the tree is logically consistent and compatible with the known facts. In some instances, there may be inconsistencies and application of the fact/hypothesis matrix will be appropriate. This powerful tool is described in more detail later in this chapter. Inconsistencies found at this point require further tree development or rearrangement.

Once the logic tree structure appears to be consistent, the first of three quality assurance tests is applied by examining the overall logic tree structure for completeness. The logic in each branch of the tree should be tested to determine if it is necessary and sufficient. (Details and tips for testing the logic are discussed in Section 9.6.2.) If the tree appears to be complete, the next quality assurance test is initiated. If the tree is incomplete, then the fact or logic problem is identified and the entire process is repeated. This is called an iterative loop .

If the logic tree appears to be complete, then the second quality control test is applied by asking the question, "Are the causes that have been identified actually related to management systems?" If the answer is yes, then the investigation proceeds to the third quality control test—the final overall review. If management system causes have not been found, then the *iterative loop* is used.

It is important to note that not all management system causes may be located at the extreme bottom points on the logic tree. Some of the management systems-related causes can be and often are located in the upper or middle portions of the logic tree diagram. Some causes can also be identified by the logic tree structure itself. For example, an overview of the entire tree structure may indicate significant gaps or overlaps in responsibilities, or it may disclose conflicting activities or procedures. These insights may be overlooked if the investigators limit their cause search to only the bottom level of the structure and fail to review the entire tree and the interrelationships between branches.

If the test for systems-related causes is satisfactory, then the third and final quality assurance test is applied. This is an overall review of the logic tree as a whole for both facts and logic. A conscientious review of each branch should be made to look for possible conflicts or inconsistencies. It is a pause to focus on the "forest" from an overall perspective, not just each "tree." The final logic diagram should be thoroughly checked against the final time-line to ensure that these two are in complete agreement. The team should also verify that none of the facts is in conflict with the tree. If the incident investigation team is satisfied with the causes identified, then the investigation proceeds to the recommendation stage. If a problem or some incompleteness is noted, then the iterative loop is reactivated.

After the tree is developed, and before moving on to the recommendations and deliberations, the team should ask, "Are there any *other* causes that anyone had in mind at the beginning of this meeting that are *not* included in the tree?" If additional causes are identified, the team adds them to the tree if there is logic to support them. Some team members may have specific concerns that the logic tree has not adequately resolved. This is the point at which remaining issues are surfaced and addressed.

In the deductive process of identifying root causes, known facts are assembled and used to develop and test one or more possible scenarios. The process normally requires multiple iterations of the cycle shown in Figure 9-9 until at least one plausible scenario is identified that fits all the known facts.

If the scenario is completely disproved by the known accepted facts, the reasoning is documented and the scenario is disregarded. If the scenario needs additional data in order to be proven or disproved, then the iterative loop path is followed and additional information is gathered. Sometimes this new information is very specific, precise, and limited in scope. Examples of tasks initiated by this iterative loop include:

- Follow-up witness interviews,
- Revisiting or reexamining a certain area of the incident scene, and
- Commissioning expert consultant opinions.

The next item in the loop is a decision point for possibly introducing the use of inductive reasoning methods into the deliberations. If the deductive process continues to indicate progress, then additional facts are procured or the logic tree is restructured. For example, one witness stated a particular valve was open, yet the post-incident inspection found it to be closed. The team must be careful to ensure that the valve is closed because of the actions taken prior to the incident, and not as a result of post-event response activities. The position of this particular valve may be a critical item in determining which of two scenarios is the more probable case. The incident investigation team would then initiate a short-term action item to conduct a mini-investigation to resolve this question.

If the deductive process has stalled and no further progress seems possible or likely, then the iterative loop calls for application of inductive investigation methods such as a checklist or HAZOP. The inductive methods may also benefit from use of the fact/hypothesis matrix tool described in this chapter.

9.6. Logic Trees

The logic tree is a systematic mechanism for organizing and analyzing the elements of the incident scenario. It is a deductive approach, looking backward in time to examine preceding events necessary to produce a specified result. This section illustrates building a logic tree using a simplified fault tree approach. Other similar methods (such as causal tree) will be somewhat different in terms of symbols and the look of the tree, but the basic concepts are the same.

Standard symbols from systems theory are often used to construct the logic tree diagram. The diagram often takes the form of a qualitative fault tree, showing the incident as the top event and the various branches using conventional AND- and OR-gates. Some investigators have simplified development of the logic tree by not distinguishing between AND- conditions and OR-conditions on the first pass through the tree. Instead, they use a "universal gate" and determine its status as the investigation progresses. Other techniques use only AND-gates.

The trees in this section will be drawn from top to bottom. Some similar techniques are drawn from left to right or right to left.

Many deductive investigation techniques use logic tree diagrams. A partial list of these methods includes fault tree analysis (FTA), causal tree method (CTM), multiple-cause systems-oriented incident investigation (MCSOII), and Why Tree. These methods are described in Chapter 4

Logic tree diagrams are developed after the incident investigation team has assembled the initial facts and has established a timeline. Logic tree development is an iterative activity extending across the life of the investigation, as additional information becomes available. Thus, in a systematic way, the logic tree provides a structure for thoroughly considering possible multiple causes. Each of the succeeding lower levels is developed by repeatedly asking "Why?" until a level is reached that allows examination of a management system or a small segment of it. The particular management system would then be scrutinized for deficiencies that caused or contributed to the incident. Identifying deficiencies provides a foundation for recommended improvements and preventive action.

9.6.1. Choosing the Top Event

Choosing the top event for the logic tree may sound like a simple task, but it is often more difficult then expected. In the Flixborough incident, the top event could be the fatalities, the potential fatalities in the office building, the fire, or the initial chemical release. If the chemical release is chosen as the top event, the discussion of why the people were in the area or why the office building was so close to the plant might never occur. As you review the example trees in the following sections, look at what might be left out, or included, if the top events were chosen differently.

It is also important and appropriate to consider the question "What could have happened?" When dealing with a near miss, there can be differences of opinion among the team members as to the credible negative consequences of the incident.

> ***TEAM MEMBERS SHOULD REMEMBER:***
>
> ***SEVERITY IS OFTEN A MATTER OF CHANCE.***

The incident investigation team must evaluate potential effects of an incident on all the stakeholders interested in a facility's continued safe operation. Public perception and good will are very important. Stakeholders include:

- Employees
- Visitors
- Community
- Contractors
- Regulators

- Customers
- Others

The top event chosen for a near miss might be a credibly possible outcome such as an injury, fire, or chemical release.

The team must also consider the likelihood of the incident being repeated, which is a major reason for pursuing near-miss incident investigations to their conclusion and learning value. This reminder should be given at the beginning of the multiple cause determination meeting and repeated as necessary.

9.6.2. Logic Tree Basics

To put it simply, a logic tree is developed by repeatedly asking "Why?" and organizing the results of the answers.

A generic logic tree for a fire incident is shown below in Figure 9-10. The top event is defined as the unwanted fire, with fuel, oxygen, and ignition depicted in the three branch conditions leading to the top event. Each of the three branches would then be examined, developed, and expanded into further detail as the investigation progresses.

The diagram can be developed from the top downward and can model a system, subsystem, or any individual component. For each level, a set of *necessary* and *sufficient* lower-order conditions or events is identified.

The basis for logic tree construction lies in the application of logic gates (other symbols are used to explain the overall system structure and analysis boundaries). The most important logic gates are the OR-gate and the AND-gate (other gates are used occasionally).

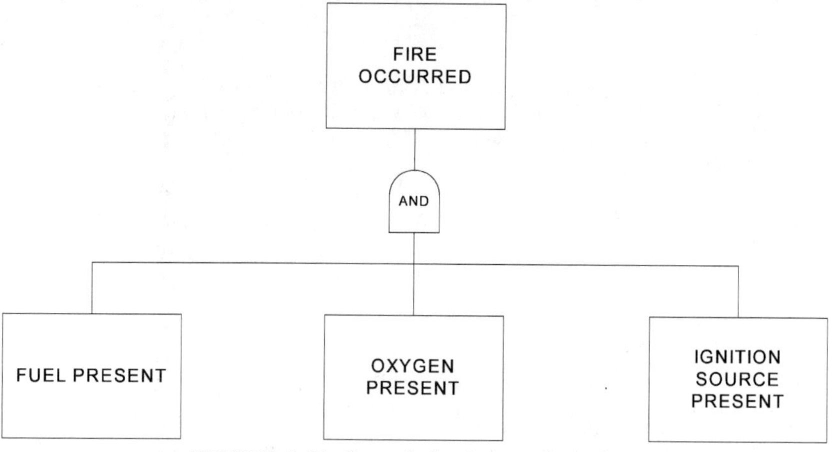

FIGURE 9-10. **Generic logic tree for a fire.**

The AND-gate is such that the *output occurs only if **all** the input events occur*. This **and** this **and** this must occur for the output event to happen.

Figure 9-11 illustrates an AND-gate: fuel, oxygen, and an ignition source must be present for a fire to occur. If any of these components were missing, the fire would not occur. These conditions are necessary and sufficient for the fire to occur.

Figure 9-12 illustrates an OR-gate. The OR-gate is such that the *output event occurs if any **one or more** of the input events occur.* This **or** this **or**

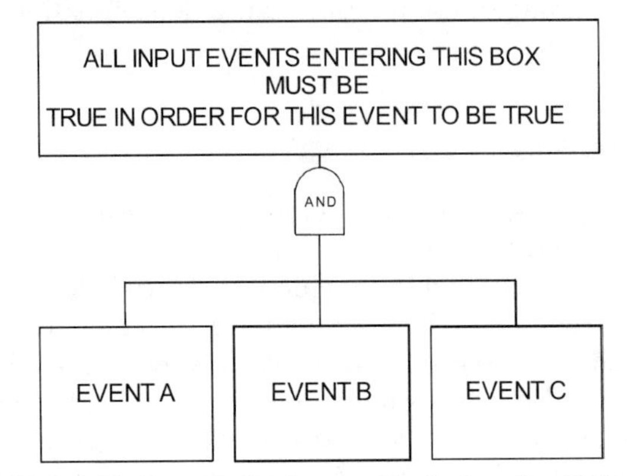

FIGURE 9-11. **Generic logic tree displaying the AND-gate.**

this . . . must occur for the event to happen. A good example of an OR-gate is an ignition source as shown in Figure 9-13.

The symbols for AND- and OR-gates are often omitted. Instead, the words are written above the connecting lines.

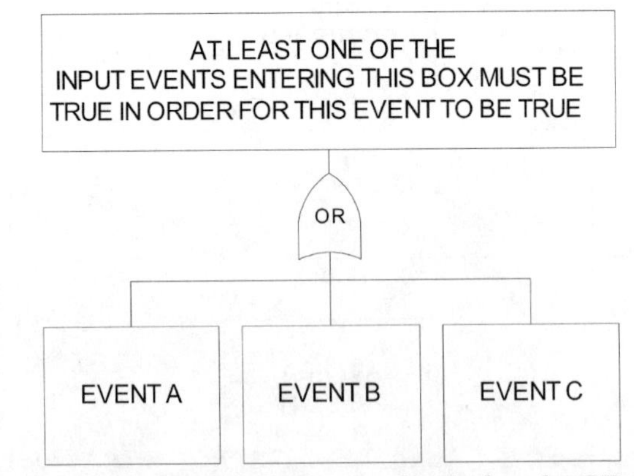

FIGURE 9-12. **Generic logic tree displaying the OR-gate.**

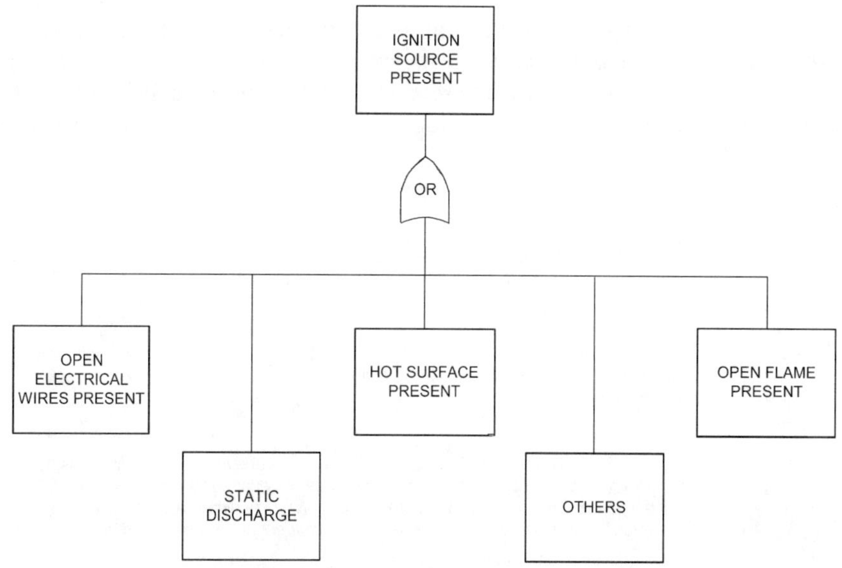

FIGURE 9-13. **Logic tree using the OR-gate to establish an ignition source.**

The tree logic should be checked every time a new event is added to the tree. If an event below the OR-gate is not sufficient on its own to cause the event above it, it needs to be joined with an AND-gate with the other necessary event. If an event below an AND-gate can cause the event above it on its own, then the event should be moved out from under the AND-gate and connected to the event by an OR-gate.

The investigation team uses an iterative process and begins to prove (accept, confirm, verify) or disprove (refute, reject) each of the OR-branches. Keeping the "OTHERS" box on the chart until very late in the tree development will help prevent the team from drawing premature conclusions.

Determining whether you have an AND- or an OR-gate becomes important when testing the tree logic, because the types of gates are tested in different ways. The type of gate is also important when developing recommendations. The recommendations will help reduce the frequency of an event when implementing the recommendations will add an AND-gate to the tree. A recommendation that eliminates only one branch of an OR-gate will be less effective (for instance, eliminating one ignition source out of many).

Some investigation techniques do not use OR-gates. If the team cannot figure out which input led to a top event, they stop tree development at that point. Speculation is not allowed. This is probably the best technique when investigating a human error incident that may involve arbitration or other potential legal issues.

Advanced investigators can use the following frequently used logic tree symbols; however, adequate logic trees can be developed without using them.

Small steps should be taken in developing the tree. One technique

BASIC Event: The BASIC event represents a basic equipment fault or failure that requires no further development into more basic faults or human failures.

INTERMEDIATE Event: The INTERMEDIATE event represents a fault event that results from the interactions of other fault events that are developed through logic gates such as those defined above.

UNDEVELOPED Event: The UNDEVELOPED event represents a fault event that is not examined further because information is unavailable or because its consequence is judged insignificant. In this book, the diamond with "Why?" inside it is used to indicate events which were not developed in the examples, but should be developed in actual investigations.

TRANSFER Symbol: As a tree is developed, identical sub-branches may appear in several branches. These duplicated structures are shown fully developed usually in only one place on the tree, and then handled by transfer gates. The symbols are labeled using numbers or a code system to ensure that they can be differentiated.

EXTERNAL or HOUSE Event: The EXTERNAL or HOUSE event represents a condition or an event that is assumed to exist as a boundary condition for the fault tree. The HOUSE symbol also implies that the event has a high probability of occurring.

available to help the team take small steps is to determine whether input blocks are active or passive. Active blocks are factors that change (examples: an ignition occurs, a valve is opened). In each AND-gate, there should only be one active event. The rest of the blocks in the AND-gate describe passive or existing conditions (examples: system contains pressure, people in control room). At the time the active event occurs, the gate happens.

Consider the incident scenario discussed previously:

A worker was walking on a concrete walkway in the process unit. There was some lube oil on the pad. He stepped into the oil, slipped, and fell. It was a sunny day; the worker was not carrying anything, was not distracted, and was not doing any particular urgent task.

The top portion of the logic tree may look something like the tree shown in Figure 9-15.

Logic Tree Tips

- Use a poster size piece of paper to start your tree. Use sticky notes (at least 2.5" x 2.5") as your event blocks. You can rearrange the blocks easily.
- To test an AND-gate: Take each input block one at a time and make it NOT true. If the event belongs in your AND-gate, the top event will not occur. Example: If the ignition source were NOT present, would the fire have occurred? No, the fire would not have occurred, so the ignition source is necessary for this AND-gate.
- To test an OR-gate: Each event by itself must be able to cause the top event.
- In an investigation meeting, test the logic tree with the group using these simple techniques. Why test as a group? It eliminates people getting defensive or upset if something they proposed is taken out.
- Remind the rest of the group that just because something doesn't belong in one level of the tree, doesn't mean it won't fit in another area of the tree.
- When you find a human error event, the type of gate is not particularly important, because while several factors may have combined to cause the error, a different person may have made the same error with only one factor present.

FIGURE 9-14. **Logic tree tips.**

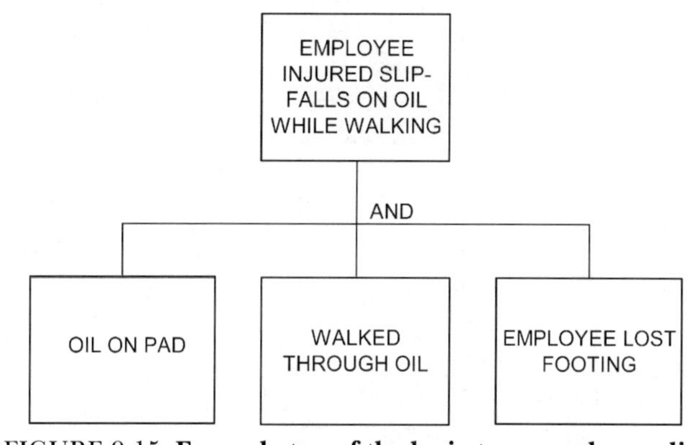

FIGURE 9-15. **Example top of the logic tree, employee slip.**

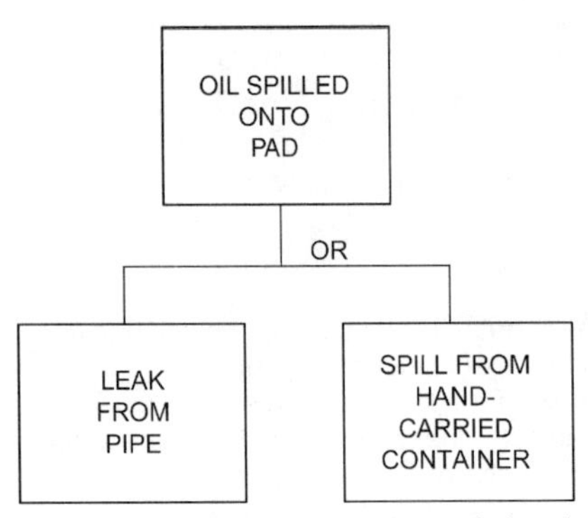

FIGURE 9-16. **Example logic tree branch level, oil spill.**

Each of the succeeding lower level events is further developed by repeatedly asking the question, "Why did this event occur?" Pursuing just one branch, for example the *Oil Spilled on Pad* branch would lead to at least two possible sources: *Leak from Pipe* and/or *Hand Carried Containers.*

Going a little farther down the tree and further developing just one of these sources, a spill from a hand-carried container would yield additional possible causes.

Each of these four subcategories can now be examined individually. Selecting one (*Right style container—but defective*) and returning to the concept of management systems leads to the following considerations.

- What is the management system involved in inspecting, repairing, or replacing the containers?
- Is the management system properly designed and arranged to achieve the desired output?
- Is the management system clearly understood and consistently enforced?

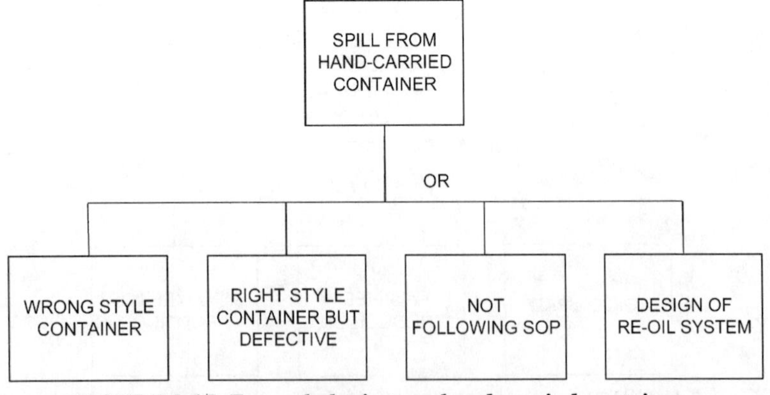

FIGURE 9-17. **Example logic tree, hand-carried containers.**

In this example, the team examined the pad surface and the employees' shoes and found both acceptable for the working conditions. Therefore, they decided "Employee lost footing" was a boundary event. The team decided to pursue the "Recognized Hazard but Walked Through it Anyway" further. Each of these items could also be pursued further.

A larger version of the tree is shown as Figure 9-18, but it is not completely developed.

9.6.3. Example—Chemical Spray Injury

Consider the following typical incident.

> *An employee is sprayed with an organic acid when loosening the handles on a filter hat in preparation for changing the filter. The employee is burned although the acid takes several minutes of contact before a burn occurs. It took the employee several minutes to get to the safety shower, because a pallet blocked the path to the closest shower. The employee went to another shower, which was farther away.*
>
> *The employee statements include: "The filter was already blocked in. I opened the drain and only a small amount of material came out, so I figured the last shift had already drained it. I can't believe that someone put that pallet there and blocked the shower access."*
>
> *A check of the records indicated the pallet had been delivered several days earlier.*

The top part of the tree looks like the one shown in Figure 9-19.

The team decides not to pursue the left branch any further, because the employee had to be in that location to open the filter.

Now, the team pursues the middle branch, shown in Figure 9-20.

The team continues asking, "Why?" and develops the branches. A more complete tree is shown in Figure 9-21, but is not fully developed. The team still needs to delve into the reasons the pallet was put in the path in the first place, and why it was allowed to remain there for several days.

They should answer questions like:

- Was there a procedure for where the pallet should have been spotted? If so, was the procedure followed?
- Do operations personnel approve locations for spotting materials?
- Do operations personnel receive any training in hazard recognition?
- Should prejob safety reviews be conducted?

Notice the event labeled "Drain/vent valve plugged" appears on two branches of the tree. This illustrates the concept of *common cause* failure,

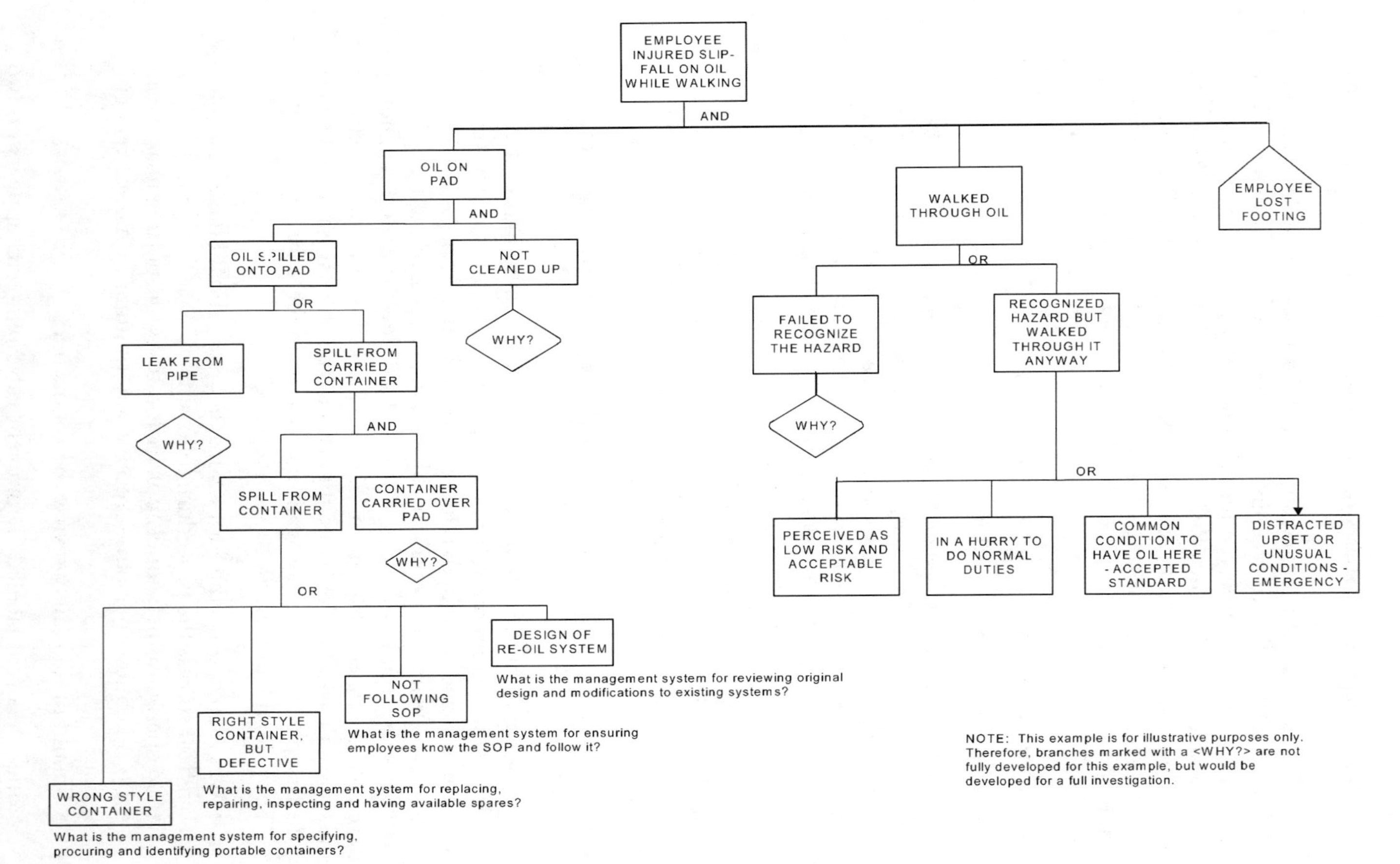

FIGURE 9-18. **Logic tree, slip/trip/fall incident.**

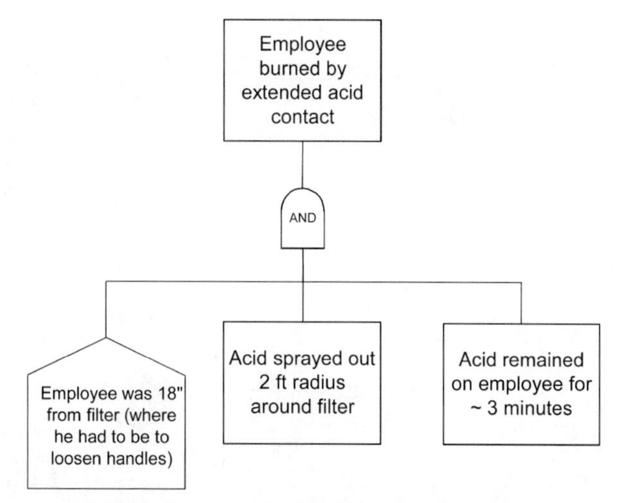

FIGURE 9-19. **Logic tree top, employee burn.**

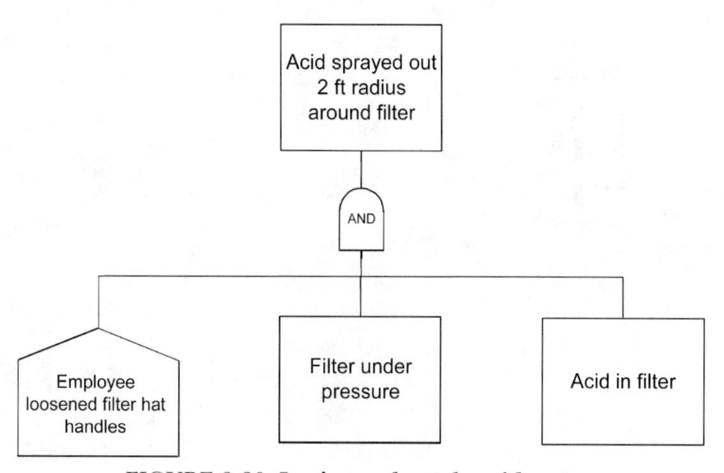

FIGURE 9-20. **Logic tree branch, acid spray.**

when the same cause appears in more than one place on the tree. In this case, the liquid drain and the vent valve were one and the same, located on the bottom of the filter. In addition, there was no way for the employee to tell if the pressure was still on the filter. In this case, the investigation team recommended that a pressure indicator and a separate vent valve be added to the filter. The pressure gauge may get plugged as well.

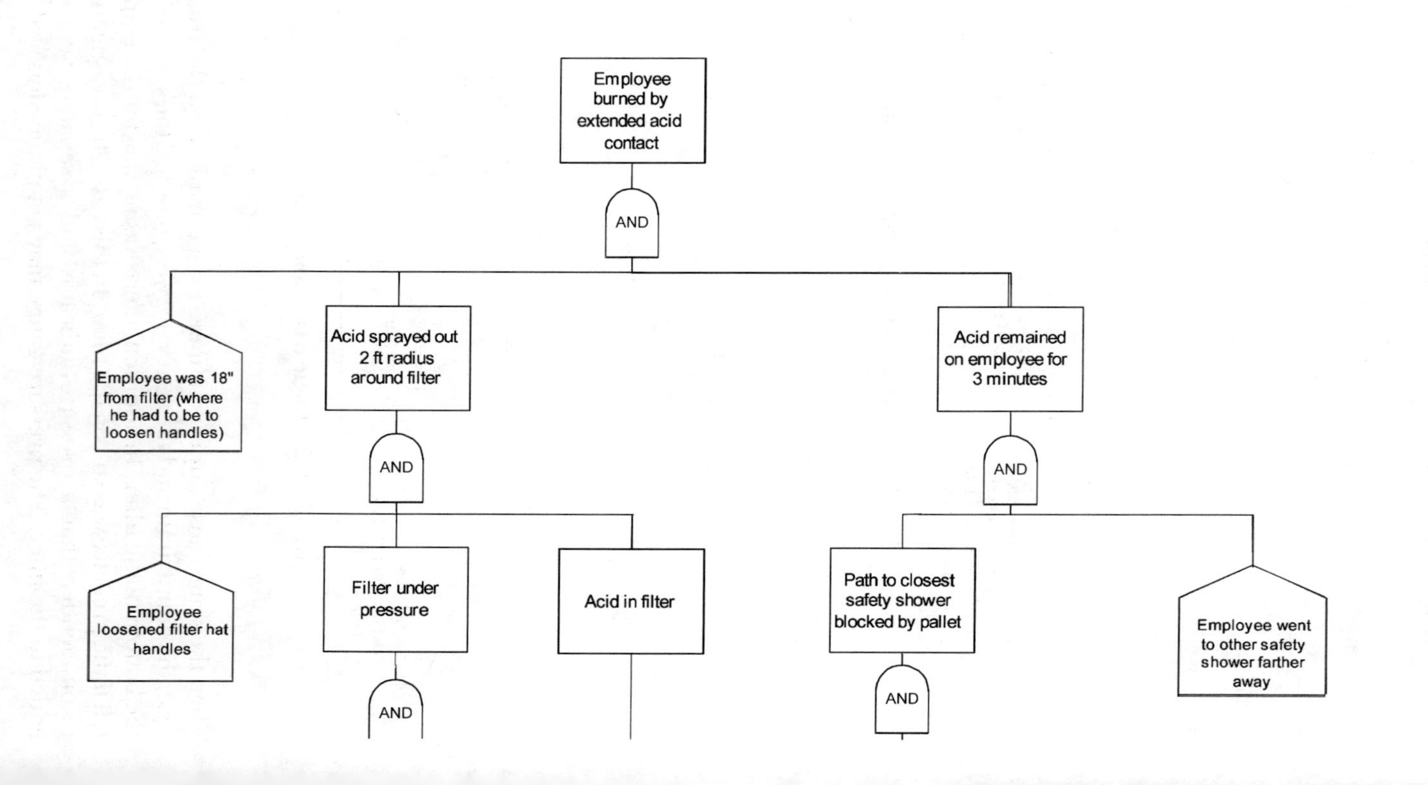

Employee burned by extended acid contact
AND
Employee was 18" from filter (where he had to be to loosen handles)
Acid sprayed out 2 ft radius around filter
Acid remained on employee for 3 minutes
AND
AND
Employee loosened filter hat handles
Filter under pressure
Acid in filter
Path to closest safety shower blocked by pallet
Employee went to other safety shower farther away
AND
AND

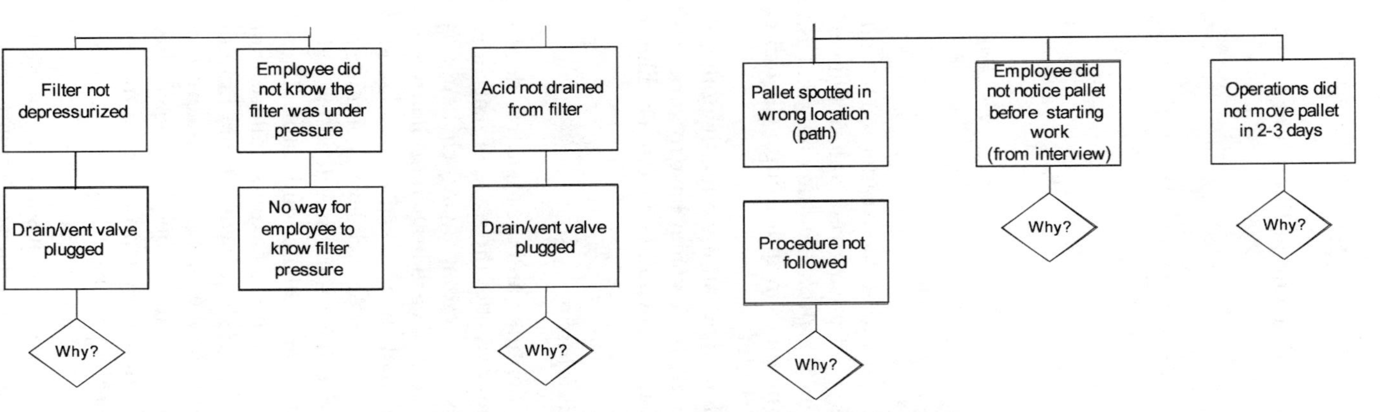

FIGURE 9-21. **Expanded logic tree sample, employee burn.**

The team should also be asking questions about the filter design, such as:

- Was the design of this filter reviewed?
- Has the filter design changed since installation? Were any such changes reviewed?

9.6.4. What to Do If the Process Stalls

The deductive process can stall for two major reasons:

1. There is no hypothesis for an event.
2. There are too many hypotheses for an event.

If there is no hypothesis for the event, use an inductive method to find potential scenarios. Inductive methods speculate a given fault or failure, then look forward in time to determine the probable outcome, that is, "What would happen if . . . ?" Inductive methods include using a Checklist or a Hazard and Operability Analysis (HAZOP).

If there are too many hypotheses for an event, use a fact-hypothesis matrix to help figure out which one it might be. You may have to work several branches of the logic tree if more than one hypothesis seems credible.

9.6.5. Guidelines for Stopping Tree Development

After the most likely scenario has been identified and the logic tree developed, the incident investigation team now reaches the stage of searching out the system-related multiple causes. An accompanying challenge is deciding when to stop further development of each branch of the tree.

Perhaps the most common mistake made by root cause investigation teams is to mistake a symptom for a root cause. At each level, continue to ask, "Why?" If you can easily answer, you have not gone deep enough. Management system deficiencies tend to be a reliable indicator. These deficiencies can include breakdowns, oversights, weakness, failures not anticipated, audits not performed, or changes not incorporated into all related systems. If a management system change is required to correct a deficiency, then the item is a strong candidate for being a root cause. Judgment is needed to determine a realistic stopping point for downward tree development. It is usually theoretically possible to develop another lower level for any event, but it may not be of any benefit.

A common intermediate level finding may be that someone failed to follow established procedure. (See Chapter 6 for details on human factors considerations.) Stopping at this point would be a mistake since *"failure to follow established procedure"* is rarely a root cause. A root cause approach would look further into the reason(s) for the operator failing to follow the procedure. Examples of possible reasons are given below.

- The procedure was unclear, hard to follow, out-of-date, sequence or facts wrong, or the situation was not covered.
- Employee perceived that a hazard was not significant.
- Enforcement or monitoring of procedures was inconsistent.
- The employee was in a hurry due to task overload (temporary or chronic).
- Some tool or supply was missing so the employee improvised.
- The employee was rewarded for violating the procedure in the past.

The management issues involved in these failures could include policies, standards, administrative controls, supervisory practices, or training.

Consider the case of a component failure in a physical system, such as a bolt or a gasket. When an unexpected failure occurs, it is for one of two reasons:

1. Something changed while the component was operating and an increased load was imposed on the component.
2. The strength of the component had degraded, but this degradation had gone undetected and/or uncorrected.

Investigators should keep developing the tree until they find issues such as:

- What was the management system involved in this failure?
- Why did the management system for plant surveillance, test, or inspection programs fail to detect the incipient failure?
- Why did the preventive maintenance program at the plant not prevent the failure?
- If the failure resulted directly from a human error, what was the underlying reason for this error?

For components or devices supplied by outside manufacturers, usually the downward progression is stopped at the component level, unless the device is normally opened, repaired, calibrated, adjusted, or inspected by in-house personnel. Electronic black boxes (similar to those under the hood of our automobiles) are good examples. We as owners may have occasion to manipulate the connection points (wires, attachment, and securing brackets), but we do not open nor usually attempt to diagnose an internal malfunction of these devices. For work on our automobile electronics black boxes, we turn to specially trained experts who use special equipment.

In a similar fashion, certain systems are assembled and maintained by operators of chemical plants. For example, for a control valve system, various components may be purchased separately and then assembled and configured by plant personnel. The incident investigation team would investigate possible accident causes associated with the methods of integration, assembly, maintenance, inspection, and calibration of the control valve system. Nevertheless, if a malfunction of a factory-sealed subcomponent were involved, the incident investigation team would seek out the appropriate expertise. The team would usually *not* attempt to analyze any factory-supplied components that normally remain sealed, without additional help. If the malfunction contributed to the incident, it should be investigated until it is understood, especially if similar components are in use elsewhere.

Another guideline is to stop the development of the tree when the events become external to the point that they can no longer be controlled by your organization. There are significant differences in the ability to control *internal* events as opposed to *external* events. Company "A" may experience a massive explosion and toxic vapor release that injures employees at the adjacent plant of Company "B." Investigators and managers at Company "B" may not be able to change systems within Company "A" to prevent a repeat explosion and release. Therefore, Company "B" may have to limit the focus to those internal actions that *they can* undertake to mitigate the effects of the release. Mitigating activities such as alerts, alarms, evacuations, shelter-in-place procedures, training, personal protective equipment, or other emergency preparedness and emergency response actions all represent internal actions that Company "B" *can* reasonably address. Each investigation team would stop its tree development at the point where they no longer had control of the events.

For serious incidents, the investigation team should document the truncation of branches and alternate cause scenarios that were disproved by the team. For major consequence occurrences or for occurrences where litigation is expected, the investigation team may have to defend its decision to reject certain potential cause scenarios. This defense may occur many months after the conclusion of the investigation team activities, so adequate documentation is critical.

9.7. Fact/Hypothesis Matrix

The fact/hypothesis matrix is another valuable tool available to the incident investigation team. This tool is used to compare the known facts against the various hypothetical scenarios. The matrix can help clarify thinking. It makes it easier to determine the most likely scenario and to refute other proposed scenarios based on the available facts. Using a matrix can help the team avoid jumping to conclusions and selecting a most likely scenario too early. The fact/hypothesis matrix can be used in the deductive stage or the inductive stage.

Real life scenarios for many process safety incidents are often rather complex. The fact/hypothesis matrix technique has proven to be useful in sorting, analyzing, and comparing information. One side of the matrix lists each plausible scenario, and the other side (usually across the top) lists the known facts, conditions, and stipulations. Each intersecting box is then examined for compatibility, known truthfulness (yes, no or unknown), and its logical fit into the particular scenario hypothesis. The degree of complexity of this matrix can vary depending on the nature of the incident, from a simple [YES/NO/?] to a variety of categories and sub-categories. A more complex set of matrix conditions might take the following form:

- **+** the fact supports the scenario
- **–** the fact refutes the scenario
- **NA** this fact apparently has no relation to this hypothesis, it neither supports or refutes the scenario
- **?** not enough information is currently available to decide on this fact

A sample appears in Figure 9-22 on page 218 for a fictitious incident. Developing the matrix is not a one-time exercise, but is usually prepared over the course of the investigation. Gradually, some hypothesis will emerge as more likely and others will become less probable. It is very helpful to others to keep unlikely scenarios on the matrix and document why the scenario was ruled out. Seeing why their pet theory was ruled out can help people accept the team's conclusions.

When determining root causes, one of the early tasks is to determine the most likely scenario and then identify the causes that led to that ultimate occurrence. It may be helpful to the investigation team to consider trying to disprove scenarios in addition to the traditional approach of proving a speculated cause scenario (hypothesis).

In addition to proving the speculated cause scenario, the team should also make a conscientious effort to disprove the scenario that has been selected as the most likely cause scenario. Most investigation teams are competent at proving the selected scenario, but it is a recognized fact that we are not as capable at disproving a favored hypothesis (scenario). In the field of critical and logical thinking, there is a concept of *falsifiability* where a specific effort is made to disprove a speculated hypothesis, in addition to the efforts made to prove the hypothesis.

It is our normal pattern to very quickly (and automatically) form a hypothesis and then begin to seek confirming evidence. We do not inherently place emphasis on seeking evidence that might disprove our hypothesis. We tend to stick to (and vigorously defend) our original hypothesis even when faced with conflicting evidence that might disprove our desired hypothesis. Investigators therefore should make a strong and conscientious effort to operate with an open, unbiased approach, especially during the early phases of an investigation. This tool can be useful with both Method A and Method B as presented in this chapter.

9.7.1. Application of Fact/Hypothesis Matrix

Example case:

Explosion and fire in an anhydrous pachydermguanadiene reduction (APGR) unit. (*Note*: This is a fictional material.)

Background:

- Tank exploded at 1:30 A.M., 3 hours after a batch product transfer was made.
- Maintenance had recently replaced a gasket on the transfer pump.
- There was a power outage shortly before the incident.

Status:

The investigation is currently incomplete. The incident investigation team is in the second day of investigation, has accumulated some evidence, and has begun to compare known facts against possible scenarios.

APGR Explosion

Fact or Condition → Hypothesis↓	Power Tripped Out 4:09 PM	Operator Added Component "A" to Batch at 10:30 PM	Storage Tank Transfer Made on Evening Shift 7:30 PM	Maintenance Changed Gasket P120B	Top of Tank Found On East Side of Warehouse	Lab Analysis Showed Zero Water in Residue	Comment
Contaminated Batch of Incoming Raw Materials	?	+	?	+	+	-	
New SOP Not Followed	?	?	+	+	+	?	
Engineering May Have Designed or Installed Wrong Gasket	NA	NA	+	+	-	NA	
Oxygen Entered Nitrogen Header from Back Flow-Preventer Device Failure	?	+	?	NA	+	-	
Oxygen entered the system during maintenance	NA	?	?	+	+	-	

Legend: (+) - the fact supports the scenario; (-) - the fact refutes the scenario; (NA) - this fact apparently is not related to this hypothesis, neither supports or refutes; (?) - not enough information is currently available to decide on this fact

FIGURE 9-22. **Example fact/hypothesis matrix.**

The matrix in Figure 9-22 is the result of their initial deliberations and is being used to develop action items and priorities, such as which direction the information gathering and cause analysis should proceed.

9.8. Case Histories and Example Applications

9.8.1. Fire and Explosion Incident—Fault Tree

The fictitious process safety incident contained in Appendix D can be used to illustrate the application of how a fact/hypothesis matrix can be used during logic tree development. Extensive details of the incident appear in the appendix but a basic summary would be:

A major fire and explosion occurred in a polyethylene manufacturing facility resulting in one fatality, five personnel injuries, and extensive damage. The fire originated in the catalyst area when a vessel was overfilled and the exit piping ruptured, releasing isopentane, a flammable material, and aluminum alkyl, a pyrophoric material. The first fireball, at approximately 11:10 AM, caused an operator fatality and a contractor injury. Emergency response was impaired because the firewater pumps were inoperable, which contributed to the severity of the consequences. The fire spread to the vertical catalyst storage tank. A subsequent explosion of an adjacent catalyst storage tank resulted in the injury of four firefighters. The local fire department and plant fire brigade extinguished the fire at 12:10.

For this illustration, the first event will be considered. The top portion of the tree for the operator fatality is developed in Figure 9-23.

The pool fire branch is further developed in Figure 9-24.

At this point, the investigation team reaches a stage where they have more than one hypothesis for the reason the isopentane line cracked. The pressure could have exceeded the design pressure for the pipe or the pipe could have failed at a point below the design pressure.

The team could use a simple fact-hypothesis matrix to decide which branch to pursue. An example matrix is shown as Figure 9-25.

In this example, assume the team obtained pipe samples of some of the remaining pipe and finds evidence of external corrosion. The team concluded that the feed line failed due to higher than normal pressure combined with corrosion of the piping system (an AND-gate). These relationships are shown in Figure 9-26.

What if the team was not able to obtain any physical evidence? They could use the absence of any corrosion inspection records plus knowledge

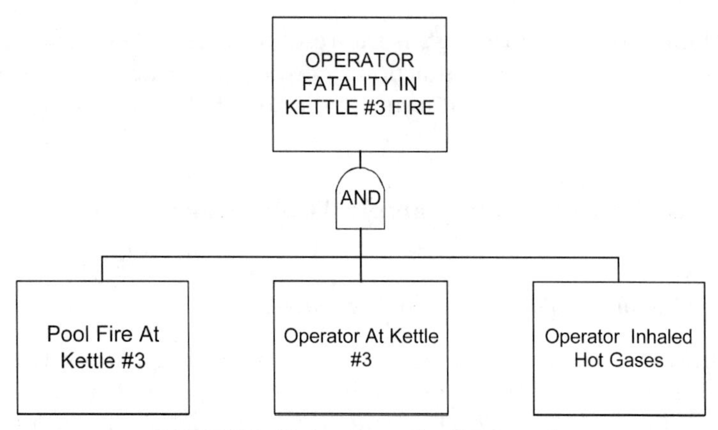

FIGURE 9-23. **Operator fatality branch.**

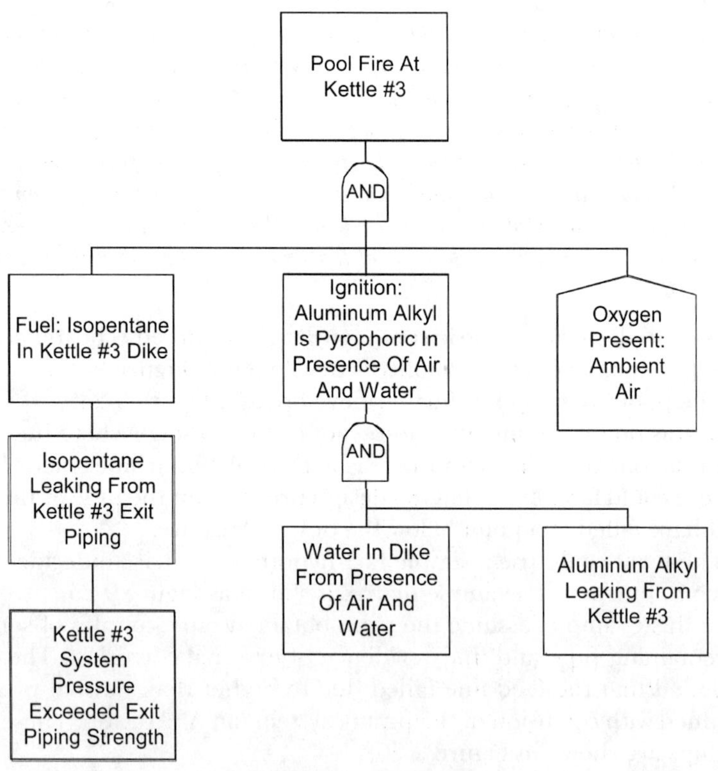

FIGURE 9-24. **Fire branch.**

Known facts→ Possible Scenarios↓	Isopentane Pipe samples show external corrosion especially in heat affected zones	Pressure indicator on system went up to 120 psig.	Pipe samples found with split running along pipe	Maintenance records indicate correct material and schedule used for repairs	Area has clearly understood gasket chart
Design pressure of 150 psig vessel exceeded	**NA**	–	**NA**	**NA**	**NA**
Pipe/vessel failed below design pressure due to corrosion	+	+	+	**NA**	**NA**
Pipe/vessel failed below design pressure due to flange gasket failure	**NA**	+	**NA**	**NA**	–
Pipe/vessel failed below design pressure due to wrong material installed	**NA**	+	– samples were correct material	–	**NA**

Legend: (+) - the fact supports the scenario; (NA) - this fact apparently has no relation to this hypothesis, it neither supports nor refutes the scenario; (-) - the fact refutes the scenario; (?) - not enough information is currently available to decide on this fact

FIGURE 9-25. **Fact/hypothesis matrix for the kettle exit piping failure.**

of the expected corrosion (internal and external) of the system as an indicator of whether corrosion was a credible possibility.

With no evidence at all, the team might develop each hypothesis as a separate branch of the tree and try to address potential causes of corrosion, improper choice of materials, flange failure, or other items.

After collecting and analyzing the available evidence, the incident investigation team constructed the logic tree diagrams shown in Appendix D. These diagrams present, in a logical and systematic format, the sequence of events and conditions that ultimately resulted in the major incident. The simplified qualitative fault-tree indicates various events and conditions that could have contributed to the incident causation and progression. Some of these sequences acted with direct impact on the trigger event, the pipe failing and initial fire, while others acted to increase the severity.

The incident investigation team's complete report is attached in the Appendix D and details of the root causes are discussed. The root causes of the incident relate to several process safety management areas:

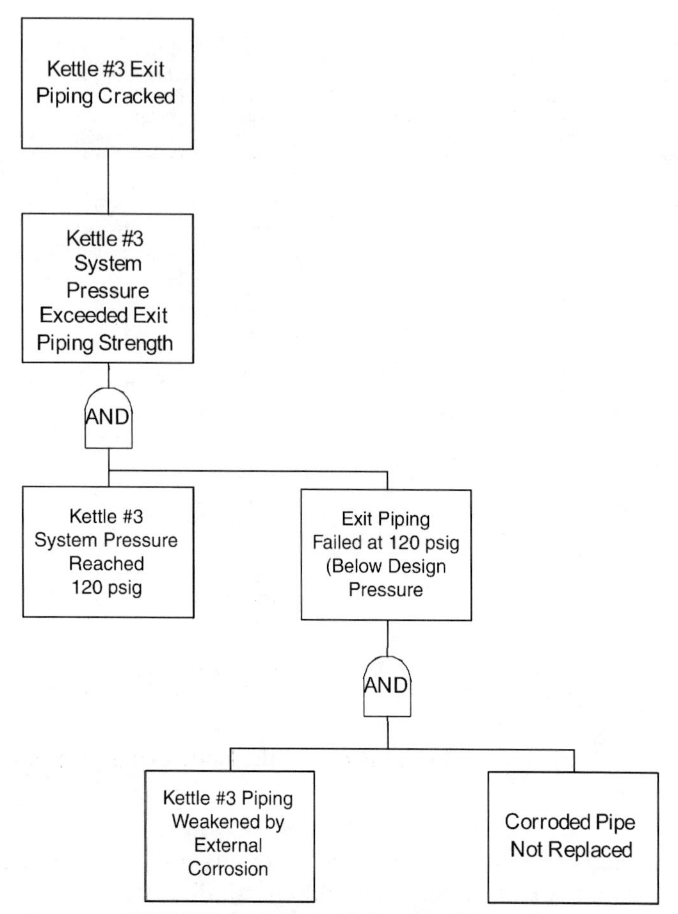

FIGURE 9-26. **Exit piping crack branch.**

- Mechanical integrity
- Contractors
- Emergency planning and response
- Process hazards analysis
- Management of change

Take the time to review the complete example in the Appendix. Look at the trees and think about what the root causes might have been if the chosen top event had been the release of isopentane. Would the team have made a recommendation about escorting and training contractors?

9.8.2. Data Driven Cause Analysis

A specialized approach is to use historical data to infer or identify potential causes. In this case, the investigators use past experience to look for patterns consistent with failure hypotheses. The technique is only as good as the records, and if data have not been put in the files, or are in error, then misleading inferences may result. In addition, if this type of event has not occurred before, the approach will not work. Failure data for the system under investigation are presented in a timeline that can be correlated with overall plant history. Two things are sought:

1. Evidence for correlation with plant state, plant condition, or external environmental effects.
2. Evidence that indicates a failure pattern that may correlate with maintenance activities.

CASE STUDY:

In a plant with a shaft-driven boiler feedwater pump, problems had historically occurred with failure of the bearing in the hydraulic coupling. There was no apparent cause, but throughout the 12-year life of the equipment, the failure occurred about once every 1 or 2 years and resulted in an outage of about 3 weeks.

Plant data did not indicate a cause, other than bearing failure—repaired/replaced. A detailed root cause investigation had never been performed. When an investigation was conducted, the equipment was operable and detailed evidence from the last failure was lost during the repairs.

A timeline for the failures was developed and patterns sought. The first thing that became clear was that failures seemed to occur predominantly following an outage in the winter. This immediately led to the thought that temperature was an important influence. The equipment is in a heated building so all components *should* have been at room temperature. If wintertime temperature was contributing to the problem, it was most likely a result of overcooling by one of the cooling systems, or normal lubricating oil systems operating at too low a temperature. The written reports of the previous failures stated that "bearing wipe" was the cause. This fact indicated that lubrication failure was a likely candidate.

When operational characteristics of the oil systems were examined in detail, it was found that the oil supply came from the main turbine lube oil system. The operators said that after a start in cold weather they had trouble maintaining anything but the minimum lube oil temperature of 120°F (49°C) until the turbine was at power. The feed pump hydraulic coupling specifications indicated that a minimum temperature of 140–160°F

(60–71°C) was required for proper operation. A reasonable failure hypothesis was that, during a start, oil temperature was too low (and therefore viscosity too high) to provide adequate flow and lubrication of the pilot bearings. This allowed excessive frictional contact and resulted in a "wiped" bearing. The corrective action was to heat the oil feed to the coupling; that reduced the number of failures dramatically.

This example is intended to show that all of the elements of the cause determination process were used, but not quite so formally as the process implies. This is important, because different techniques achieve the intent of the process via specific but different approaches. The investigator must understand the functional objectives that provide the foundation of the multiple cause determination. Without this understanding, a "shotgun approach" is often used; there is neither rigor nor a search for completeness. The first identified potential cause often becomes the cause, and the investigation terminates. This is one reason that failures recur although remedial action was taken after an earlier failure. There is also a tendency to stop the process at the intermediate causes level. In the case study, the general cause of bearing wiping was lubrication failure. Suggested cures were proposed for bearing redesign, new materials, vibration monitoring, etc. Even modified bearings would be prone to continued failure following winter outages until the underlying cause—low temperature—was identified and corrected.

9.9. Root Cause Determination Using Predefined Trees—Method B

As discussed in Chapter 4, many CCPS member companies report that they generally use one of two main methodologies to determine root causes. The first involves timeline construction followed by logic tree development as discussed above. The second main methodology involves timeline construction, identification of causal factors, followed by the use of predefined trees or checklists. This latter approach is discussed in detail below.

The following section presents a systematic discussion of the concepts and actions depicted in Figure 9-27.

The initial tasks are similar to those of the logic trees previously described:

- The accumulation of facts, information, observations, insights, questions, and preliminary speculations gained from the evidence collection activities.
- The development of a chronology of events leading to the incident based on the available known times and sequences using a *timeline*.

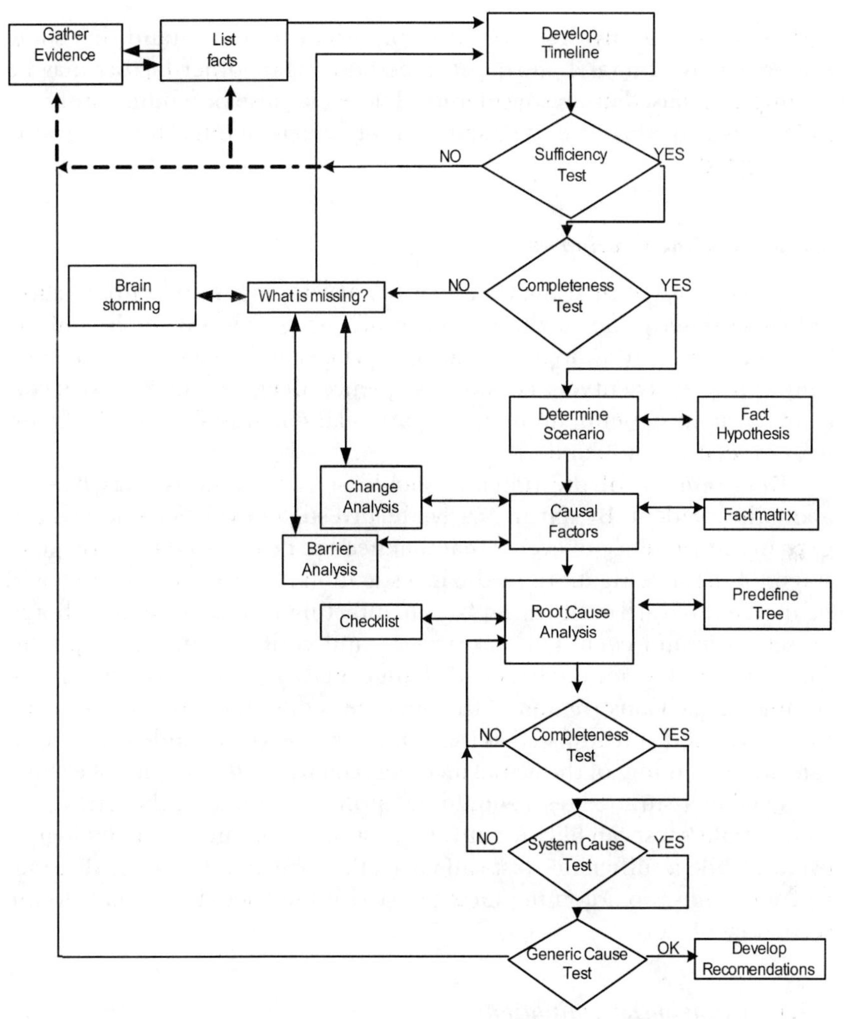

FIGURE 9-27. **Flowchart for root cause determination—predefined tree/checklist.**

These tasks have been discussed previously. The following represents a brief summary, but more detail about organizing data with a timeline is provided in Section 9.3.

9.9.1. Evidence Gathering

The first phase of process safety incident investigation involves gathering all the pertinent facts from the collected evidence, whether derived from interviews, site and equipment inspections, or document reviews. At this early stage of the investigation, it is important to understand that some evidence may ultimately turn out to be false, while other factors may be assumptions based upon speculation. There may also be a number of possible scenarios for the incident, all of which should be rigorously investigated.

9.9.2. Timeline Development

The next phase of investigation involves developing a preliminary chronological description of the sequence of events that led to the failure. Timelines can be developed in various formats and levels of detail, from simple lists of events to complex sequence diagrams or causal factor charts, usually dependent upon the particular circumstances of the investigation being conducted.

Development of the timeline should start as soon as facts emerge about the incident. By starting early, the investigator will become aware of gaps in the sequence of events that that need to be resolved in subsequent stages of the investigation. As the investigation progresses, efforts should be directed toward confirming facts (conflicting or false evidence should be set aside and eventually discarded), and verifying any assumptions. Tools such as barrier analysis and change analysis may be used to ensure completeness. Construction of the timeline is therefore an iterative activity; it is refined and adjusted as the team gains a more complete and accurate understanding of the actual incident scenario and sequence of events.

Quality assurance tests should be applied to enhance the process. A completeness test can help identify gaps or omissions in the chronology of events, while a sufficiency test can check the logic in a sequence diagram by comparing two adjoining facts to verify if one fact always leads to the second fact.

9.3.3. Scenario Determination

If necessary to assist in determining the actual incident scenario from two or more possible scenarios, a fact/hypothesis matrix may be employed to help resolve conflicting facts. This is most efficiently performed prior to identifying the causal factors.

9.9.4. Causal Factors

Once the timeline or sequence diagram based upon the actual scenario has been developed, the next phase of the investigation involves identifying the causal factors. Causal factors involve human errors and equipment failures that led to the incident, but can also be undesirable conditions and failed barriers. The causal factors are the negative events and actions that made a major contribution to the incident. They can be identified by asking whether the incident would have occurred if each event on the timeline had not existed.

The process of evidence gathering, timeline development, scenario determination, and causal factor identification is somewhat iterative, and therefore some of the tools and quality tests previously described may assist in causal factor identification. More specifically, barrier analysis and change analysis, together with a completeness test, can ensure that all valid causal factors are identified.

9.9.5. Predefined Tree

The causal factors need to be examined further to determine why those factors existed. The investigation team may use a predefined tree to examine each causal factor individually. The first causal factor is analyzed starting at the top of the tree, and then working down all of the branches as far as the facts permit. When an appropriate subcategory on one of the branches is identified, it is recorded as a root cause. The remaining branches are checked as one causal factor may have multiple root causes. The procedure is then repeated for each causal factor in turn.

Several quality assurance tests should be applied when using predefined trees. Firstly, predefined trees are designed to capture most root causes, but may not be comprehensive. A completeness check should be conducted on each branch of the tree to see if there are other root causes associated with the category of that branch that are not listed on the tree.

Some predefined trees do fully reach down to the root cause level. A system test should be applied to each identified root cause to ensure that it relates to a management system failure. By applying the 5-Whys tool to each cause identified at the end of the relevant branches of the tree, the investigator can determine if another underlying cause can be identified.

After the predefined tree has been used, a final generic cause test should be applied. The plant operating history, especially previous incidents, is considered to indicate if other generic management system problems exist. For example, repetitive failures may indicate generic causes that would not be apparent by only investigating the current incident. It is also an opportunity for a final overall review of the investigation to focus on the big picture, not just individual facts or causal factors. The team should ask, *"Are there any other causes that anyone has in mind that have not been included?"*

If the incident investigation team is satisfied with the root causes identified, then the investigation proceeds to the recommendation stage. If a problem or some incompleteness is noted, then an iterative loop is followed.

9.10. Causal Factor Identification

Once the evidence has been collected, a timeline or sequence diagram developed, and the actual scenario confirmed, the investigation can proceed to the next stage, the identification of causal factors. These causal factors are the negative events and actions that made a major contribution to the incident.

Causal Factor*—a major unplanned, unintended contributor to the incident (a negative event or undesirable condition), that if eliminated would have either prevented the occurrence, or reduced its severity or frequency.*

Causal factors involve human errors and equipment failures that led to the incident, but they can also be undesirable conditions, failed barriers (layers of protection, such as process controls or operating procedures), and energy flows.

In practice, the initial stages of the investigation are an iterative process with a significant degree of overlap. For example, the events and causal factors on a preliminary timeline or sequence diagram may be revised many times as new facts about the failure mode and sequence of events emerge, and the investigator seeks evidence that confirms the actual incident scenario. Sometimes it may be necessary to progress the investigation along more than one scenario until the team gains a complete understanding of the actual incident scenario. Because of this iterative process, the investigator's initial perspective of causal factors may change, and it is therefore important to remain objective and not jump to conclusions too early.

Causal factor identification is relatively easy to learn and apply to simple incidents. For more complex incidents with complicated timelines, one or more causal factors can easily be overlooked, however, which inevitably will result in failure to identify their root causes. There are a number of tools, such as Barrier Analysis, Change Analysis, and Fault Tree Analysis, that can assist with bridging gaps in data and the identification of causal factors. Each of these tools has merits that can assist the investigator in understanding *what happened* and *how it happened.*

9.10.1. Identifying Causal Factors

The simplest technique for identifying causal factors involves reviewing each event or condition on the timeline. The investigator repeatedly asks the following question:

> *Would the result have been significantly different if the event or condition had not existed at the time of the incident?*

If the answer is YES, that is, the incident would have been prevented or mitigated, and it is a negative event or undesirable condition, then the fact is a causal factor. Generally, process safety incidents involve multiple causal factors. This technique is equivalent to step #15 in Figure 9-7.[6] Once identified, the causal factors become the candidates to undergo root cause analysis.

The investigator may streamline this technique by focusing upon each unplanned, unintended, and/or adverse fact (negative event or undesirable condition) on the timeline. It is also important to recognize those items that are still speculative and based on an assumption, as these will need to be tested later to verify if they are accurate facts.

It is critically important that the wording or the phrasing of each causal factor accurately and clearly describes the factor. Teams will struggle with cause analysis if the causal factor is not crystal clear to all. In the case of an incident arising from work on a pump that has not been adequately isolated from energy sources, an investigation team may say one causal factor is 'no lockout/tagout'. However, this short statement can be interpreted in a number of ways, depending upon individual team members' views of the evidence and personal biases.

For example, "no lockout/tagout" can mean:

- No procedures for LO/TO exist
- Procedures exist but the employees involved had no knowledge of them
- An attempt was made to perform LO/TO, but no locks were available
- No effort was made, a rules violation.

When different team members approach an issue from different directions, it is not unusual to see much churning in the causation discussions. It is better to resolve what the team believes the evidential issue is before starting the root cause analysis.

In the same example, if the team has the evidence to support that there were adequate procedures, training and equipment in place, and the failure involved a technician circumventing the rules, the causal factor should be worded along the lines of, "Technician failed to install the required locks and tags on the pump." This provides little room for differing interpretations.

If the team is not settled on what the evidence tells them, it is indicative that more investigating needs to be done—cause analysis is premature at this point.

Some investigators review each of the causal factors to determine the immediate causes (unsafe acts or unsafe conditions) of the incident, as an intermediate step before proceeding to determining the root causes. Causal factors can be immediate causes.

The following tools can assist with the identification of causal factors for complex incidents with complicated timelines.

9.10.2. Barrier Analysis

The design of most process plants relies on redundant safety features or layers of protection, such that multiple layers must fail before a serious incident occurs. Barrier analysis[9] (also called Hazard–Barrier–Target Analysis, HBTA) can assist the identification of causal factors by identifying which safety feature(s) failed to function as desired and allowed the sequence of events to occur. These safety features or barriers are anything that is used to protect a system or person from a hazard including both physical and administrative layers of protection. The concepts of the hazard–barrier–target theory of incident causation are encompassed in this tool. (See Chapter 3.)

The term *barrier* encompasses a wide range of safeguards and preventative measures. Some examples of barriers are:

Physical
- Closed valve
- Blast/fire wall
- Electrical insulation
- PPE

Natural
- Distance
- Time
- Laws of nature

Administrative
- Standard Operating Procedure (SOP)
- Pre-Startup Checklists
- Lockout/Tagout Procedure
- Design Standard

Human Action
- Supervision
- Manually Controlling Process
- Monitoring Process Parameters
- Taking Process Samples

Barrier Analysis may be performed by asking a series of questions while studying the timeline or sequence diagram. Typical questions are:

- What physical, natural, human action, and administrative controls are in place as barriers to prevent the incident?
- Where in the sequence of events would these barriers prevent the incident?
- Which barriers failed to work?
- Which barriers worked successfully?
- What other physical, natural, human action and administrative controls might have prevented the incident if they had been in place?

The tool helps the investigator to understand and focus on the failed barriers, which are normally identified as causal factors. These failed barriers may need to be strengthened, replaced, or supplemented, especially where weak administrative controls are highlighted. Even successful barriers that prevented more serious consequences may require reinforcement. Therefore, barrier analysis can give the investigator valuable insights into how the incident happened and some of the multiple causes that need corrective action to prevent recurrence.

9.10.3. Change Analysis

Change analysis[10] (also known as Change Evaluation/Analysis, CE/A) is another tool that can assist the identification of causal factors. It is useful for brainstorming about what has changed since conditions were safe, or perceived as safe. It may also be used for hypothesizing potential contributory factors to a hazardous condition or action.

A basic premise of change analysis is that "change signals trouble." If a system performs satisfactorily for a period and then suddenly fails, the failure will be due to a change(s) in the system. By identifying what has changed, it should be possible to determine the factors that led to that failure arising. Six investigative steps characterize the technique.

1. Examine the incident situation.
2. Consider similar, but incident-free situations.
3. Compare the two situations.
4. Write down all the differences between the two situations, whether they appear relevant or not.
5. Analyze the differences for effect on causing the incident.
6. Integrate the differences into incident causal factors.

The practical application of these basic ideas has been incorporated into a worksheet. Potential influencing factors are listed on the left column of the worksheet, and then the influence of each of these factors on the incident is tabulated in terms of:

- the present situation,
- prior comparable situations,
- the differences that exist between the two situations, and
- affective changes that have taken place.

The potential influencing factors are selected based on whatever is relevant to the task or process, and will likely include factors such as working conditions, administrative controls, communication, supervision, training, and equipment design, operation, and maintenance. The inclusion of questions such as *What? Where? When?* and *Who?* may also be helpful. The worksheet may be developed using a flip chart on which the input from the investigation team, aided by a facilitator, is documented.

Any change that caused an adverse impact should be designated as a causal factor and added to any causal factors identified by other means. The investigator may also wish to review the local management of change system and identify why a change(s) that had an undesirable outcome was allowed to occur.

9.10.4. Quality Assurance

There are a number of quality assurance checks that should be considered *before* identifying the final list of causal factors. It is important to test for *sufficiency* of the information when compiling a sequence diagram. This test for sufficiency may be performed by asking one or more of the following questions, when comparing two adjoining facts in the sequence of events:

- Will (insert the complete statement of Fact B here) always lead to (insert the complete statement of Fact A here)?
- Every time Fact B occurs, does Fact A have to follow?
- Just because Fact B occurs, will Fact A always follow?
- Are there any layers of protection that should have prevented Fact B from progressing to Fact A?
- Does anything else have to occur or does any other condition have to be satisfied for Fact B to lead to Fact A?
- Are there any other potential causes of Fact A other than Fact B?

The use of this sufficiency test is described in Figure 9-7 on page 195.

Once the timeline or sequence diagram has been compiled, a test for *completeness* should be performed by reviewing the entire chronology for any omissions or gaps. The investigation should then focus on gathering evidence on any identified gaps and adding new data to the sequence diagram. Barrier analysis and change analysis may be used in addition to brainstorming to assist this test. Any new data added to the diagram should be subjected to the sufficiency test above.

The entire timeline or sequence diagram should also be reviewed to identify any conflicting facts. The aim should be to determine a single scenario of events that caused the incident, although on occasion it may not be possible to distinguish between potential scenarios. The fact/hypothesis matrix approach should be used to resolve any conflicting facts and determine the most likely scenario. It may not be necessary to tabulate the data in a matrix, but the same logic should be applied in comparing all of the information.

9.10.5. Causal Factor Summary

The identification of causal factors points us to the key areas that need to be examined further for why that factor existed. It acts as a filter to limit the number of areas that are subjected to further analysis to determine root causes. This critical activity must be performed diligently and systematically to identify every causal factor applicable to the specific incident. If a causal factor is missed, one or more root causes will likely be omitted as well, which could lead to similar incidents in the future.

9.11. Predefined Trees

Once the actual incident scenario is understood and its multiple causal factors identified, this information may be used to determine the incident's root causes. One means of performing root cause analysis involves the use of ready-made, predefined trees. A predefined tree provides a systematic approach for analyzing and selecting the relevant elements of the incident scenario. It is a deductive approach, looking backward in time to examine preceding events necessary to produce the specified incident.

Predefined trees arrange a relatively complete list of potential root causes organized by subject matter, such as human error and equipment failure, into various categories and subcategories in a hierarchy of branches and sub-branches. Although the trees do not display any logic symbology, each of the nodes between branches and sub-branches represents an OR-gate. An example of a section from a proprietary predefined tree[7] is shown in Figure 9-28.

Unlike the procedure followed in developing logic trees, the investigation team does not construct the tree. Rather they apply each causal factor to each branch of the predefined tree in turn, and those branches that are not relevant to the incident are discarded. This prescriptive approach offers consistency and repeatability by presenting different investigators with the same standard set of possible root causes for each incident.

The consistency offered by predefined trees with standard categories and subcategories of root causes also facilitates statistical trend analysis. This allows an organization to more easily collect and analyze data from the investigation of incidents and near misses over a period of time to determine any trends not apparent from single incidents. Some organizations deliberately structure the root cause categories and subcategories along the lines of their management system in order to focus on common system issues. Incident trend analysis is discussed further in Chapter 13.

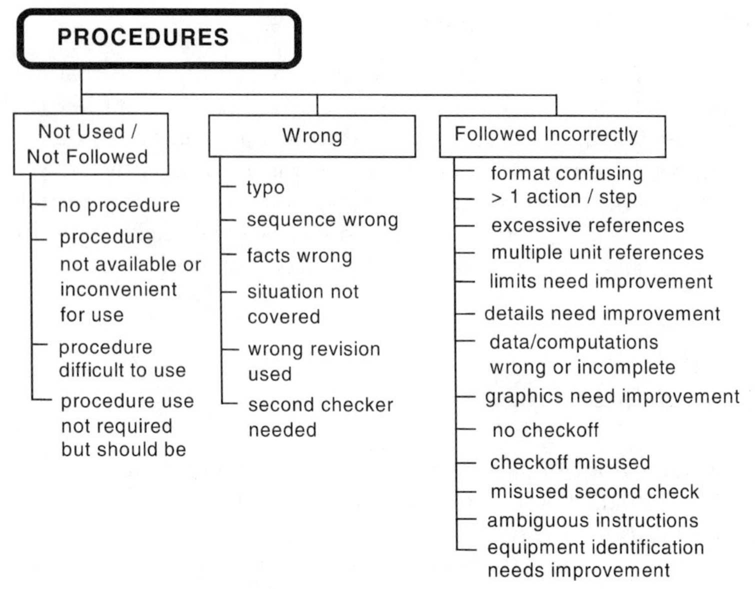

FIGURE 9-28. **Example of root causes arranged hierarchically within a section of a predefined tree.**[7]

While the use of predefined trees does not directly challenge the investigation team to think laterally of other possible causes, many predefined trees present a wide range of causal factors, some of which the team may not have otherwise considered. It is therefore possible, but unlikely, that the incident could involve a novel root cause that was not previously experienced by those who developed the predefined tree. The addition of a final test based on another tool, such as brainstorming, can overcome this apparent weakness.

9.11.1. Background—MORT

The Management Oversight Risk Tree (MORT) [8,11] was developed by the System Safety Development Center (SSDC) of the Department of Energy (DOE) for the investigation of occupational incidents at DOE sites. Although MORT is loosely based on fault tree analysis (FTA) logic, it represents one of the earliest predefined trees. The branches of MORT were modeled after safety management system best practices. Many of the process safety incident investigation tools used in the chemical and allied industries today, which are described elsewhere in this book, are based upon the original MORT concept.

The barrier and energy concepts are central to the MORT model. An accident is defined as an event for which a barrier to an unwanted energy flow is inadequate or fails with adverse consequences. Wherever there is a risk that persons or objects may encounter an energy flow or an environmental condition that could cause harm, it is necessary to isolate the energy flow or the environmental condition.

Application of the MORT technique is based on a predefined tree structure laid out vertically (top down) in eight interconnecting trees. The structure is quite complex, but contains 98 generic problem areas and up to 1500 possible causes. A user's manual [(12)] provides detailed instructions on how to use the tree.

The MORT diagram starts with the incident, which is equivalent to the top event in FTA. The second step consists of an OR-gate, and the investigator must choose between *assumed risk* or *management oversight and omissions*. Assumed Risk is defined as having been identified, analyzed and accepted at the appropriate management level, whereas unanalyzed or unknown risks fall under Oversight and Omissions by default. The next decision point, another OR-gate, separates *what happened* from *why* it happened. The *what happened* category addresses the controls that should be in place, while *why* considers general management system factors. Eventually the tree breaks down each of these factors until root causes are reached, which could take up to 13 levels of the tree. An example of a segment of the Oversights and Omissions portion of the MORT tree is shown in Figure 9-29. A simpler Mini-MORT variation has been developed to reduce complexity.[(1)]

The MORT technique has received domestic and international recognition, and has been applied to a wide range of projects from investigation of occupational incidents to hazards identification. It is supported by detailed documentation, and has been subjected to continued development efforts since it was originally introduced. Today there are several predefined trees available from public and proprietary sources that are based, at least in part, upon the MORT tool.

9.11.2. Using Predefined Trees

Although there are differences between various predefined trees, the basic method to perform a root cause analysis using the trees is similar whichever tree is used. The following basic steps apply:

1. First, it is necessary to identify the multiple causal factors of the incident. The procedures in Causal Factor Identification may be used to identify the causal factors from a timeline or sequence diagram (including a causal factor chart).
2. The first causal factor is then analyzed starting at the top of the predefined tree and working down the branches as far as the facts

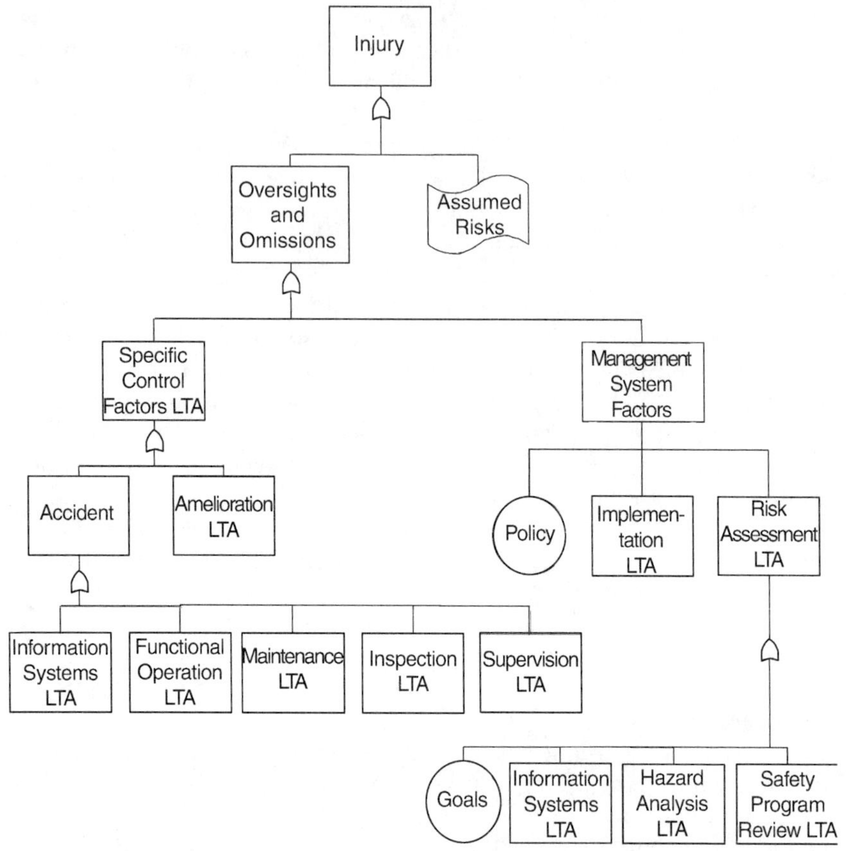

FIGURE 9-29. **Top portion of the generic MORT tree.**

permit. If the category of a particular branch appears to be an appropriate cause of the incident, the branch is followed to successively lower levels until a subcategory matches an appropriate root cause. (*Note:* In some circumstances, the facts may not allow root causes to be identified without further investigation.)

3. All branches and sub-branches should be considered because an individual causal factor can have more than one root cause.
4. As each branch is considered, the investigator should ask if there are other root causes associated with that category that are not listed on the tree. The team should ask, "Are there any *other* causes that anyone has in mind that have *not* been identified?" (Predefined trees are designed to capture most, but not necessarily all, root causes.)

5. The procedure (steps 2 through 4) is then repeated for each causal factor in turn.
6. When all the root causes have been identified from the tree, the investigator should ask *why* to each one in turn as a test to ensure that they are really root causes. If it is possible to identify a lower level cause, this should be recorded as the root cause. (*Note:* This is analogous to the 5-Whys.)
7. Finally, the investigator should consider other generic causes of the incident that are not identified by the predefined tree categories. The investigator should consider the plant operating history. Other incidents may indicate repetitive failures that may indicate generic management system problems.

Predefined trees are relatively easy to use and generally require less training and effort to conduct root cause analysis than logic trees.

9.11.3. Example—Environmental Incident

The following is an example of the use of a predefined tree to analyze an environmental incident. While the structure (number of branches and levels) and terminology of predefined trees vary, this example demonstrates the basic method.

During a normal night shift at a process plant, a temporary water treatment unit, operated by contract personnel, overheated and released hot, low pH water to one of the plant's outfalls, which ultimately resulted in fish being killed in the local river. The overheating occurred when a firewater hose, providing cooling water to the temporary water treatment unit, ruptured. The plant was provided with an automatic trip that apparently failed to work, and an alarm to which the operator did not respond.

The sequence of events is shown in Figure 9-30.

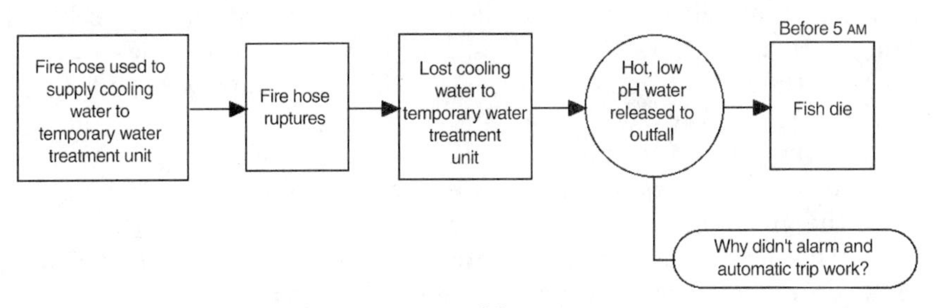

FIGURE 9-30. **Incident sequence.**

The investigation team interviewed all contract operators and their supervisor, the temporary water treatment unit vendor's engineers, plant personnel at the process plant unit, procurement personnel, and operations management.

Causal Factor Identification

After the interviews and other evidence gathering activities are complete, a more detailed causal factor chart can be developed. Causal factors are indicated by black triangles.

Four causal factors are identified. (Note that each causal factor includes all of the information attached to it in the causal factor chart.)

1. Contract operator falls asleep
2. Fire hose ruptures
3. Automatic shut-off jumpered
4. Contract operator can't hear alarm due to noise

A more complex incident may require the use of barrier analysis or change analysis to assist in developing the causal factor chart.

Each of the causal factors can now be analyzed for its specific root causes using a predefined tree, as shown in Figure 9-31.

Analyzing a Causal Factor

The following is an analysis of one of these causal factors: contractor operator (CO) falls asleep. The basic technique works with any of the predefined trees commonly used within the process industry. However, for the purposes of this example, a proprietary tool [7] has been selected, and therefore the structure of the tree and the terminology used is specific to that tree.

To analyze the causal factor, the investigator starts at the top of the tree and works down the tree through a process of selection and elimination. The investigator asks and answers questions to identify the specific root causes for the causal factor.

In this case, the causal factor (contract operator falls asleep) is identified as a Human Performance Difficulty (one of the four major problem categories at the top of the tree), and the other three categories are discarded. (Different predefined trees use different terminology and structure, but generally cover similar choices.)

The investigator then follows the Human Performance Difficulty category through a series of questions (or subcategories). These questions help the investigator identify which of several human performance related branches (sometimes known as basic causes) to investigate further. (Some predefined trees use statements rather than questions, but the selection process is similar.)

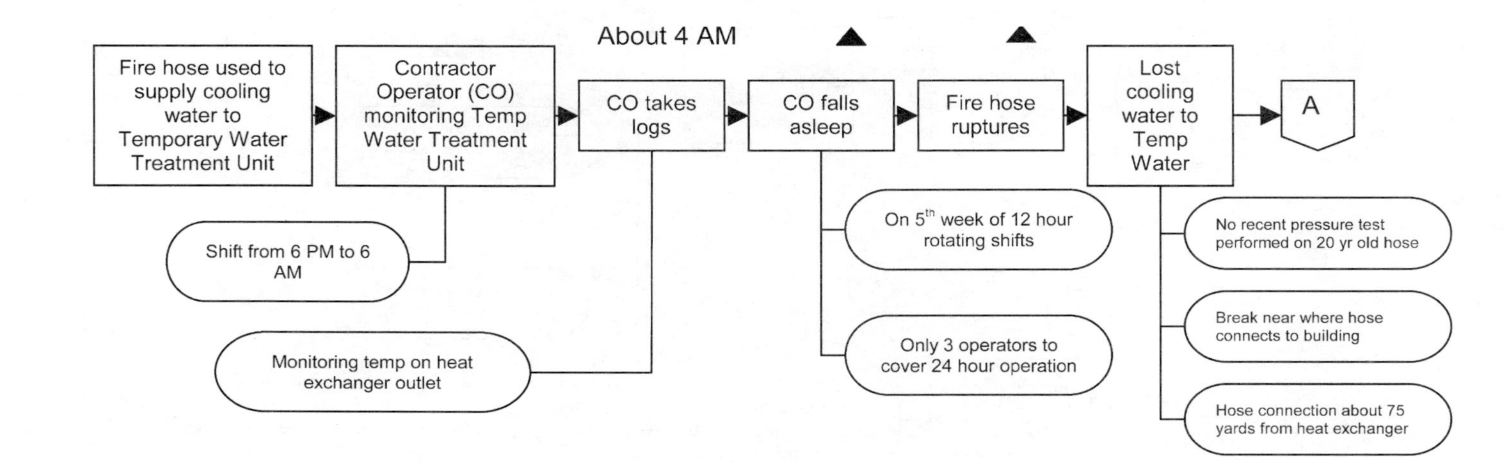

FIGURE 9-31. **Complete causal factor chart for fish kill incident (*continued on next page*).**

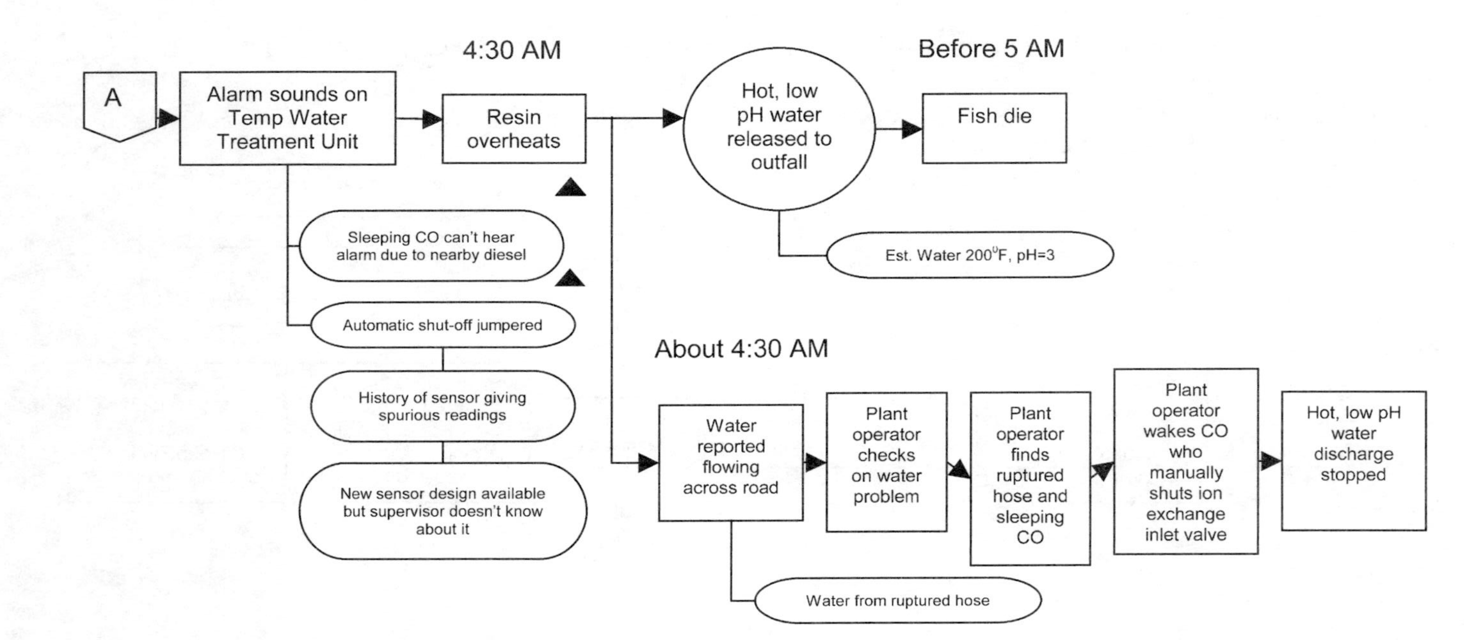

FIGURE 9-31 (*continued*). **Complete causal factor chart for fish kill incident.**

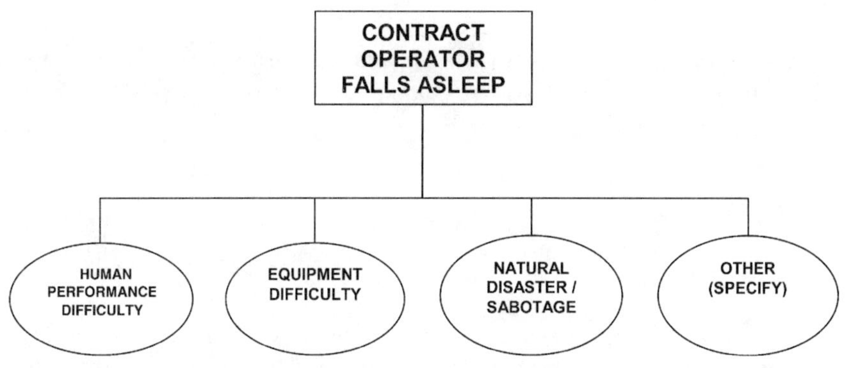

FIGURE 9-32. **Top of the predefined tree.**

The human performance related branches are:

- Procedures
- Training
- Quality Control
- Communications
- Management System
- Human Engineering
- Work Direction

Each branch is investigated further to see if it is relevant, that is, one or more related root causes contributed to the problem, or if it can be eliminated.

In the case of the fish kill incident, the first of the questions, shown in Figure 9-33, is answered YES because the contract operator was considered to be both fatigued and bored. This indicates that the cause may be related to Human Engineering and/or Work Direction.

A different predefined tree may express this question as one or more simple statements, such as:

- Rest/sleep less than adequate (fatigue)
- Attitude/attention less than adequate

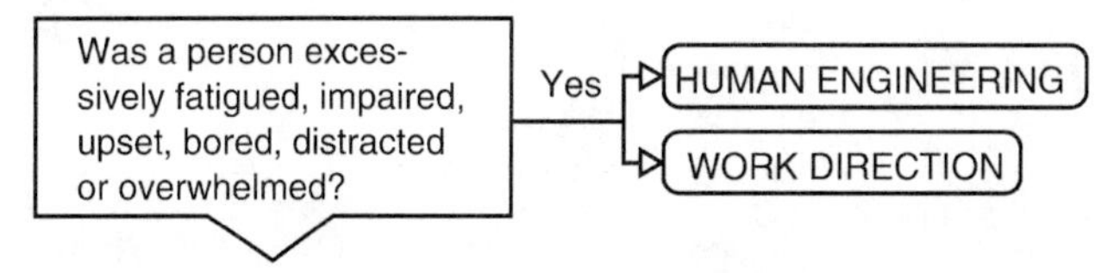

FIGURE 9-33. **First question of the human performance difficulty category.**

However, the basic method is identical.

When all the questions on the Human Performance Difficulty category are answered, the following branches of the tree are indicated for more investigation:

- Human Engineering
- Work Direction
- Management System
- Procedures

Figure 9-34 illustrates one of these branches, Human Engineering, showing three levels of the tree, designated as *basic cause*, *near-root cause*, and *root cause*. (Note that other trees may use different terminology for these levels, although "root cause" is a common term.)

Each lower sub-branch (near-root cause) is then considered in turn to determine if any of the potential root causes on that sub-branch is a valid reason for why the causal factor existed at the time of the incident. Valid root causes are recorded and invalid causes are eliminated.

In the Fish Kill example, the completed analysis of the Human Engineering branch is shown in Figure 9-35. Under the Human-Machine Interface sub-branch, *monitoring alertness needs improvement* is selected as a valid root cause, and the remaining subcategories have all been discarded.

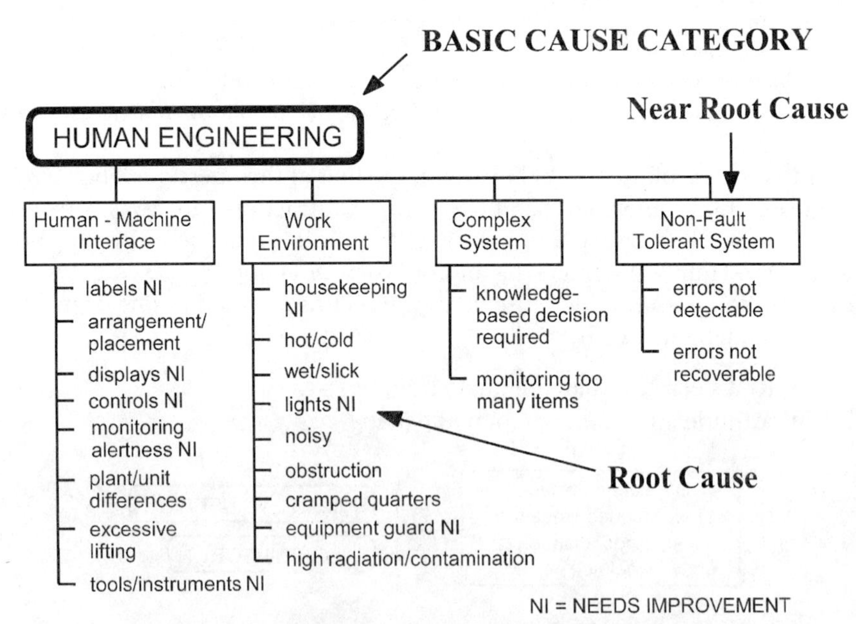

FIGURE 9-34. **Human Engineering branch of the tree.**

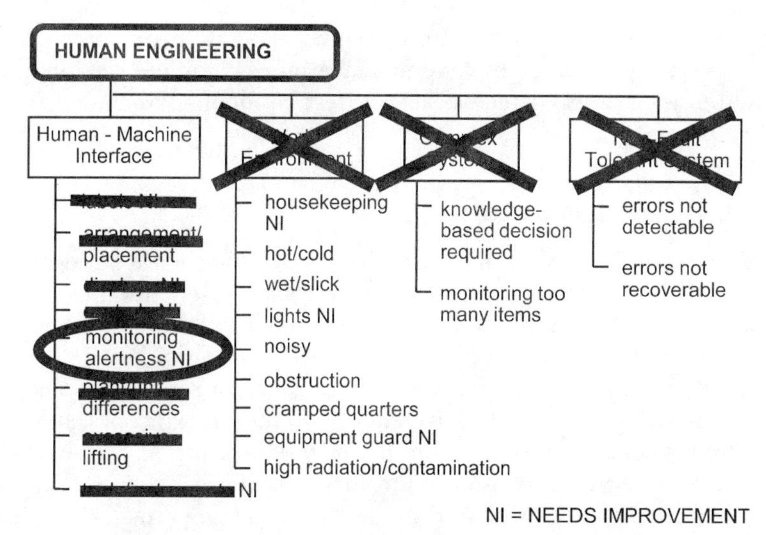

FIGURE 9-35. **Analysis of the Human Engineering Branch.**

When the first causal factor is analyzed using the remaining applicable branches (i.e., Work Direction, Procedures, and Management System), the following root causes are identified:

1. Monitoring alertness needs improvement.
2. Shift scheduling needs improvement.
3. Selection of fatigued worker.
4. The "no sleeping on the job" policy needs to have a practical way to make it so that people can comply with it.

The investigation team then repeats the process by considering the remaining causal factors one at a time:

- Fire hose ruptures
- Automatic shut-off jumpered
- Contract operator can't hear alarm due to noise

Finally, the investigation team considers generic causes that pertain to the overall management system for the process plant by considering the operating history and any other incidents that may have related causes.

Developing Corrective Actions

Once all of the root causes are identified, the investigator is ready to develop the corrective actions, as described in Chapter 10.

Some of the figures and text in this example appear courtesy of System Improvements Inc. and are copyrighted material. They are used here by permission of System Improvements, Inc. Reproduction without permission is prohibited by federal law. A more complete version of this example, including corrective action development, is included on the CD-ROM.

9.11.4. Quality Assurance

There are a number of quality assurance checks that should be considered when conducting an incident investigation using predefined trees. Most of these checks have already been discussed, but will now be summarized in relation to the corresponding phase of the investigation.

Predefined trees are designed to capture most root causes, but may not necessarily be comprehensive enough to identify all root causes. It is therefore necessary to conduct another *completeness* test. As each branch of the predefined tree is considered in turn, the investigator should ask if there are other root causes associated with that category that are not listed on the tree.

The root causes identified by applying the causal factors to a predefined tree should be subjected to a *management system* test to ensure that they are management system failures. Some predefined trees are quite detailed, while others do not necessarily fully reach the root cause level. The system test essentially applies the 5-Whys tool to each cause identified at the end of the relevant branches of the predefined tree.

The 5-Whys tool adds some structure to group brainstorming by taking a logic tree approach without actually drawing the logic tree diagram. [13] The team asks why an event occurred and/or condition existed with the goal of asking *"why?"* enough times to reach a management system deficiency. Sometimes this takes two *"whys"*—sometimes five or more. Typically, the team needs to ask *"why?"* five times to reach root causes hence the name.

Just like the guidance for stopping logic tree development, judgment is needed to use the 5-Whys effectively. If it is easy to answer why at a level, then the analysis has not gone deeply enough and should continue to ask, *"Why?"* Similarly, if it is possible to take a potential root cause identified from a predefined tree branch, ask *"why?"* and answer easily with another underlying cause, then the original cause is not a root cause and should be replaced by the new underlying root cause. This new root cause should also be tested by asking *"why?"* again.

After the root causes have been identified from the predefined tree, a *generic* cause test should be applied. By considering the plant operating history, especially other incidents that may indicate repetitive failures, the investigator may identify other generic management system problems.

These generic causes would not necessarily be apparent from investigating the latest incident alone.

9.11.5. Predefined Tree Summary

Predefined trees are a convenient means of identifying root causes. Providing all of the causal factors have been determined, use of a comprehensive predefined tree should ensure that most, if not all, root causes are identified. Several quality assurance tests should help identify any remaining root causes.

Varieties of public and proprietary predefined trees are available for use, although most owe some allegiance to MORT. The comprehensiveness of the different trees varies from some that may not fully reach root causes, to others that are very detailed with numerous categories and sub-categories. Several of these trees are listed in the appendix and examples of TapRooT® [7], SOURCE™/Root Cause Map [6] and Comprehensive List of Causes (CLC) [14] are included on the CD-ROM.

9.12. Checklists

Checklists of varying content and detail are used in incident investigation methodologies as a user-friendly tool to assist root cause analysis. Sometimes a comprehensive checklist may be used as the primary root cause analysis tool, or alternatively a checklist may be simply used to supplement another primary tool.

Another situation where checklists can be very helpful is when the investigation team has no hypothesis as to what caused an occurrence. The checklist is an example of an inductive approach that can be used to get off top dead center.

Checklists used for process safety incident investigation share many similarities with predefined trees. They can comprise a series of questions or statements related to root causes based upon experience of safety management systems. Some checklists need care in use because the statements that they contain can infer blame to the casual observer, rather than discourage blame seeking. Checklists also offer consistency and repeatability by presenting different investigators with the same standard set of potential root causes for each incident. This consistency facilitates statistical trend analysis of multiple incidents involving recurring problems within an organization.

While a checklist may not encourage the investigation team to think laterally of other potential causes, it can overcome a lack of experience within the team and present causes that the team would not have otherwise considered.

9.12.1. Use of Checklists

The use of checklists as a primary root cause analysis tool is virtually identical to the use of predefined trees. This is hardly surprising as most predefined trees are really a succession of checklists organized by subject matter (category) into an arrangement of branches within the tree.

A timeline or sequence diagram is first developed, and then causal factors identified. Care should be taken to ensure that the checklist is not used too early. Be sure to determine *what* happened and *how* it happened before determining *why* it happened. Otherwise, the team will think that they have identified the right root cause(s), when in reality only one or two of several multiple causes have been determined. The causal factors are then applied one at a time to each page of the checklist(s) to identify relevant root causes. Those pages that are not relevant to the particular incident of interest are discarded. Similar quality assurance checks should be applied as those described for predefined trees.

The use of checklists to supplement another root cause analysis method can be a very powerful technique, for example, human factors checklist(s) may be used in conjunction with logic trees. The checklist may be used as a guide during development of a logic tree, or as a check after the tree has been developed. The checklist essentially acts as a memory jogger to direct the investigation team. This is especially helpful if the team lacks previous experience in the subject matter.

Checklists may also be used in combination with structured brainstorming tools, such as What If/Checklist Analysis. [15]

9.12.2. Checklist Summary

Checklists represent a root cause analysis tool that has similar advantages to, and ease of use as, predefined trees.

A variety of public and proprietary checklists are available that vary in comprehensiveness. There is no reason for an organization to start from scratch in developing a checklist. A human factors checklist and tables are included in Chapter 6. The Systematic Cause Analysis Technique (SCAT)[16, 17] is an example of a proprietary checklist. The accompanying CD-ROM also contains examples of checklists which can be modified for the readers use.

9.13. Human Factors Applications

Investigators are discovering that an increasing number of failure causes are related to inadequately addressing human factors or the relationship of the human to the machine. There is an opportunity to improve process safety management performance by improving human performance and human reliability. Although technology advances have resulted in increasingly complex and highly automated processes, the facilities do not run themselves. Proper operation requires periodic and sometimes constant intervention from humans. System designers are now realizing this and are considering the expectations placed on the operator by management and the physical systems. Human factors are a discipline concerned with matching the system to human capabilities and limitations. A mismatch leads to human performance deficiencies that often result in repeated incidents.

Example

If a component fails because of a human error, "counseling" the worker may prevent him or her from performing the same error again, but what of the other members of the operating crew? Conditions that led to the original failure remain, so others are still prone to committing the same error! Many repeat occurrences could be avoided if the correct information and reasons for those errors are uncovered by the investigation team and (1) corrected and/or (2) communicated to others who might also be at risk of committing them.

Structured root cause analysis uncovers the underlying reasons for human error and consequently provides guidance on suitable corrective actions. Humans make errors. Our task is to design systems that detect and correct an error before it leads to a serious consequence. Chapter 6 provides extensive information applicable during root cause analysis.

9.14. Conclusion

With use of the Method A flowchart sequence in Figure 9-9 or the use of the Method B flowchart sequence in Figure 9-27, the underlying multiple root causes present in almost every process safety incident can be found. By applying the iterative loop, testing the facts, and systematically applying quality control tests, incident investigators will be able to uncover those underlying multiple causes that could result in future incidents.

The success of the cause analysis is a direct function of the quality of available and discovered information as well as the perceptiveness of the incident investigation team. The goal of the cause analysis is to find the information needed to determine cost effective and practical preventive measures.

Endnotes

1. Ferry, T.S. *Modern Accident Investigation and Analysis.* 2nd[nd] Edition. New York: John Wiley & Sons, 1988.
2. Benner, L. Jr., "Accident Investigations: Multilinear Events Sequencing Methods." *Journal of Safety Research,* 7(2):67–73, 1975.
3. Benner, L. Jr. 10 MES Investigation Guides. Oakton, VA: Starline Software Ltd., 2000.
4. Hendrick, K. and Benner, L. Jr. *Investigating Accidents with S-T-E-P.* New York: Marcel Dekker, 1987.
5. Buys, R.J. and Clark, J.L. *Events and Causal Factors Charting.* DOE 76-45/14, (SSDC-14) Revision 1. Idaho Falls, ID: System Safety Development Center, Idaho National Engineering Laboratory, 1978.
6. Bridges, W.G. and Vanden Heuvel, L.N. *Finding Root Causes: A Proven Approach to Incident Investigation.* Knoxville, TN: ABS Consulting, 2001.
7. Paradies, M. and Unger, L., *TapRooT® - The System for Root Cause Analysis, Problem Investigation, and Proactive Improvement,* Knoxville, TN: System Improvements, Inc., 2000.
8. Johnson, W.G. *MORT, Safety Assurances Systems.* New York: Marcel Dekker, 1980.
9. Trost, W.A. and Nertney, R.J. *Barrier Analysis.* DOE/SSDC 76-45/29. Idaho Falls, ID: System Safety Development Center, Idaho National Engineering Laboratory, 1985.
10. Kepner, C.H. and Tregoe, B.B. *The Rational Manager.* 2nd Edition. Princeton, NJ: Kepner-Tregoe, Inc., 1976.
11. Department of Energy, *Accident/Incident Investigation Manual.* 2nd Edition, DOE/SSDC 76-45/27. Idaho Falls, ID: System Safety Development Center, Idaho National Engineering Laboratory, 1985.
12. Buys, R.J. *Standardization Guide for Construction and Use of MORT-Type Analytical Trees.* ERDA 76-45/8. Idaho Falls, ID: System Safety Development Center, Idaho National Engineering Laboratory, 1977.
13. ABS Consulting. *Introduction to Reliability Management: An Overview.* ABS Group Inc., Knoxville, TN, 2000.
14. BP (formerly BP Amoco). *Incident Investigation. Root Cause Analysis Training. Comprehensive List of Causes.* London, 1999.
15. CCPS, *Guidelines for Hazard Evaluation Procedures, 2nd Edition with Worked Examples.* New York: AIChE, 1992.

16. Bird, F. E., Jr. and Germain, G. L. *Practical Loss Control Leadership*. Loganville, GA: International Loss Control Institute (ILCI), 1985.
17. International Loss Control Institute. *SCAT—Systematic Cause Analysis Technique* Loganville, GA: Det Norske Veritas ,1990.

Additional References

Dowell, A. M., Philley, J., and Pearson, K. "Structured Root Cause Investigation." Training course presented at Texas Chemical Council Safety Seminar, 1999.

Feynman, R. P. *What Do You Care What Other People Think?* New York: Norton, 1988.

Gano, Dean L. *Apollo Root Cause Analysis—A New Way of Thinking*. Apollonian Publications, 1994.

Gilovich, Thomas. *How We Know What Isn't So, The Fallibility of Human Reason in Everyday Life*. New York: Macmillian, 1991.

CCPS. *Guidelines for Technical Management of Chemical Process Safety*. New York: American Institute of Chemical Engineers, Center for Chemical Process Safety, 1989.

Hon. Lord Cullen. "The Public Inquiry into the Piper Alpha Disaster." London: UK Department of Energy, HMSO, 1990.

Jones, B. "A Process Quality Management Tool, Prevention of Unwanted Events, Training and Text," 1991.

Kletz, T. A. "Accident Investigation: How Far Should We Go?" Paper 11d presented at the American Institute of Chemical Engineers Loss Prevention Symposium, 1983.

Livingston, A. D., Jackson, and Priestly. *Root Causes Analysis: Literature review*. WS Atkins Consultants Ltd., Birchwood, Warrington: WS Atkins House 2001.

Lucas, D. A., and Embrey, D. E. "Human Reliability Data Collection for Qualitative Modeling and Quantitative Assessment." In Colombari, V. (Ed.), *Reliability Data Collection and Use in Availability Assessment*. New York: Springer Verlag, 1989.

Norman, D. A. *The Design of Everyday Things*. New York: Basic Books, 1988.

Philley, J. "Root Cause Incident Investigation Can Be Tricky." Presented at International System Safety Conference, 2001.

Sanders and McCormic. *Human Factors in Engineering and Design*, 6th Ed. New York: McGraw-Hill, 1987.

Stern, A. and Keller, R. "Human Error and Equipment Design in the Chemical Industry" *Professional Safety*. May 1991.

Winsor, D. A. "Challenger: A Case of Failure to Communicate." *Chemtech*. September 1989.

10

Developing Effective Recommendations

Using structured approaches such as those presented in the preceding chapter, an investigation team identifies the multiple system-related incident causes. These approaches provide the mechanism for understanding the interaction and impact of management system deficiencies. When the investigators understand what happened, how it happened, and why it happened, they can develop recommendations to correct immediate, contributing, and root causes.

Effective recommendations identify changes in management systems that will eliminate the multiple management system-related causes of the incident.

Effective recommendations can reduce risk by improving the process technology, upgrading the operating or maintenance procedures or practices, and most critically, upgrading the management systems. Recommendations that correct management system failures should eliminate or substantially reduce the risk of recurrence of the incident and other similar incidents. This critical step is necessary to achieve the ultimate benefits from an incident investigation.

This chapter develops characteristics of high quality recommendations necessary for achieving successful implementation. The first section is a presentation of the major concepts related to recommendations, such as attributes of good recommendations, management of change, and inherent safety. The second section expands on the attributes and presents a systematic discussion of the flowchart for recommendations.

10.1. Major Issues

Figure 10-1 presents an overview of the activities in this chapter beginning with the multiple system-related causes already identified. Each system

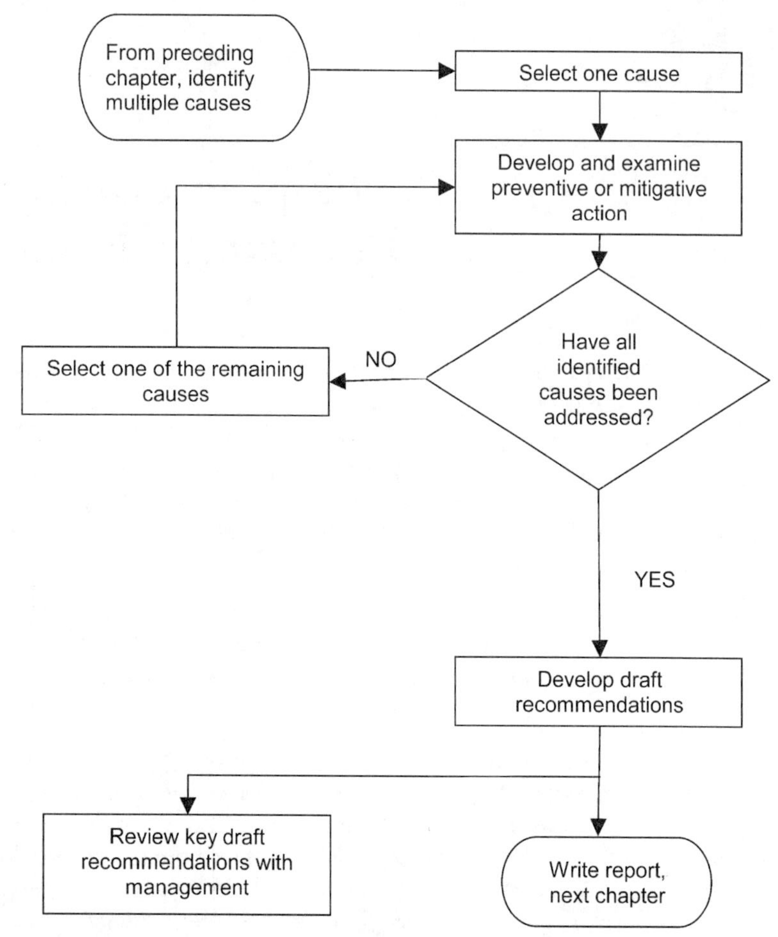

FIGURE 10-1. **Incident investigation recommendation flowchart.**

cause selected should be addressed by a recommended preventive or mitigating action item or comment. Each cause is examined individually. Proposed actions to prevent a recurrence of the incident are evaluated for technical merit. The incident investigation team develops recommended actions, and then presents these recommendations to the management team responsible for accepting (or rejecting) these recommendations.

Recommendations should be written clearly and should be practical to implement. Recommendations that target improvements to the inherent safety of the process design are preferred to those that merely add extra protection features. All recommendations will potentially change management systems; therefore, it is necessary for the team to consider the implications of those changes when developing recommendations.

Recommendations should undergo a management of change evaluation to identify and consider the impact and risks associated with any changes. It is imperative that the management of change evaluation be completed *before* implementing recommendations.

10.2. Developing Effective Recommendations

10.2.1. Team Responsibilities

The incident investigation team is responsible for developing practical recommendations and submitting them to management. It is then the responsibility of management to approve, modify, reject, communicate, implement, and follow-up on the recommendations, including:

- allocating sufficient personnel and capital resources for timely completion and
- implementing changes and following up with those affected by the changes to assure measures are working as expected.

10.2.2. Attributes of Good Recommendations

A well-written recommendation specifically describes and defines successful completion in clear and measurable terms. Satisfactory completion of the recommended action should be easy to recognize, that is, you should clearly know when the corrective action is finished. Vague or ambiguous terms do not belong in a written recommendation. Terms such as "appropriate safeguards" or "improve the quality of training" should be avoided unless they are expanded upon and specifically define performance terms.

Recommendations should be written in such a way that they transfer full responsibility and ownership to the receiving department. They should engage the reader to think about the recommendation before merely acting on it. Recommendation writers should avoid assertive or prescriptive statements that simply direct the recipient to carry out an activity. For example, instead of "*Rewrite the operating procedures*" a recommendation might suggest

- *Conduct a step-by-step review of the reactor charging operating procedures with a multidisciplined team and update the procedures as necessary. The incident investigation team identified several steps in which details seem to be missing: purging, blocking in reactant A, and disconnecting trailer.*

The team may also make appropriate higher-level recommendations such as,

- *Ensure there is a system to review and update procedures on an annual basis and after changes occur, or*
- *Ensure a system is in place to enforce existing and modified procedures in the field.*

Clearly written recommendations allow little opportunity for confusion. The owner is fully responsible for understanding and managing the follow-up. A good practice is to word the recommendation as a stand-alone thought that includes an explanation of why it was made, specifying the consequences to be avoided. Ambiguously worded recommendations are subjective. They could be open to different interpretations and obscure the intent of the writer.

Recommendations concerning restart of operations are often developed before the final report is published. This is a management decision, but may or may not be part of the incident investigation scope. Nevertheless, these recommendations should be clearly written, understood by all, and be approved and accepted by the management organization responsible for safe operation of the facility. Restart criteria deserve special attention and are addressed in further detail later in the chapter.

Appendixes D and E contain two reports—Appendix D is one on a fictitious fire and explosion incident and Appendix E is one from an actual incident. The recommendations contained in the reports are typical of the range of detail that can be found in recommendations that address changes in management systems.

Five attributes of a successful recommendation are

1. It addresses a root cause of the incident (that is, it fixes the problem).
2. It clearly states the intended action, why it is needed, and what it should accomplish.
3. It is practical, feasible, and achievable.
4. It adds or strengthens a layer of protection.
5. It eliminates or decreases risks, consequences, or the probability of either one.

Sometimes several specific causes or recommendations related to each other will be consolidated into a general statement. The preventive action items would appear as individual distinct recommendations, yet a general description may appear as an observation or conclusion.

For clarity, there is a distinction between findings and conclusions.[8] Findings are restricted to known facts (information, data). Conclusions state judgments derived from evaluations made by the investigators.

10.3. Types of Recommendations

Successful recommendations decrease the risk of an occurrence by reducing the frequency or decreasing or eliminating the consequences of an occurrence. The following sections describe methods to decrease risk and give examples. There are several different approaches to categorizing recommendations, for example:

- Recommendations targeted at reducing the probability of a given incident (for example, increasing preventive maintenance inspection programs to reduce the probability of simultaneous failure of critical circulating pumps and back-up pumps)
- Recommendations for minimizing personnel exposures (for example, minimizing duration of exposure or relocating noncritical groups of workers to areas remote from potential blast zones)
- Recommendations to change OR-gates to AND-gates. These AND-gate recommendations will result in a lower frequency of occurrences than OR-gates. (For example, if an overflow was caused by the failure of a single level controller—an OR-gate with transmitter failure, valve failure, leads plugged—a second level transmitter could be used to shut a remote valve on high level. This would add an AND-gate to the tree so that it would take a failure of both systems to result in an overflow.)
- Recommendations intended to eliminate or reduce the consequences of a given incident (for example, reducing inventories of hazardous materials)

10.3.1. Inherent Safety

Recommendations that lead to inherently safer designs are preferred to those limited to adding-on extra mitigation or prevention features. [1] Inherently safer designs limit reliance on human performance, equipment reliability, or properly functioning preventive maintenance programs for successful prevention of an incident. Inherently safer design changes yield greater economic benefits if they are implemented during the early design phases. Nevertheless, the incident investigation team should consider them when developing recommendations.

Intrinsic and passive safety design features are by nature more reliable than design features that depend on a consistent performance of a physical or human system.[2] Passive designs reduce the probability of an occurrence occurring or minimize the severity of the adverse consequences without the requirement for successful operation of devices, controls, procedures, or interlocks. Some examples of general strategies [3] for increasing inherent safety are:

- **Reduction of Inventories:** Advancements in process control and changing acceptable risk standards may have removed the initial justification for large inventories of hazardous raw materials or products. For example, tight quality control of on-time deliveries of hazardous raw materials may allow for a one or two day supply on hand versus a one- or two-week supply.
- **Substitution:** Sometimes substitution of a less hazardous material is feasible. For example, many chlorinating systems for water purification have recently converted from pressurized cylinders of liquid chlorine to a pelletized, hypochlorite salt.
- **Intensification:** Sometimes it is possible to achieve significant reductions in reactor size (and inventory) with improved mixing technology. Another example of intensification is changing from a batch operation to a smaller scale continuous operation.
- **Change:** It may be possible to use a totally different process or method to accomplish the same objectives.

The incident investigation team should evaluate the inherent safety features of recommendations. Changes can be either beneficial or detrimental, so investigators should be alert for inherently unsafe features in recommendations. Two common examples of design changes that can *increase* overall risks are the use of flexible joints and the use of glass (rotameters, bulls eyes, sight glasses, or additional control room windows).[4] Seal-less pumps are generally considered to be inherently safer than pumps with mechanical seals. The failure mode of any recommended new valves should be carefully considered. Most valves should be designed to fail to a closed position, although there are some exceptions, such as cooling water control valves.

10.3.2. Hierarchies and Layers of Recommendations

The concept of multiple layers of protection (barriers) has widespread support throughout the chemical processing industry. By providing sufficient layers of protection against an accident scenario, the potential risk associated with that accident can be avoided or at least reduced. For a given scenario, only one layer must work successfully for the consequence to be prevented. However, since no layer is perfectly effective, multiple layers of protection must be provided to render the risk of the accident tolerable.

Well-known process safety expert Trevor Kletz identifies three layers of barrier considerations for recommendations as follows[5]:

- **First layer remedies use immediate technical recommendations targeted to prevent a particular incident.** Consider the case where an employee is injured by inhalation exposure while taking a liquid chlorine process sample. First-layer recommendations would address such items as changes to the sampling procedure, refresher training, and selection and use of personal protective respiratory equipment.
- **Second layer recommendations focus on avoiding the hazard.** A deeper and broader perspective is used for this second layer, and often the focus is on improving the normal barrier measures placed between the person and the hazard. Typical remedies for the above chlorine incident might include modifications to the sampling apparatus, sampling at a different location, or perhaps an in-line analyzer, which would eliminate the need for manual sampling.
- **Root causes are addressed in the third layer by identifying changes in the management systems**. These third-layer recommendations act to prevent not only this particular incident, but also similar ones. Preventive measures, which result in changes in management systems, are in theory more consistent and enduring. For example, if an incident occurred due to an out-of-date operating procedure on a particular process plant, updating that procedure will only prevent the specific recurrence at that plant. However, if the management system is changed to ensure that all operating procedures on all plants are accurate and up-to-date, then similar incidents at other plants will be prevented as would dissimilar occurrences cause by out-of-date procedures.

A general sequence of safety layers has also been proposed, as shown in Figure 10-2.[6] This sequence starts with soft then hard engineering, progresses through maintenance, inspection, and operating practices, to instrumentation, overpressure protection, and emergency response. Each layer is a management system in itself and may contain latent failures.

As another example, in the chlorine exposure case the investigators would consider such items as:

1. **Improvements in the methods in which process sampling is established.** (Who participates in the decision? What are the criteria for determining location method and devices? Who authorizes? Is there a periodic audit or reevaluation?)
2. **Improvements in the management system for establishing, evaluating, and monitoring standard operating procedures.** (Are the procedures adequate, understood, and consistently performed? Is the task still necessary?)

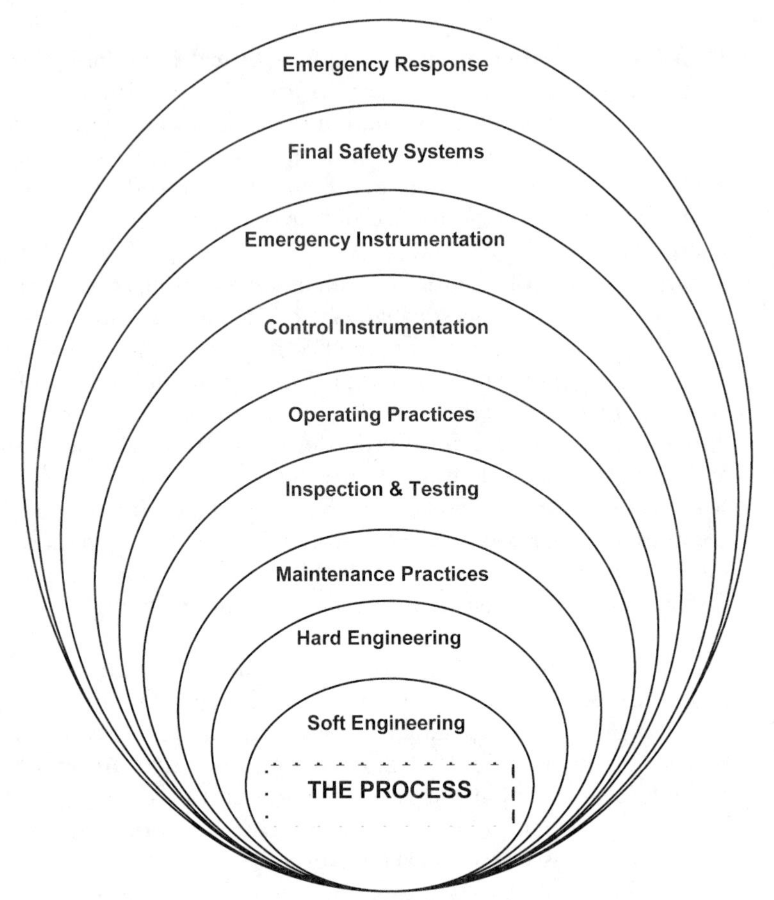

FIGURE 10-2. **Layers of safety.**[6]

3. **Is there a routine mechanism such as Job Safety Analysis (JSA) in which tasks such as this are systematically reviewed for potential hazards?** JSA is a procedure that systematically identifies: (1) job steps, (2) specific hazards associated with each job step, and (3) safe job procedures associated with each step to minimize accident potential.

Multiple layers of protection are a concept incorporated in the American Chemistry Council Process Safety Code of Management Practices.[7] Management Practice number 15 endorses "sufficient layers of protection through technology, facilities, and personnel to prevent escalation from a single failure to a catastrophic occurrence." This approach can be applied to multiple system root causes when the investigation team evaluates potential remedies. These layers offer guidance in selecting practical recommendations.

The key issue is that causes addressed in recommendations must address system-based issues, and the team must always look for multiple causes and corrections.

10.3.3. Commendation /Disciplinary Action

Whenever an investigation reveals an employee action worthy of commendation, the incident investigation team should point this out in their comments. Cool rational actions in the middle of an emergency often limit the consequences of an incident. Those employees who demonstrate this type of response should be recognized and commended.

Incident investigations should very rarely result in disciplinary actions. The team should assume that disciplinary actions are not part of the investigative outcome. Even the perceived threat of disciplinary action has detrimental effect on an investigation and may discourage cooperation during interviews.

10.3.4. The "No-Action" Recommendation

There may be rare occasions where a particular cause is identified but no accompanying recommendation is submitted. After thorough analysis and deliberation of available existing information, the incident investigation team may decide not to issue a recommendation addressing a particular cause. This decision may be based on a valid and documented quantitative probability assessment verifying that the risk of recurrence is too low to warrant further action, or the cause may be outside of the control of the organization. An example would be a part falling off an aircraft flying at 20,000 feet and striking a tank of flammable material.

Even in the case of a no-action decision, the fault and the decision to take no action should be documented. The logic of the decision-making should be documented, along with any facts and reference documents.

10.3.5. The Incompletely Worded Recommendation

Another special case is the incompletely worded recommendation, where the investigation team recommends evaluation of an existing safeguard. It is common for an investigation team to generate a recommendation to confirm that an existing physical system (or administrative measure such as written procedures or training program) does provide adequate protection. In these instances, the team should specify what action should be taken if the safeguard is found to be inadequate. If the team only specifies an action such as "review the startup procedure," then the implemented action may or may not meet the team's intentions. The team should include instructions to describe what specific action is needed if the existing safeguard is found to be inadequate.

10.4. The Recommendation Process

10.4.1. Select One Cause

The process of developing recommendations is summarized in Figure 10-1. Starting with the set of multiple causes determined previously in Chapter 9, each cause is evaluated individually to consider actions that would prevent (or satisfactorily mitigate) a recurrence. Ideally, each recommendation should cover just one item and spell out precisely what action is recommended.[9]

10.4.2. Develop and Examine Preventive Actions

The incident investigation team should strive to find recommendations that:

- prevent a recurrence (actually fix the problem),
- are reasonably achievable,
- are compatible with other organizational objectives such as profitability, protecting the community and environment, and
- are measurable, clearly worded, and describe specific outcomes.

Recommendations can focus on changes to improve:

- physical systems, such as hardware, equipment, and tools,
- software systems, such as procedures, methods, training, or
- overall management systems.

A hierarchical perspective can be applied. Preventive measures that increase inherent safety are generally considered the most effective approach. See examples for assessing the effectiveness of individual recommendations in Figure 10-3.

10.4.3. Perform a Completeness Test

The next activity in the sequence is to check for completeness, such as, "Have all the identified causes been addressed?" The incident investigation team should remember that the multiple causes are not all necessarily located at the bottom of the logic tree structure. Sometimes causes may be

Example Recommendation	How Its Effectiveness May Be Assessed
Revise the procedure to change the warning to an action step	Review the revised procedure to determine that all action steps are included and all warnings and cautions do not contain action steps. No inappropriate steps, cautions, or warnings should be found in the procedure
Review and revise all existing procedures to ensure proper format is used for information contained in the procedure (steps, cautions, warnings, notes, etc.)	Periodically review existing procedures to determine format errors. No inappropriate steps, cautions, warnings, notes, should be found
Revise the procurement specifications for the gaskets used in this line	Verify that the revised procurement standard contains the specification for the proper gasket
Perform a review of other gaskets in the lines carrying the material to determine the proper gaskets are being specified. This review should start with at least a 10% sample of current applications	Verify that existing procurement standards contain the specification for the proper gasket Review maintenance work orders to determine the number of gasket replacements that are performed as a result of the inappropriate gasket materials
Revise the design standard used to design lines carrying this material	Periodically review the design of equipment carrying this material to ensure that the proper gaskets are being specified Review maintenance work orders to determine the number of gasket replacements that are performed as a result of the inappropriate gasket materials
Train the engineering staff on the revised standard	Periodically review the design of equipment carrying this material to ensure that the proper gaskets are being specified Review maintenance work orders to determine the number of gasket replacements that are performed as a result of the inappropriate gasket materials
Provide an additional communication system, such as portable radios, that maintenance personnel can use to communicate with supervisors and planners. This should allow the current paging system to be used exclusively by operations	Periodically monitor communications on the paging system to determine if other groups are using the system Assess the traffic load on the new maintenance communication system to ensure that it has adequate capacity Periodically survey the maintenance personnel to determine if they are having difficulty with the communication equipment Periodically survey operations personnel to determine if they have problems with other groups using the paging system
Modify the overtime policy to limit the number of hours of overtime that can be worked in a given time period	Monitor the number of overtime hours being worked. Determine if individuals are exceeding the guidelines that the new policy outlines

FIGURE 10-3. **Example recommendations and assessment strategies.**[10]

identified from occurrences or combinations of events from upper levels of the logic tree structure.

10.4.4. Establish Criteria to Resume Operations

The decision authority and responsibility for restart rests with site and corporate management. Depending upon the scope of the investigation, the incident investigation team may be requested to assist by identifying any recommendations that should be implemented before resumption of operations. If so, then the team should clearly specify the minimum acceptable criteria recommended for safe resumption of operations. This list should be in writing and may contain a special checklist of items to be verified and completed before attempted restart. If a specific authorization or waiver of constraint is needed, this fact should be clearly indicated and communicated.

For most process safety related incidents, the remaining or adjacent sections of the facility will be in operable condition. Thus, the question of restart criteria may be an early task for the team. The topic is often addressed in stages. Initially there may be many restrictions because there will be many unanswered questions. Initial restrictions can be gradually lifted as information becomes more available. Some restrictions on operating parameters (such as flow rates, pressures or temperatures) may continue for extended periods.

There will be occasions where restart authorization, prior notification, or coordination with other organizations is required. Regulatory agencies such as federal and state health, safety, and environmental regulators, as well as some municipal agencies (such as fire departments, hazardous material response (HAZMAT) teams, building officials, local emergency planning committees, and others) are examples of outside organizations that may have specific jurisdiction authority over restart. These communications and coordination activities are usually sensitive and can sometimes require more time than originally expected. It is a good practice to verify and document communications of this nature. It is also a good practice to develop a specific checklist of outside parties that should be notified regarding the restart activities. This cooperation and coordination is worth the extra effort.

If the team identifies a specific change that must be made before restart, then the recommended change should be evaluated using the site's management of change procedures before restart begins for potential hazards and impacts on the system. These immediate recommendations might receive less than the required amount of scrutiny because of the circumstances, especially if most of the facility is intact and apparently ready to restart. There may be a tendency for short-range thinking, such as considering resuming operations just until the next planned outage (that may be in the very near future). The team should maintain consistent standards of diligence and discipline when these pressures arise and insist that all changes be evaluated using the site's management of change procedures.

Restart criteria recommendations should receive the same degree of scrutiny as the final and formal recommendations. The management-of-change and pre-startup safety review implications of each restart recommendation should be considered and evaluated for possible adverse consequences. Some common changes encountered in restart criteria recommendations include:

- changes in operating procedures or instructions (should be in writing and specific),
- changes in control instrumentation set points for alarms, and
- changes in expected emergency response actions by on-duty operations personnel.

Team participation in startup activities also offers a special opportunity to verify that completion of the recommended action items was done as intended and that those employees operating the unit understand the implications.

10.4.5. Prepare to Present Recommendations

While preparing to present the recommendations to management, the incident investigation team should perform some analysis. Related recommendations may be grouped by:

- Priority
- Systems affected
- Time frame for implementation
- Cost (or level of approval required)
- Need for outside resources (such as further research or special expertise)

Identify and anticipate challenges to successful implementation. Although the impact of each recommendation should have been largely identified during the course of recommendation development, look at this aspect again. The team should research and resolve expected questions and concerns from line management. If information is available to compare the recommendations to other similar operating processes, then this information should be taken into consideration when preparing to present the recommendations to management.

10.4.6. Review Recommendations with Management

The next activity as shown on Figure 10-1 is a presentation and review of the recommendations with the members of line management who have responsibility for operation of the affected unit. As discussed previously, management has the responsibility to approve, modify, reject, or implement the recommendations.

At this stage of review, before writing the full incident investigation report, only the essential recommendations may have been developed. However, line management are invariably more interested in these key recommendations, and the other less critical recommendations may not be reviewed with management until the report has been drafted.

If a recommendation is submitted to correct a safety problem exposed during an investigation, a legal concept sometimes referred to as "increased knowledge" can then be applied. The concept is based on the idea that additional responsibility accompanies this additional knowledge. Management has a legal duty to respond to all known hazards. This is particularly applicable in the case where a previously unrecognized or underestimated potential hazard becomes evident (or further understood) by the investigation. Failure to resolve the recommendation in a reasonable manner can result in increased liability. Most members of upper and middle management understand this concept. However, others, such as first line supervisors, may not fully realize the possible impact of a decision not to act on a formal safety recommendation. Company legal representatives should be consulted when the management system is established to address these concepts, before a situation of "increased knowledge" arises.

10.5. Reports and Communications

The recommendations as accepted for implementation are now ready to be documented in formal written reports and communicated, as addressed in Chapter 11.

Internal corporate legal departments may suggest separating the report from the recommendations to simplify and streamline potential litigation. The Center for Chemical Process Safety endorses a policy of including the recommendations in the formal report. A single document is more likely to be consistently applied. A single document should communicate the story and causes of the incident better, and thus achieve improved prevention results.

If the team conducting the investigation has been chosen for their experience, technical knowledge, and skills, they are best placed to develop recommendations to prevent recurrence. In some governmental organizations, however, a separate group, rather than the evidence gathering team, develops the actual recommendations and generates a separate report. This is a common practice for many U.S. governmental agencies, such as the Department of Energy, National Transportation Safety Board, Department of Labor, and Nuclear Regulatory Commission. One basis for this practice is that the evidence-gathering and analysis team is often staffed with highly trained professionals who specialize in physical evidence analysis. These specialists can have a relatively narrowly focused scope of knowledge and may not always be in the best position to formulate recommendations that include changes in management systems. In this case, the recommendations may be separated from the rest of the report. Another reason for separating the recommendations from the rest of the report recognizes the practical aspects of the implementation process in governmental agencies. It is common for higher levels within a federal agency to add to or strengthen the original recommendations.

This approach is not recommended for process safety incident investigation. If the team conducting the incident investigation has been chosen for their experience, technical knowledge, and skills, they are best placed to develop the recommendations to prevent a recurrence.

Endnotes

1. Kletz, T. A. "Make Plants Inherently Safe" *Hydrocarbon Processing Magazine*. Houston, TX: Gulf Publishing Company, September 1985.
2. Hendershot, D. C. "Design of Inherently Safer Chemical Processing Facilities" Presentation Paper at Texas Chemical Council Safety Seminar, Process Safety Management Program, 1991.
3. Knowlton, E. R. *An Introduction to Hazard and Operability Studies—A Guideword Approach,* 7th Printing. Vancouver, British Columbia: Chemetics International Co., 1989.
4. Englund, S. M. "Design and Operate Plants for Inherent Safety" *Chemical Engineering Progress*, Part 1, March 1991; Part 2 May 1991.
5. Kletz, T. A. *Learning from Accidents in Industry,* Boston-London: Butterworths Publishers, 1988.
6. Csengery, L. V., "The Time for Process Hazard Analysis Has Arrived." Paper 68d, AIChE Loss Prevention Symposium, 1991.
7. American Chemistry Council. *Resource Guide for Implementing the Process Safety Code of Management Practices.* Washington, DC, 1990.
8. U.S. Nuclear Regulatory Commission. NUREG-1303, *Incident Investigation Manual*. Washington, DC, 1988.

9. Ferry, T. S. *Modern Accident Investigation and Analysis*. 2nd Edition. New York: John Wiley & Sons, 1988.
10. ABS Consulting. *Incident Investigation/Root Cause Analysis Training: Results Trending and Assessment*. Knoxville, TN: ABS Consulting, 2001.

Additional Reference

Center for Chemical Process Safety (CCPS). *Inherently Safer Chemical Processes: A Life-Cycle Approach*. New York: American Institute of Chemical Engineers, 1996.

11

Communication Issues and Preparing the Final Report

With information gathered, multiple system causes analyzed, and recommendations formulated, the incident investigation team sets about the task of preparing the formal written incident report. What are the attributes of high-quality incident investigation reports? How do they differ from other communications such as interim reports? This chapter describes practical considerations for formal, written incident reports. Attributes of quality reports are presented with special focus on the report reader or user. A generic report format is presented along with a discussion of avoidable common mistakes. Figure 11-1 presents a flowchart of the activities and concepts addressed in this chapter.

11.1. Interim Reports

Some process safety investigations will be extensive in duration, particularly where serious or high-potential incidents are involved. In these situations, some form of interim report may be necessary. Additionally, incidents that are more serious may warrant investigation by a team, which includes a safety professional and others with specific experience and knowledge as deemed appropriate. In such cases, an investigation team leader—independent to the incident—should lead the incident investigation and report writing. The interim report content should be as accurate and thorough as possible, yet should remain flexible and responsive to new information. It should communicate where the investigation is at that point in time, what is known, and what is not. It should not speculate. Each report issued by the incident investigation team should be retained, and its distribution documented. The team leader should coordinate all such interim reporting activity. Someone should serve as the appointed liaison

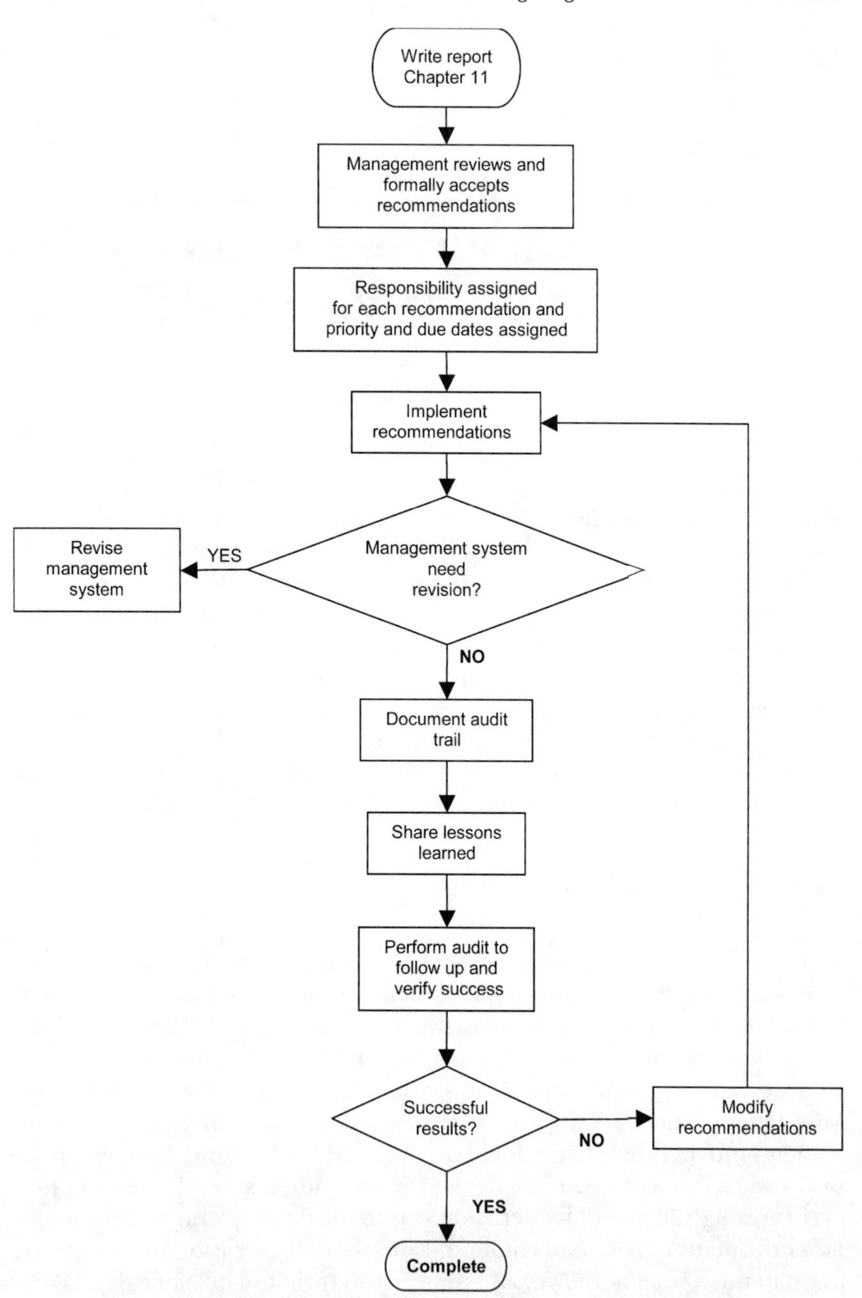

FIGURE 11-1. **Flowchart—formal reports and communications.**

between the incident investigation team, management, and external organizations. This is often the team leader, but others with special training may also fill this function. A single communications channel is especially helpful when team members must deal with external regulatory agencies.

For complex incidents or causes, confirmation of all scenarios or causes may require a lengthy period—perhaps years. Rather than waiting for this, the team should write a complete report, indicating open research action points. As the open research items become resolved, this status should be documented. If new information indicates that conclusions on recommendations should be changed, the report document should be updated or annotated as necessary.

11.2. Writing the Formal Report

11.2.1. General Guidance

The formal written incident investigation report is the official vehicle for documenting and communicating the investigation results. Previous experience indicates that reports come from a variety of backgrounds and cover a variety of topics, but unless a report is laid out in a certain manner, the impact of its presentation is not as efficient as it could be. A quality report can be extremely useful, leading to process safety improvements and enhancing, summarizing, and extending the impact of the team's investigation. Likewise, a poorly prepared report can be nearly useless, negating weeks or months of productive investigation.

A mechanism for capturing and applying the results of the investigation should be an integral part of the management system for process incident investigation. *Guidelines for Technical Management of Chemical Process Safety,*(1) page 118, states that, "The lessons learned from an incident investigation are limited in usefulness unless they are reported in an appropriate manner." The American Chemistry Council recognizes this need by including it as one of the twenty-two management practices in the Responsible Care®, Process Safety Code of Management Practices.(2) The formal written report should convey the findings and recommendations of the investigation clearly and succinctly.

One key to success is to ensure that the report is written for its intended audience.

Identify the users and ensure the report contributors understand the needs of the entire audience. Although it may be unreasonable to expect that all the needs will be met completely, considering them during the writing phase will help approach that goal. The large variation in the readers' technical backgrounds, the need to include technical information and the need to be reasonably concise may limit the usefulness of a single report. Every final report represents a balanced trade-off of content, details, quantity of information, and the expected needs of the readers and users. It is reasonable to expect that the report user has some general knowledge of chemical process technology and hazards. It is also reasonable to expect that the readers have some genuine interest and a desire to gain from understanding and applying the lessons available. The report should not only document and communicate the findings and recommendations, it should be a tool to motivate or inspire action.

Carper, in Chapter 11 of his book *Forensic Engineering*,[3] recognizes a primary and a secondary audience and acknowledges the reality that the report should not be expected to reach all audiences and satisfy all questions. *Professional Accident Investigation* by Kuhlman[4] develops the concept that different levels of management have different needs and priorities. The report should anticipate and address questions the intended user might be expected to ask.

The American National Standards Institute (ANSI) has established a voluntary consensus standard for recording basic facts relating to the nature and occurrence of work injuries, ANSI Z16.2.[5] This standard has been in use for more than 30 years and provides a consistent baseline for many report formats. The standard focuses on injury to personnel and presents standard classification categories. If alternate codes are used in incident reports, special effort and notation must be made to avoid confusing the reader.

Although it is the most important single document, the final published report is only a portion of the overall record of the investigation. A full and complete set of documents should be maintained for future reference. This systematic documentation package is sometimes referred to as the *audit trail*. It provides subsequent reviewers and investigators with the opportunity to understand the team's decisions and analysis more completely. The document set should contain lists of relevant files. All documents associated with the investigation should be preserved according to the company records retention plan.

An exceptional investigation report:

- explains the technical elements and issues associated with the incident,
- describes the management systems that should have prevented the occurrence,
- details the system root causes associated with the human errors and other system deficiencies involved with the incident.

The incident investigation team should constantly perform some kind of self-examination to ensure objectivity and independence. Sometimes after being immersed in the details for a while, the team can lose its sense of perspective.

Incident reports will often include information that may be sensitive or controversial. The report should include relevant information even if it is not yet completely understood. If there is a reasonable doubt as to the certainty of an event sequence, this information should be addressed in the report.

As discussed, the major purpose of the investigation is prevention. The tone and choice of words used in the formal report should reflect the attitude of preventing a repeat of a similar incident rather than affixing the blame.

Some suggestions on **wording:**

- **Brevity:** Be brief where possible.
- **Facts:** In presenting the information, stick to the facts. State the facts clearly and concisely. Any opinions should be so identified.
- **Findings:** As with facts, state whether a finding is opinion or factual. Weigh the evidence for your conclusions or recommendations carefully. If in doubt, provide the reasons behind your findings.
- **Causes:** For each finding, carefully determine the causes, which could be split into immediate, contributing, enabling, and root causes.
- **Recommendations:** Recommendations require careful wording. Before finalizing recommendations, review the report's intended and potential audiences, company protocols, expectations, and past practices.
- **General:** Many reports are presented in a satisfactory manner, with the conclusions and recommendations plainly stated and readily found. Some reports, however, although containing voluminous text, tend to be more of a discussion document and without proper formalization of positive recommendations. This makes it difficult for other facilities to derive optimum benefit without researching the document at some length. Other documents have clear and positive recommendations and conclusions, but they are buried in a dialogue of discussion notes. Some reports are not dated. This simple omission makes the report highly ambiguous if a case history is being researched on a particular facility or operation over a few years.

It takes careful reading of such documents to extract and collate actionable information, and the effectiveness or impact of such reports is much reduced. For a report to have maximum effect, essential information should be readily identifiable.

Avoid creating an emotional effect to what you are saying, such as by using superlatives. Consider replacing *Procedures were insufficient and totally inadequate* with *Procedures must be improved to prevent similar future incidents*. Avoid suggesting negligence or blame; rather, provide constructive recommendations that improvements or additions are necessary.

11.3. Sample Report Format

Because there is no single universal report that simultaneously satisfies all needs of all potential readers and users, a variety in degree of detail and circumstance has been included to give readers the opportunity to select the format best suited to their needs. Most sample reports answer the basic questions of:

- What happened?
- How did it happen?
- Why did it happen?
- What were the multiple management system-related root causes?
- What can be done to prevent a repeat or lower risk?

The subject matter of the report may influence aspects of a report's layout. However, the guidance below presents a logical sequence of events surrounding an incident and would allow someone reading the report to get a better understanding of the circumstances. Some organizations mandate a format for minor occurrences.

11.3.1. Executive Summary

The executive summary should be a brief summary describing the occurrence and key (or main) consequences, root causes, any significant findings, and recommendations. It is usually very helpful to present a highly simplified summary in the first one or two sentences. The purpose is to send a mental image of the occurrence. For most formal reports, this synopsis is written after completing the other sections. Some experts stress limiting this summary to one page including all headers, footers, legal disclaimers and other legal information. They emphasize that the executive summary is a summary—not a separate version of the report itself. The single page layout can make a good document for sharing results through internal web pages, bulletin boards, safety meeting bulletins, and training manuals.

TABLE 11-1
Typical Sections of an Incident Investigation Report

Section	Content
Title Page	Include the date of the report.
Table of Contents	
Introduction	General summary, terms of reference, conduct of the investigation, team description.
Executive Summary	Summary of occurrence, consequences, causes, and recommendations.
Background	Process description, purpose, and scope of investigation, conditions preceding the incident. Historically significant issues may be discussed.
Sequence of Events and Description of Incident	Description of the occurrence scenario, sequence, consequences, and response summary.
Evidence and Cause Analysis	Identification and discussion of the root (primary), contributing, and immediate (secondary) system causes of the incident.
Findings	Factual findings.
Recommendations	Recommended preventive actions.
Noncontributory Factors	Discussion of particular factors that were found to be in no way responsible for the incident.
Attachments or Appendices	Miscellaneous back-up information such as: discussion of rejected or less-probable scenarios, documents of special interest or value, method and conduct of the investigation and team membership, photographs, diagrams, calculations, lab reports, references, noncontributory factors, terms of reference.
References	All documents that the team reviewed and/or used should be listed.

11.3.2. Introduction

The body of the report may include an introduction summarizing the occurrence or incident, the incident investigation team members, and the scope of the investigation. Background information for the facility (unit) includes such items as history, age, size, expansions, major occurrences, and technical sophistication as necessary to understand the occurrence.

11.3.3. Background

This section presents an overview of the process and occurrences leading up to the incident. It should reconfirm the purpose of the investigation, and clarify any limitations on the scope of the team activities. If a particular program such as a periodic inspection program or job safety analysis is involved in the incident, it is referenced in the background section. Information about the existing management systems, procedures, and policies is normally included in this segment, as are any unusual external occurrences such as labor relations issues, personnel turnover rate (if applicable), maintenance shutdowns or turnarounds, and interruptions or distractions (for example, a power dip).

The background of the individuals involved should be addressed, such as experience level, qualifications, experience with the particular task involved, time in the position, years of experience, time at that task that day, whether working overtime or rotating shift.

The environmental conditions present at the time can be particularly significant. Those may include time of day, temperature, lighting, and weather conditions such as rain, fog, ice, snow, or wind conditions.

Significant process conditions preceding the incident should be identified, especially if the process is a batch operation or if there was any known deviation from normal conditions of sequences, flows, pressures, concentrations, temperatures, pH, or other process parameters. Often it is helpful to separate the background conditions into several distinct periods. One category may be normal conditions, a second category may be the time period from 48 hours up to 1 hour before the occurrence, and a third section may address the background immediately (1-20 minutes) before the occurrence.

11.3.4. Sequence of Events and Description of the Incident

In this section of the report, the scenario is described (usually in chronological order) and the results are identified. This is the WHO–WHAT–WHEN–WHERE portion of the report as well as the progress through the incident of the action taken to deal with the situation. It should give precise and specific information using identification numbers and location of equipment. The military 24-hour clock method of recording time is especially applicable to this section. Many chemical processes operate using this format already, as it is less ambiguous. When this section is done successfully, the reader has a sense of witnessing the occurrence as it unfolds. The extent of injury, details of the damage, and an estimated out-of-production time are included in this section. Anticipated questions and items of special interest should be clearly addressed. It should be remembered that diagrams are often more useful than long paragraphs. If a timeline has been developed, it should be included in the report. The observations should be backed up with statements from those involved. Wherever possible, supporting documentation in the form of drawings, photographs, flow diagrams and calculations should be included. The investigation should comment on the adequacy of the emergency response to the incident, and highlight any shortcomings in emergency plans, resources, logistics, etc.

11.3.5. Evidence and Cause Analysis

Root (or primary) causes, immediate (or secondary) causes, and contributory factors are identified, analyzed, and discussed in this section of the report. As described in Chapter 9, process safety incidents are the result of many factors, and therefore singling out one cause is rarely the proper approach. Some experts indicate that if a fault tree or causal factor chart was developed as part of the investigation it should be incorporated to facilitate understanding.

It may be appropriate to review the various types of evidence:

- People (interviews)
- Physical (for example, equipment, machinery, and parts)
- Electronic (for example, operating data recorded by a control system, both current and historical, and controller set points)
- Positions (people and equipment)
- Paper (for example, procedures and process data)

It is often helpful to discuss how the various items of evidence relate to events before, during, and after the incident. This should help explain how and why the incident occurred leading to the root causes.

This section can also summarize any specialized studies or analysis that may have been commissioned to explain the circumstances of the incident. For example, studies such as metallurgical analysis of components, chemical reactivity, and supporting documentation could be included in the report appendices.

11.3.6. Findings and Recommendations

The attributes of successful recommendations are addressed in detail in Chapter 10. The incident investigation team has the responsibility to *develop and submit* the recommendations. It is management's responsibility to act on and resolve the recommendations. Management usually determines target dates and assignment of responsibility.

Incident investigation reports for the chemical process industry sometimes differ from governmental agency reports concerning recommendations. For many government reports, the official recommendations appear in a separate document. The CCPS strongly endorses a policy of including the recommendations in a single incident report document.

Principal or main findings and recommendations should be discussed at length to highlight their importance. This subsection should precede a tabulated format of all findings, causes, and recommendations. As a rule, up to five findings should be regarded as principal.

Each finding should be tabulated as a separate item so that the individual subject matter under discussion is sufficiently and clearly separated from other points. The list of recommendations derived from these findings should be cross-referenced, allowing reasons supporting a particular recommendation to be readily found. This is most easily done in a table format. One example of a tabulated format is shown in Table 11-2; for each finding, root and secondary causes may be shown together with contributory factors (if appropriate). However, various tabulated formats are possible. Each recommendation should be individually numbered to facilitate subsequent follow up and tracking.

In the course of the investigation, the team will invariably become aware of a number of minor points relative to their particular interest. These minor observations should be specifically mentioned within the table as one or more findings, and a formal, positive recommendation submitted along with the more important recommendations.

The report becomes clumsy if negative comments or those suggesting improvement are not progressed from findings into recommendations.

This table should include:

- The team's conclusions as to the significant findings of the incident.
- The team's opinion as to the probable factual cause(s) of the incident. Where the cause of the incident cannot be definitively ascertained, the team's opinion as to the probable cause should be given. The causes given should be clearly stated and where applicable include the root or underlying causes and as well as secondary causes, plus any possible contributory factors and explanations.
- The team's recommendations (immediate and long term) as to steps that might be taken to prevent any recurrence of incidents of a similar nature. These recommendations should not be confined to such matters as plant modification, but should, if appropriate, embrace training (retraining), assessment of operational procedures, and strengthening of safety measures. Recommendations that improve work systems and conditions make it harder for people to make mistakes. The team should also indicate what

TABLE 11-2
Table of Findings, Causes, and Recommendations

	Finding	Cause	Recommendation
1.	The end cap of the filter pot apparently blew off and struck the operator on the head.	**Root Cause**—The threads of the top cap union were badly worn and insufficient to contain the applied air pressure. **Secondary Cause**—There was no secondary retaining device in the event of thread failure. **Contributory Factor**—The operator closed a valve deadheading and pressuring the system with compressed air.	1.1. Issue a safety bulletin to all other facilities to inspect the threads on filter pots of similar design. 1.2. Issue a safety bulletin to all other facilities to raise awareness of the hazards of LP air purging. 1.3. The Engineering Department should investigate the practicality of installing secondary retaining devices.
2.	The purging operation was performed within the design pressure rating of the injection system. However, the mechanical condition (worn threads) of the filter pot and end cap was insufficient to contain the applied air pressure.	**Root Cause—**There is no formal inspection and maintenance program for the filter pots. **Secondary Cause—**There was excessive wear of the top cap union threads, and corrosion was evident on the inside of the cap and on the threads. **Contributory Factor—**The lower threads were full of dirt, and the cap may not have been fitted tight enough and/or cross-threaded.	2.1. A maintenance and inspection program should be developed for the injection pump filter pots. 2.2. A maintenance and inspection program should also be developed for other critical equipment, especially pressure containing items.
3.	The end cap of a similar filter pot had been dislodged on at least one previous occasion, but without personnel injury. No action appears to have been taken following this incident.	**Root Cause—**There is no effective system to record all near-miss incidents. **Secondary Cause—**The incident was not reported to senior management. **Contributory Factor—**The cap was reportedly only dislodged 6 to 8 inches in the air and no personnel injury resulted.	3.1. The facility should implement an effective system to ensure that all near misses are reported and followed up.

follow-up measures are necessary to ensure that the recommended actions have been effected.

Each recommendation should be brief (two or three lines only), sufficient to identify a particular topic, and individually numbered to facilitate management of follow-up and resolution. If appropriate, each recommendation should have appended a cross-reference number to enable a fuller explanation, description, or argument to be extracted from other sections of the report in support of the recommendations. Some investigators recommend adding a column to the table entitled 'Evidence' which allows the team to tie in important pieces of evidence to the findings and identified causes. In effect, this column answers the questions, "How do we know this? Why do we believe this is a cause?"

11.3.7. Noncontributory Factors

Record those factors, both human and system-based, that were analyzed and found not to be relevant. For example, concern frequently centers on personnel competency issues, and if this is not a factor in the incident, it should be stated. Other noncontributory factors should also be recorded.

11.3.8. Attachments or Appendices

The remaining contents of a formal written incident report can vary significantly depending on circumstances. A collection of data and additional reference information that some, but not all, readers may need is often included as an appendix. Typical supplemental information might include:

- Flow sheets
- Diagrams
- Photographs
- Material safety data sheets (MSDS)
- A full list of reference materials consulted
- Scope and objectives of the investigation, including a list of team members
- A description of investigative methods and approaches used
- A glossary of terms and acronyms
- A bibliography
- Log sheets
- Computer printouts
- Pertinent extracts from witness interviews
- Maps
- Initial incident reports

- Copies of work permits
- Detailed injury information
- Detailed equipment damage information

If a map is used, it should focus on the area of interest, and show only the minimum amount of nonessential information.

Medical evidence is usually omitted from formal incident reports due to the need to respect medical doctor–patient confidentiality. Names of injured and other participants are also frequently omitted for privacy reasons. Descriptions such as Operator 1, 2, or 3 can be substituted.

Although it is not required by existing regulations, it is a good practice to include the reasons for eliminating other possible causes and alternate scenarios. This can be extremely useful and enlightening to subsequent investigators or analysts who may follow 5 or 10 years later.

All members of the incident investigation team should review and reach consensus on the content of the report before it is finalized and published. It may be appropriate for them to sign the final report depending on local practice. This is an indication of personal endorsement of the team consensus.

11.3.9. Criteria for Restart

If conditions or restrictions were imposed on restart of operations, these items should be included in the report. The discussion should include the basis for placing the restrictions as well as the reasons for removing them.

11.4. Capturing Lessons Learned

11.4.1. Internal

Lessons learned from one facility's incident often have applicability to other facilities within the same organization or similar business.[1] A management system should be in place to ensure that the understanding of the lessons learned is not restricted to a single location. Another way to express these thoughts was presented in Kletz's paper, "Organizations Have No Memory."[6] In actual practice, organizations may find it extremely difficult to maintain continuity of the lessons learned from an incident. This challenge remains regardless of the quality of the initial investigation. A well-designed and operated incident investigation management system should set up a mechanism to communicate lessons learned to all appropriate company groups and, where appropriate, other companies with similar technology.[2] This includes maintaining an incident log, and ensuring that incident reports become a part of the process safety information document package.

> ***Organizations have no memory.***
>
> ***Only people have memory.***
>
> ***A proactive and sustained effort is needed so that lessons learned once will not have to be relearned.***

There should be a plan for issuing the information from the investigation. Specific responsibilities and target dates should be established. Not all groups have the same information needs, so it may be practical at times to prepare more than one information release. The personnel and contractors who work directly in the unit that experienced the incident have a need to know what happened, why, and what direct, specific changes are to be made. Personnel and contractors in adjacent units may not be exposed to the same specific hazards, and therefore may need a more generic presentation of the lessons learned. Workers in other similar facilities within the same parent organization or even at other companies may be exposed to the same hazard. Unless the company has an active communication program as part of its incident investigation system, these workers could remain uninformed and a similar incident could occur. Additional information on case histories can be found in Chapter 15.

Sometimes formal written reports represent a trade-off between quantity and quality of information. The information contained in the outgoing communication should be general enough to ensure that an incident is not regarded as an isolated phenomenon, and yet specific enough to ensure that the information is not regarded as too vague or too general to be useful. Depending on the complexity of the incident or near miss, a short, high-quality report may be all that is needed.

Some companies have found success with a periodic publication of incident abstracts. Each incident may appear as a one or two paragraph summary in a quarterly bulletin. This information is circulated widely within the corporation or is available online. If a site has a special interest in one particular incident, full details are then requested by direct contact between the two sites. The highly abbreviated summary has other uses, such as material for employee safety training meetings or bulletin board postings. This is another example of the practical trade-off between quantity of incidents publicized and the amount of detail presented for each incident.

Historical incident recording and communicating in useful form has several beneficial results. This information suggests precautions for other facilities, allows lessons learned to be taken into account in future design, and helps identify trends not apparent from single incidents. Because incidents have many causes, some causes may not be identified in the investigation of a single incident. For example, if an incident occurs on a Saturday, this may simply be coincidence, or it may be a symptom of deficiencies in management systems on weekend shifts. If a pattern of weekend incidents develops, then management can take appropriate action. Without incident recording and analysis of the record, such patterns may go unnoticed and result in even more incidents. Figures 11-2 and 11-3 show two examples of summary historical analyses.

Common causes and trend identification analyses provide the opportunity to apply deductive hindsight to past incidents. This process allows a productive allocation of resources and can maximize incident prevention and control. However, establishing relevant categories for future researchers is not as easy as it might appear. There will be hard choices between having too many small categories and too few very broad, somewhat nonspecific categories. Searchers should be able to clearly focus on commonalities. The categories should include specific system deficiencies or breakdowns such as design, training, process mechanical integrity, and equipment specifications. Other categories may be categorized by specific hazard exposure.Trend analysis can be facilitated by using the categories and subcategories in predefined trees.

Trend analysis can be confused or invalidated by a sample that is too small. If the charting or analysis is limited only to major incidents, there will often be too few within a period to arrive at meaningful conclusions. For example, a facility with one thousand employees may experience only one or two serious incidents per year, and several years' worth of data would be needed to make any meaningful statistical analysis. Minor incidents and near misses can be as useful in trend analysis and preventive prediction as major incidents. All process incidents should be reported, classified, and investigated as appropriate. **The severity of an incident is frequently more a function of chance than actual fundamental system differences among accidents and near misses.**

Computer-aided trend identification offers potential benefits, but is dependent on the quality of the input information. Expert systems and artificial intelligence are tools being tested. When successful, they may give improved insight into identifying common causes and trend analyses.

A standard form is a common and useful tool for the relatively minor process incident. Forms tend to steer the investigators' thoughts within narrow, limited channels. Forms, therefore, may not be as useful for investigating serious process-related occurrences as they are for minor process incidents. Some companies have increased the effectiveness of reporting forms by printing memory jogger information and questions on the back of the form. Forms can be helpful for capturing cause data in a uniform manner so it can be coded for computer analysis.

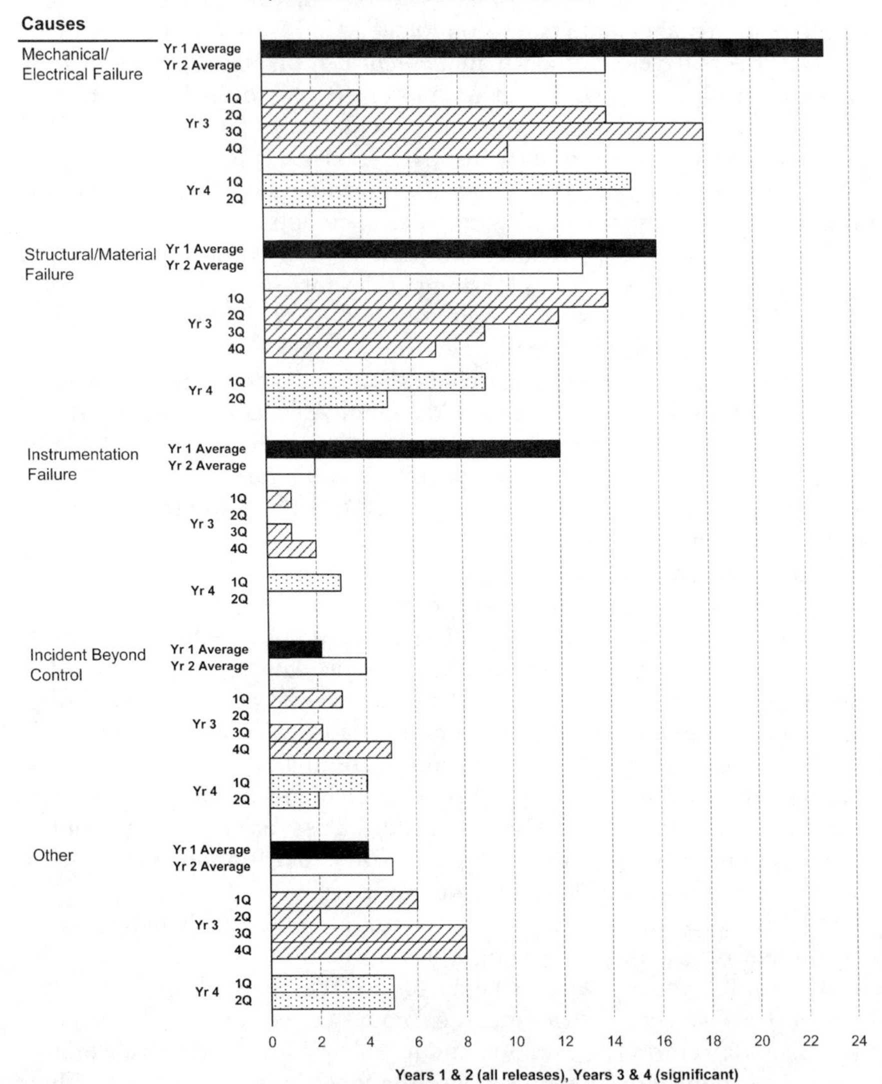

FIGURE 11-2. **Representative incident trend data.**

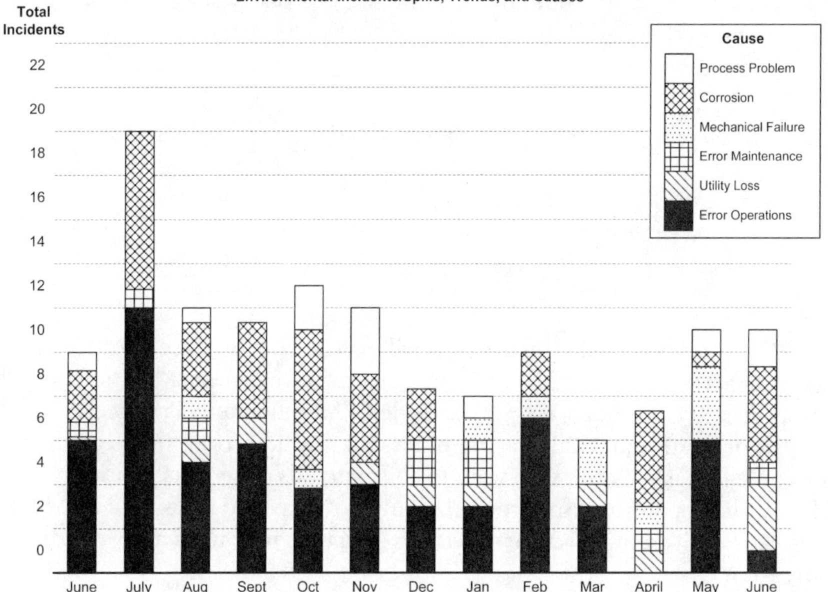

FIGURE 11-3. **Representative incident analysis data.**

Each investigation is unique and can be used as an opportunity to improve the incident investigation system. A critique of the investigation process should be conducted and should include everyone who participated on the investigation team. It should identify potential changes that would make the next investigation more successful. It should also identify those items that were successful and do not need change. Evaluation of the planning, team supplies, coordination, and communication issues often will show areas for improvement. Proposed changes to the incident investigation system should be evaluated and executed using the management of change system. Changes adopted should be implemented and tracked as any other recommended action item. Modifications to the incident investigation team-training program should be kept up to date. These types of modifications are discussed in detail in Chapter 14.

11.4.2. External

Process safety is a dynamic field. Enhancing process safety knowledge is one of the twelve key elements recognized by CCPS for successfully managing process safety.

> *"Management commitment to using all available resources for enhancing process safety knowledge of all levels in the organization is a key difference in having a minimum, adequate, or outstanding process safety program."*
>
> —*Guidelines for Technical Management of Chemical Process Safety* (1) page 137.

The American Chemistry Council recognizes this area in Management Practice #5, *Information Sharing*, as an integral part of the code of Responsible Care® Process Safety Code of Management Practices.

In practice, external sharing of lessons learned from investigations is not an easy task. Those who wish to share details of serious incidents with the public encounter numerous challenges. Many process safety incidents involve fatalities or third party activity, and thus may ultimately end up in litigation.

One method of enhancing safety knowledge is to learn from the experience of others. (See Chapter 15.) A sampling of industry databases that contain information about incidents is presented below in Table 11-3. An internet search can help the user in locating and evaluating these and other databases.

TABLE 11-3
Examples of Industry Databases

Center for Chemical Process Safety Incident Database (PSID)
PSID is a respected and valuable database of incidents that CCPS member companies have submitted. This database is searchable and has features allowing the building and customizing of reports for specific needs. The PSID is a professionally managed database with verified inputs and has the ability to pull in some incidents reported in other systems such as those listed in this table. It is available to participating membership.
Accidental Release Information Program (ARIP)
ARIP involves collecting questionnaire information from facilities that have had significant releases of hazardous substances, developing a national accidental release database, analyzing the collected information, and disseminating the results of the analysis to those involved in chemical accident prevention activities. ARIP also helps to focus industry's attention on the causes of accidental releases and the means to prevent them. The database is publicly available and covers incidents from 1986–1999.

Chemical Safety Board—Chemical Incidents Report Center (CIRC)
This is the CSB database containing incidents reported to them. It currently contains reports from 1998 onward, but there are plans to expand as time permits.

CAIT Accident Investigation Summary Matrix
For a list of accidents that EPA's Chemical Accident Investigation Team has investigated or is investigating.

(Continued on next page)

US EPA CEPPO Chemical Accident Histories
Contains data and has additional links to US and international databases of interest.

Major Hazard Incident Data Service (MHIDAS)
Published by AEA Technology on behalf of the UK Health and Safety Executive, this database is an excellent resource for querying incident information. The service is available on CD-ROM, or online.

Mary Kay O'Connor Process Safety Center at Texas A&M University
The Mary Kay O'Connor Process Safety Center is conducting a number of projects on "Development of Accident History Databases," "Analyses of Accident History Databases," and "Risk Assessment Based on Accident History Databases." One such project involves the compilation of incidents from various sources.

Major Accident Reporting System (MARS)
The Major Accident Reporting System (MARS) is a distributed information network, consisting of 15 local databases on a MS-Windows platform in each Member State of the European Union and a central UNIX-based analysis system at the European Commission's Joint Research Centre in Ispra (MAHB) that allows complex text retrieval and pattern analysis.

US Department of Energy's Occurrence Reporting and Processing System (ORPS)
Summaries and reviews of the US Department of Energy's Occurrence Reporting and Processing System (ORPS) can be found at http://www.dne.bnl.gov/etd/csc/. DOE has been tracking, trending, analyzing, and compiling chemical safety concerns at its facilities since August 1992.

Hazardous Materials Accidents—National Transportation Safety Board Reports
This database provides a list of hazardous materials accident reports prepared by the National Transportation Safety Board.

UK IChemE Accident Database
The Accident Database is a compilation of safety case studies and accident reported compiled by the (UK) Institution of Chemical Engineers (IChemE). The database is a subscription product but a time-limited trial version is available.

For years, AIChE has sponsored public presentation of actual incident descriptions with accompanying lessons learned. Some of these case histories are contained in the AIChE publications on *Ammonia Safety Symposium, Loss Prevention*, *Chemical Engineering Progress,* and *Plant Operations Progress*. The American Petroleum Institute also publishes a series of booklets on specific lessons learned. In the UK, the Institution of Chemical Engineers[7] publishes *Loss Prevention Bulletins* that contain specific case histories.

Valuable, although unofficial, case history information can often be exchanged verbally during conferences and seminars. This is especially true for those incidents in which litigation is not expected.

Other avenues for increasing process safety knowledge exist. Previously unrecognized potential hazards and latent properties of materials are written about in industry and technical journals. Peer-reviewed research work in the area of process safety is published in several international scientific journals. Manufacturers and suppliers of process equipment often publicize previously unrecognized failure modes. Additional information on case histories can be found in Chapter 15.

11.5. Tools for Assessing Report Quality

11.5.1. Checklist

The incident investigation team should review the investigation and final draft of the formal written report to ensure the intended result was achieved. An example checklist is shown in Table 11-5. Reports should have technical review for clarity and accuracy, and where appropriate, legal review.

TABLE 11-5
Example Checklist for Formal Written Reports

CHECKLIST FOR FORMAL, WRITTEN REPORTS
❑ Intended reader/user identified and technical competence level chosen
❑ Purpose of report identified
❑ Scope of investigation specified
❑ Summary/abstract no longer than one page
❑ Summary/abstract answers what happened, why, and general recommendations
❑ Background—describes process, investigation scope
❑ Narrative—clearly describes what happened
❑ Root Causes—identifies multiple and underlying causes
❑ Recommendations—describes specific action for follow-up
❑ Other—necessary charts, exhibits, information
❑ Content agreed to by team members
❑ Distribution identified

11.5.2. Avoiding Common Mistakes

For improved quality of formal written incident reports, the incident investigation team should follow these guidelines:

1. Avoid jargon specific to the process that the intended reader may not understand. One good guide is to try to have the written report intelligible to the technical-minded layperson.
2. For increased readability and comprehension, reduce the use of abbreviations and acronyms. Most of these can be avoided.
3. Decide on the selected reader's level of technical competence and then be consistent in writing to that level. Assume the reader has a certain minimum knowledge of chemical process safety.
4. Avoid intermixing conclusions when presenting the factual findings.
5. Be sure to include a complete list of reference materials used during the investigation. Subsequent investigators or analysts who review the report several years later should be able to substantiate the conclusions reached by the investigation team.
6. Identify multiple system-related root causes determined by the investigation team after a systematic analysis. Root causes reported should not be limited to the opinion or judgment of a participant or witness of the incident.
7. Include specific equipment identification, because omission can cause problems for readers in other process units or facilities who may have the same equipment and remain unaware of its hazards.
8. Avoid downplaying human performance factors when drafting the report. There is a natural hesitancy to criticize or address normally encountered performance limitations or errors. Effects of fatigue from working excessive overtime are not always addressed in the written report. If these human performance factors are neglected, the error may be repeated. All facts of the incident must be considered relevant.
9. Publish only one official version of the report. Sometimes the team may release a preliminary draft copy that differs from the final version. This can sometimes cause unnecessary and avoidable confusion if the interim report is not handled carefully. If a preliminary draft is published, the team should ensure that all copies of the preliminary version are replaced with the final edition. The preliminary report should be conspicuously identified as a draft. Consider printing a message on each page's footer or header such as *Draft—not a final report—subject to change.*
10. Delay writing the abstract/summary until after the main body of the report has been drafted.

11. Start writing the report as soon as the investigation begins. Focusing on the result can help keep the team focused on the process and the product.

In summary, the report is the vehicle for transmitting and documenting the investigation. It is a direct reflection of the quality and professionalism of the incident investigation team. The entire team should review the report before release.

Endnotes

1. Center for Chemical Process Safety. *Guidelines for Technical Management of Chemical Process Safety*. New York: AIChE, 1989.
2. American Chemistry Council. *Responsible Care®, A Resource Guide for the Process Safety Code of Management Practices*. Washington, DC: 1990.
3. Carper, K. *Forensic Engineering*. New York: Elsevier, 1989.
4. Kuhlman, R. *Professional Accident Investigation*. Loganville, GA: Institute Press, International Loss Control Institute.
5. American National Standards Institute. *ANSI Z16.2 Method of Recording Basic Facts Relating to the Nature and Occurrence of Work Injuries*. New York.
6. Kletz, T. A. "Organizations Have No Memory." Paper presented at the AIChE Loss Prevention Symposium, April 1979.
7. *Loss Prevention Bulletin #061*. Rugby, England: Institution of Chemical Engineers, February 1985.

12

Legal Issues and Considerations

The potential for legal liability following any incident where there is injury, death, environmental damage, or even business interruption comes in many different forms. Such an incident could result in civil litigation, criminal prosecution, or both. After incidents involving health implications, companies often face situations where its employees are litigants on both sides of a case—as plaintiffs if claiming injury or illness—and defendants if named as a manager of the area or activity.

Legal access to an incident investigation team's documentation of deliberations and findings is an increasingly complex issue. It is sensible and in the public's best interest for companies to investigate an accident and determine root causes to prevent future incidents. This is complicated when companies are concerned about the threat of discovery and the use of collected information in litigation. An incident investigation team can best address these complicated issues and considerations with open access to legal counsel throughout the investigation. In fact, some companies are assigning legal counsel to an investigation team before their first meeting.

The conduct and conclusions of the incident investigation team, the documents developed, and records of the investigation will become a focal point if legal actions ensue. It is critical that legal counsel be available to the investigation team from the very beginning in order to identify potential legal liabilities for management or for individuals in cases where conduct could be considered criminal under today's broad statutory authorities. In addition to providing advice, legal counsel can also review documents in order to avoid unnecessary legal exposure.

12.1. Seeking Legal Guidance in Preparing Documentation

12.1.1. Use and Limits of Attorney–Client Privilege

Some documents created by an incident investigation team may be subject to disclosure to:

- government agencies under their inspection authorities, and
- plaintiffs' lawyers under the rules of discovery that govern litigation.

The appropriate use of the attorney–client privilege during an investigation, however, can help promote frank and open communication between the incident investigation team and legal counsel, and through legal counsel to management. When documents are prepared at the request of counsel or when communications are transmitted to counsel in order to obtain legal advice, these communications may generally be protected from disclosure. The attorney–client privilege exists so clients can communicate frankly with their attorney. Usually, the attorney can provide sound representation without the substance of those communications becoming public. In most European countries, however, the concept of privilege is extremely narrow and in the United States, judges may apply privileges sparingly.(1)

If a document is considered privileged information, the organization must severely restrict access to that document to maintain that privilege. An organization must choose if it wants to share information or restrict it. Some progressive organizations have stopped seeking privilege, as they believe the open ability to share (both internally and externally) is more important.

Because there are many attacks on the use of the attorney–client privilege today, each investigation team member must treat any note, email, report or communication as if it would become a public document available to the press, government or the public in general, including competitors. Legislative requirements indicate that reports on process safety incidents that are covered by the US OSHA process safety management (PSM) standard and the US EPA Risk Management Program (RMP) must be shared with workers. The American Chemistry Council's Responsible Care® guidance(2) promotes the sharing of such information. Companies may want to share the results of the investigation with their stakeholders and others in industry that may have similar situations. The primary advantage of the attorney–client privilege is to allow the preliminary fact-finding process and legal analysis of the situation to be protected. If outside experts are needed to assist in the investigation, legal counsel will be responsible for retaining the expert. The experts may then assist counsel in the defense of any legal actions that may follow.

Under certain circumstances, the courts will deny attorney-client privilege and order the disclosure of investigative reports. Other protections that may apply include *The Work Product Doctrine* and *The Self-Critical Analysis Privilege*. (1)

The work product doctrine was created to protect materials prepared in anticipation of litigation from discovery. Although technically speaking a lawyer does not have to be involved for material to acquire work product protection, attorneys might need to be involved for several reasons. First, some courts believe the involvement of a lawyer is required. Second, involving a lawyer suggests the matter should not be considered ordinary course of business. Third, the lawyer's involvement emphasizes that the work is being done in anticipation of litigation.

The self-critical analysis privilege is not widely recognized. Counsel should be consulted about the possible use of this privilege.

To help protect the confidentiality of an incident investigation, a request should be made by the company either to a lawyer, in-house counsel or outside counsel, for legal advice on matters arising from or related to the incident. The lawyer should then direct the activities to be undertaken, making it clear that the information is to be provided to counsel for the purpose of providing legal advice to the company or preparing for litigation, or both.

12.1.2. Recording the Facts

There may be a perceived conflict between the need of the investigation team to gather information quickly and record observations versus the legal risk the company could face from hastily prepared notes or erroneous preliminary conclusions. In fact, the investigation team and legal counsel have the same basic goal: helping the organization to avoid loss. (2) Haste in making notes without clearly distinguishing between factual observations and speculation can cause unnecessary legal risk to the company. The company will end up spending a great deal of time and money trying to explain the hasty notes in litigation or enforcement actions. The investigation team should take accurate notes and record only facts. Any opinions or speculation should be clearly noted as such. Facts cannot be altered, but conclusions can change as the investigation continues. In some cases, the legal counsel should review documents that are prepared by the investigation team for outside distribution as well as the final official reports as they are drafted. The guidance by legal counsel can help to limit unnecessary liability. Typical guidance to investigators may include:

- Involving legal counsel as soon as possible if the incident appears to have potential liability for the company
- Using header and footer designations to identify official incident team internal documents. Legal counsel may recommend adding statements such as, "Privileged and Confidential—Attorney–Client Privileged Information" or other designators on each page of certain documents
- Refraining from use of inflammatory statements such as *disaster, lethal, nearly electrocuted,* and *catastrophe*
- Refraining from use of judgmental words with special legal meanings such *as negligent, deficient,* or *intentional*
- Refraining from assigning or implying blame
- Refraining from offering opinions on contract rights, obligations, or warranty issues
- Refraining from making broad conclusions that cannot be supported by the facts of this investigation
- Avoiding unsupported opinions, perceptions, and speculations
- Refraining from *overly* prescriptive recommendations, that is, allowing for alternative resolutions of the problems and weaknesses found
- Following through on each recommendation and documenting the final resolutions, including why a recommendation was rejected or modified if that is the final resolution
- Reporting, investigating, and documenting near misses as well as incidents to demonstrate the company's commitment to learning where there are weaknesses and to improving risk control

Chapter 11 provides additional information on preparing documents.

12.2. The Importance of Document Management

Of equal importance to documenting investigation activities appropriately, is properly managing all documents and evidence developed by the investigation team. The team needs to develop a document management system to track all documentation and evidence. A log should be developed and every piece of evidence or documentation given an identifier and entered into the log. Legal counsel should also be consulted on the scope of distribution lists of documents that are prepared by the team. If the investigation is being conducted under attorney–client privilege, the counsel will determine the scope of those who need to be on distribution lists. It is important to keep control of preliminary copies and draft reports issued for team review and comment. A good practice is to include a full distribution list on each copy, so that receivers of the document know who else has been copied. This is especially important on sensitive documents related to serious incidents. In addition to the use of headers and footers noting confidentiality, expert investigators include DO NOT COPY on some documents and always use the pagination style that notes the identification *this is page x of y* markers on certain documents. It is important to end up with only *one final report* that everyone agrees is the consensus of the team and then make sure all previous versions are destroyed.

A chain of custody must be maintained for all evidence. It is likely that items could be sent for testing by a specialist or examination by a regulatory agency. It is essential to preserve the condition and quality of the evidence as well as knowing precisely where it is at any given time.

Incident investigation document retention is another important issue to consider. Lawyers and engineers are likely to disagree about which documents to keep and how long to keep them. Lawyers know that each document retained may cause, at best, additional legal expense and, at worst, increased legal liability. On the other hand, engineers may believe certain documents must be retained to maintain corporate memory. Lessons learned may otherwise be lost. The document retention controversy cannot be resolved here. Each organization must develop and implement its own policy. Several universal principles must be recognized. In the United States, EPA and OSHA have established certain retention requirements for those facilities subject to process safety management regulations. Under these regulations, incident investigation reports must be retained for five years. Refer to 40 C.F.R. §68.81(g); 29 C.F.R. §1910.119(m) [7]. Note that these regulations do not require the retention of related documents.

12.3. Communications and Credibility

A company's reputation and credibility are fragile assets, taking years to establish. A single statement, action, decision, or occurrence can destroy both. Accordingly, any statement must be carefully reviewed to be certain that it is soundly based on fact and not hopeful speculation. Any statement made by the investigation team can become the subject of enforcement actions and litigation. Failure to exercise care in written documents and communications can result in years of legal costs in explaining what the team really meant or defending the factual stance that the statement was only preliminary.

While legal counsel can provide valuable assistance, the training and experience of the team are essential to the credibility of the investigation's results. A primary goal of the investigation team is to prevent recurrence. While that is also a major goal of the legal counsel, it is also his or her goal to limit liability. Limiting legal liability is important to all organizations, but the veracity of the investigation should not be compromised.

In addition to legal concerns, part of the company responsibility arising from an incident is to inform employees, contractors, neighbors, governmental agencies, trade associations, and the public, in an accurate and appropriate fashion and in the proper degree of detail. In fact, the American Chemistry Council's Responsible Care® Process Safety Code, Management Practice #5, requires that each member company have an ongoing process safety program that includes *Sharing of relevant safety knowledge and lessons learned from such incidents with industry, government and the community*. Progressive companies have abandoned the stonewalling "no-comment" response. This change recognizes practical realities. The public is entitled to a certain reasonable amount of information and the media will deliver a story regardless of the cooperation or noncooperation of the company.

Credibility becomes a concern for the process safety incident investigation team as soon as the first request for information is made, from within or outside the company. Official communications from the team to management and other stakeholders such as statements, progress reports, preliminary determinations, findings, and other communications are normally the responsibility of the team leader; however, others with special training may fill this role. Team members must follow corporate communication protocols and the guidance in the incident investigation management system. There is often an initial flood of requests for information, followed by a short hiatus, which in turn, is followed by renewed demands for the expected answers as to what happened, why, and what will be done to prevent a recurrence. Selecting the team spokesperson is a particularly important task. Guidance from management and persons experienced in corporate communication with the public should be sought, together with legal counsel, in making such a selection. The investigation team spokesperson should resist the temptation to speculate or offer possible conclusions before the investigation yields a factual basis for such conclusions.

12.4. The Challenges and Rewards of Sharing New Knowledge

The best way to avoid claims and litigation is to prevent an incident from occurring in the first place by having adequate systems and preventive measures in place. The goal of the incident investigation is to learn the root causes of an incident in order to improve management systems, and by doing so, prevent incident reoccurrence. By sharing best practices and lessons learned, the entire industry can be rewarded with continuous safety improvement.

Along with this reward is the responsibility imposed on a corporation after it has acquired additional knowledge and expertise from someone else about a hazard or remedy. Legal liabilities will be much more significant for the company if knowledge gained from incident investigations is not acted upon to prevent other incidents. In addition, many chemical processes are generic and possess common inherent hazards. There is decreasing tolerance on the part of the public toward industrial incidents; therefore, one company's poor safety performance can adversely affect the entire industry.

In today's litigious environment, sharing the details of investigations with those at similar facilities that might benefit can be problematic. In addition to concerns about litigation issues, practical logistics can make it difficult to succeed in communicating details and lessons learned consistently. These logistics are especially challenging in large corporations or even within a single large facility. Knowing who has a potential interest can be a considerable task, due to turnover of personnel and sheer size of the organization. Notwithstanding these challenges, appropriate communication of investigation findings is critical to the overall reduction of incidents. Chapter 13 addresses specifics of communicating investigation findings to all potentially affected workers, to outside organizations, and maintaining long-term continuity of the lessons learned.

12.5. Employee Interviews and Personal Liability Concerns

Because of the very real risk today under various laws for personal liability, it is important to get general guidance from legal counsel concerning employee interviews. If information develops during an interview that indicates an individual could have personal liability, legal counsel needs to be informed to assess whether or not the individual needs to obtain his or her own legal counsel. Sometimes an investigation team wants to record verbatim what the person being interviewed is saying. There are legal implications associated with this and legal counsel should consider it. Some attorneys believe the substance of the interview can be captured but the person should not be asked to sign the document. If they do, it later can be construed as the person's statement and as such, copies of these statements may be subject to disclosure to governmental agencies and opposing parties in litigation.

Legal counsel should be consulted so that employees are aware of their rights before being interviewed by any governmental authority.

Company lawyers have the company as their client, not necessarily the employees. In many situations the employees of a company may assume that the company lawyer represents their interest when in fact the lawyer may be seeking to distance the company from individual actions. This is an ethical issue. The incident investigation team and the governmental authorities will want to interview the same employees. It is important that the employee understand what his or her rights are under each circumstance. The best way to achieve this goal in advance of an incident is to include the topic of employee rights whenever the facility provides training on general incident investigation topics as described in Chapter 7. Including this topic in training can help employees understand and adopt the company's stance that the prime importance is getting the facts and preventing recurrence, not blaming those involved.

The manner of conducting witness interviews has been extensively covered in Chapter 8. The discussion here is limited to those points that deserve additional emphasis or particular legal consideration.

First, witnesses should be interviewed as promptly as possible. Human memory fades rapidly. When the incident is fresh in a witness's mind at the time of the interview, the recollection will be more accurate. Moreover, especially in the aftermath of major incidents, government investigators are likely to interview witnesses. Depending on their responsibilities, some government investigators may have a different objective than company investigators because of their compliance and enforcement role. All questioning should be objective in order to achieve accurate answers. If leading questions are asked the witness may feel encouraged to say things, or say things in a manner, that are sympathetic to the questioner's point of view. One recommendation is that the company investigator interviews the witness first.

Witnesses should be told the purpose of the interview and how the information will be used. They should not generally be advised that the information they give will be held confidential.

Witnesses should be provided documents pertinent to their statements and allowed time to review them before or during the interview. The documents may help the witness recall information that otherwise would be forgotten. Asking the witness to sort out any apparent inconsistencies in his recollection and information contained in documents before committing to a recorded version is a practice to consider. This avoids statements that he may later have to modify or retract.

The incident investigation team has responsibility for collecting facts, determining causes, and recommending mitigation measures to avoid reoccurrence. The team does not have responsibility for determining disciplinary action. The team may forward specific information regarding employee, supervisor, or management behavior. However, only management (with appropriate legal counsel) should be responsible for making decisions as to what, if any, disciplinary action is warranted. The team also has no authority to grant limited immunity in order to elicit information. If a witness perceives (correctly or incorrectly) that the investigation interview is somehow a disciplinary hearing, it will most assuredly have a negative impact on the interview. In addition to being counterproductive, incorrect perceptions by the witness will hamper the research into the sequence of occurrences and facts before causes are established. Fair and appropriate measures should be taken as dictated by established company policy and the facts of the incident. These objectives should be clearly incorporated into the charter of the team, reinforced during team mobilization, and explained to every potential witness before an interview.

12.6. Gathering and Preserving Evidence

Investigators must also be aware of legal principles relating to the preservation of evidence. Although the law varies from jurisdiction to jurisdiction, generally a duty arises to preserve all evidence relating to the subject matter of a potential lawsuit when that lawsuit is filed, threatened, or reasonably foreseeable. In any cases where there has been property damage, or workers or neighbors have been injured or subjected to potentially illness-causing chemicals, a lawsuit may be considered reasonably foreseeable. In the wake of such an incident, physical evidence and documents should be preserved. The failure to do so subjects the company to potential sanctions for "spoliation of evidence."

Some jurisdictions recognize a separate cause of action for damages for spoliation of evidence. The destruction of documents or other evidence can lead to criminal indictments for obstruction of justice.

More typically, the failure to preserve evidence can hurt the company in a lawsuit. For example, where a defense expert in a product liability case conducts destructive testing on the product, thereby rendering it unfit or unavailable to the plaintiff's expert, the court may sanction the conduct by excluding the testimony of the defense expert. Alternatively, if crucial documents are destroyed or lost, the judge may instruct the jury to assume the worst—that the documents contained information harmful to the defendant's position in the litigation.

Often physical evidence can only be inspected or tested once without permanent alteration of the data. For these reasons, the incident investigation team needs to make efforts to preserve all physical and documentary evidence in its time-of-the-incident condition. To the extent that is not possible, lawyers should be consulted to take appropriate steps to avoid spoliation of evidence claim. For example, it may be appropriate to take detailed photographs or videotape of the equipment in its condition postincident condition before it is repaired.

All destructive testing and even part disassembly can be permanently altering. When it is necessary to test such physical evidence, it is important for all interested parties to achieve a consensus on the testing plans. The investigation team should get the buy-in of other interested parties when conducting permanently altering tests or inspections. Appropriate planning and approvals can provide a testing scenario that will satisfy the requirements all. Even the appearance of impropriety should be avoided when handling evidence that can be analyzed only once. Consider this example:

> *A facility is interested in disassembling a valve to inspect it for deposits. In the process of the disassembly, some of the material inside the valve is dislodged. The valve cannot be put back together and given to the next party for examination. Another party may have wanted to test the flow rate through the valve. This can no longer be determined because it was already disassembled. If the test was conducted without the knowledge and support of all interested parties, the facility could be accused of purposely altering or destroying data.*

Because there is only one chance to test most physical evidence, it is essential to achieve consensus with interested parties prior to testing. A description of test plans for physical evidence is located in Chapter 8.

12.7. Inspection and Investigation by Regulatory and Other Agencies

In some circumstances, outside organizations that have the authority and responsibility to enforce safety or environmental legislation will carry out their own investigation. Government entities such as the Occupational Safety and Health Administration (US OSHA), the federal Environmental Protection Agency (US EPA), the United States Chemical Safety and Hazards Investigation Board (CSB) or any state or provincial agency with jurisdiction may become involved. Inspections or subsequent enforcement may be inevitable. Enforcement may be civil or criminal. This depends on incident severity and whether there are allegations regarding failure to comply with regulations. Since US OSHA and US EPA have similar process safety management regulations, a facility where a process safety incident occurs can face multiple agency inspections. In addition, the Chemical Safety Board has authority to investigate process safety incidents and may conduct its investigation at the same time that the other governmental inspections are on going. Where an incident has resulted in serious injury or death or significant environmental damage, state-prosecuting agencies, such as the local district attorney or state attorney general's office could also decide to become involved. Finally, there is also the potential for civil litigation, particularly by contractors or members of the community surrounding a facility where an incident occurs.

Companies have specific legal rights during investigations by government agencies. Legal counsel can help companies to determine their rights and obligations, and can assist them in preparing for investigations and on site inspections. The management system can include actions for companies to take when preparing for an agency inspection. Whether or not to consent to an immediate entry by government inspectors in the aftermath of an incident is a difficult question to answer in any situation. It is impossible to answer generically. Remember that the incident site and evidence will be under regulator control. Facility managers should know that the Fourth Amendment guarantee to be free of unreasonable searches and seizures does apply to industrial establishments. A government entry into and search of a facility in the wake of an incident may be unreasonable. It may be appropriate to refuse to consent to entry in some cases. In others, it may be appropriate to consent to the government entry on conditions. The conditions might include limits on the scope and duration of the inspection or specific agreements about the taking and sharing of photographs and interviews of employees. Of course, a government warrant must be obeyed if the visitors have one. To summarize, cooperation should always be considered first when faced with an agency seeking to perform an investigation. In the long term, this approach can help forge a good working relationship with the agency

Whether or not the agency is admitted to the facility on consent or under a warrant, the government's purpose should be kept in mind. That purpose is not the same as that of the company incident investigation team. The company seeks to identify the factors contributing to the incident and the underlying causes. The government seeks to identify regulatory violations and the evidence on which to bring an enforcement action. Both want to ensure that lessons are learned to prevent future incidents. Especially when an accident causing death or personal injury has occurred, government investigators are likely to assume that a preventable condition caused the incident, that the condition violated a statute or regulation, and that regulatory penalties should be imposed. This view is not a criticism, but an observation on the way the regulatory system operates.

There are many problems from the facility's perspective. Facility personnel must manage the incident and its aftermath, but also divert resources to educate agency personnel. They must cooperate, but avoid volunteering too much information. Plant staff may be asked questions that are premature.

The government will seek to interview employees. US OSHA has express statutory authority to interview employees and their (union) representatives. US EPA does not have that statutory authority, but will seek to interview employees. Unless subpoenaed to testify, employees are not required to submit to interviews. Moreover, employees are entitled to have counsel—either company counsel or their own—present during agency interviews. The company may inform employees of these rights but must be careful to avoid doing so in a way that appears to be evasive or to constitute an obstruction of the government's investigative process. Consider involving legal counsel in these situations. One best practice is to align findings and recommendations between the company and the regulator.

A regulator's approach to incident investigation has to be different than the company's as "proof beyond a reasonable doubt" is required if a criminal case is warranted. The government will also seek documents and physical evidence. Absent a warrant, government investigators are not authorized to take documents. Generally, a facility should allow reviewing and copying those documents that the facility is required to keep and make available to the agency. Such documents would include, under the US OSHA process safety management standard, for example, copies of process hazard analyses, incident investigations for the past five years, and the last two compliance audits. Requests for other documents should be accepted in writing and considered by management and counsel. Procedures should be implemented to track any documents produced.

The incident investigation team should consult legal counsel to determine other activities that will assist in preparing for facility inspections or investigations and document preparation depending on the incident.

12.8. Legal Issues Related To "Postinvestigation"

The importance of prompt follow up and implementation of recommendations cannot be overemphasized. The impact on a government investigator, or a jury, of recommendations made but ignored cannot be overstated. This is especially true if it appears that the recommendations were not implemented due to their costs. It is the company lawyer's nightmare to learn, for example, that a past audit identified a problem and years later, a personal injury-causing accident occurred because the problem was not fixed.

Concerning written recommendations in reports, consider reviewing potential recommendations orally with managers responsible for imple-

Assigning personnel to assist with the inspection or investigation

Determine how many inspectors there will be and the extent of the inspection so appropriate personnel will be prepared and available to accompany each inspector.

Reviewing document control procedures

Ensure that employees and inspectors know the policy for transmitting documents to inspectors. All document requests should go through the assigned document control manager.

Having supplies readily available (cameras, notebooks, video equipment, copies of regulations)

It is appropriate to take photographs of whatever the inspectors photograph.

Take duplicate samples or perform duplicate testing alongside inspectors.

Training employees on regulations, protocol, and their rights

Employees should be reminded to answer questions concisely, directly, and truthfully giving only the facts, not opinions or conclusions.

Having a logistical plan for the movement of the outside agency while in the facility

A plan will help ensure that the inspection does not unreasonably affect business or the inspector's time.

Ensuring that paperwork is in order

Self pre-inspect to ensure compliance with regulations.

Review self-audits, insurance audits, previous recommendations, and follow-up actions.

FIGURE 12-1. **Preparing for an inspection or investigation.**

menting them before documenting the recommendations in writing. It may be that there are problems of which the incident investigation team members are unaware that could make implementing certain types of recommendations impossible or difficult. Management should identify those problems, discuss them, and potentially resolve them before the recommendation is written and formally sent to management. It is possible that other approaches could be considered and included as alternative recommendations. For example, the engineer's ideal solution to a problem may be an engineering fix that would be extremely expensive. However, procedural changes or administrative changes could achieve similar results at far less capital cost. Rather than stating a preferred alternative, it might be more appropriate to list both, allowing management to make a decision without necessarily appearing to be choosing the cheapest alternative.

It is also extremely important to document the follow up and implementation. A true story best illustrates this principle.

In the late 1980s, a manufacturing facility that used chlorine retained an outside consultant to assist with the development of emergency planning documents required by the Emergency Planning and Community Right-to-Know Act, which had been enacted in 1986. While the consultant was onsite, someone suggested that he perform a review of the chlorine handling system. The consultant wrote a highly critical report that used faultfinding language. The report set out a series of findings which boiled down to the conclusions that the system was poorly designed, poorly constructed, poorly maintained, and operated poorly by poorly trained personnel. Management was surprised by this report, not only by the findings and recommendations, but also by the report's very existence. No one had advised management that the "audit" was to be undertaken. Nonetheless, the report was taken seriously. All of the findings and recommendations were analyzed, discussed, and, where possible (the design could not be changed), appropriate actions were taken. However, the corrective actions taken were not documented. After an incident, the report might stand as evidence, but establishing proof that corrective actions were implemented would be more difficult.

12.9. Summary

The ultimate message to the incident investigation team regarding related legal issues is that the team needs to conduct a thorough and effective investigation while minimizing legal implications. Preplanning and a well-designed incident investigation management system using some of the guidance provided above will help address legal issues that may arise.

Key points to remember are listed below.

- Be aware of potential litigation and follow up on any incident and its ensuing investigation.
- Get an attorney's advice at the beginning of the investigation and decide early if attorney-client privilege is going to be used.
- Use a good document management system and stress to the team that almost any document generated may become part of the public record.
- Have good procedures for managing the handling and chain of custody of all evidence.
- When appropriate, discuss employee interviews, tack, and potential discoveries with an attorney prior to the interview and after to properly provide legal protection.
- Have a plan in place for how to deal with and communicate with outside agencies.

Endnotes

1. Adams, Kathy B. *Accident Investigation Reports: What Are Your Lawyers Afraid Of?* Presentation for American Society of Safety Engineers Conference. Baltimore, Maryland. June 14, 1999.
2. American Chemical Council. *Responsible Care®, Process Safety Code*. Arlington, Virginia, 2000.

13

Implementing the Team's Recommendations

The ultimate objective of incident investigation is preventing recurrence of a specific incident scenario or related similar incidents. Considerable effort and resources are expended in determining an incident's root causes and identifying suggested preventive measures. Despite this effort, *the potential for a repeat occurrence remains unchanged until recommendations are implemented.* The value of the investigation is entirely dependent on the effectiveness of follow-up activities. This chapter focuses on the fourth step of incident investigation: implementing the team's recommendations and following through on their application.

Depending on the investigation team's charter, its responsibility may be complete when the recommendations were submitted in the final incident report; however, the *company's* responsibilities are far from over. Other portions of the organization may be assigned responsibility for actual implementation and follow-up of the recommendations identified by the investigation team.

Implementation of recommendations is a good and necessary business practice for a variety of reasons. Regulators may need to share lessons learned as part of wide ranging duty for personnel and public safety. In one example, a chemical manufacturer was cited for failing to apply lessons learned from a similar incident investigation.[1] There has also been an increased emphasis within progressive organizations to identify, investigate, publicize, and take action on near-miss occurrences.

This chapter addresses:

- major concepts related to implementing the recommendations,
- examples of repeat incidents where previous incidents were not followed-up adequately to prevent recurrence, and
- practical guidelines for achieving successful implementation.

13.1. Three Major Concepts

After presentation of the incident investigation report findings and recommendations to management, follow-up activities can fall into three distinct groups:

1. Initial resolution of recommendations submitted by the investigation team
2. Implementation of the accepted or modified recommendations
3. Sharing lessons learned from the investigation

Failure of any of these three components may eventually result in a repeat incident. A system to follow-up and track each recommendation or resolution until completion must be in place. The investigation should produce a work product that facilitates prediction of the results of recommended changes that are observable and capable of being monitored because they are observable. Figure 13-1 presents an overview of the activities recommended in this chapter.

The implementation and follow-up phase begins with the review and acceptance of the recommendations by management, and the assignment of implementation responsibilities. Each recommendation should have an individual assigned as personally responsible for monitoring the implementation through completion. If the original recommendation needs to be changed, postponed, or rejected for a valid reason, this decision should be fully documented. The basis for the decision should be specified, along with any new information or new options that were considered. The allocation of resources and timing of implementation will depend upon the priority placed on each action item.

Responsible managers should monitor and document progress of all actions through to completion to ensure that the corrective and preventive actions are achieving the intended results. Inevitably, some actions may result in changes to local management systems and equipment, and a strict management of change rigor should be adopted to ensure that all potential consequences of implementing the recommendations are understood and acceptable, and that all persons whose work assignments may be affected are aware of the changes. Responsible managers will also ensure that lessons learned from the investigation are widely disseminated throughout their organization and throughout the industry as recommended in the American Chemistry Council's Responsible Care® Process Safety Code of Management Practices. Auditing is an integral step to verify that the actions have been correctly implemented and that the lessons have been retained within the management system.

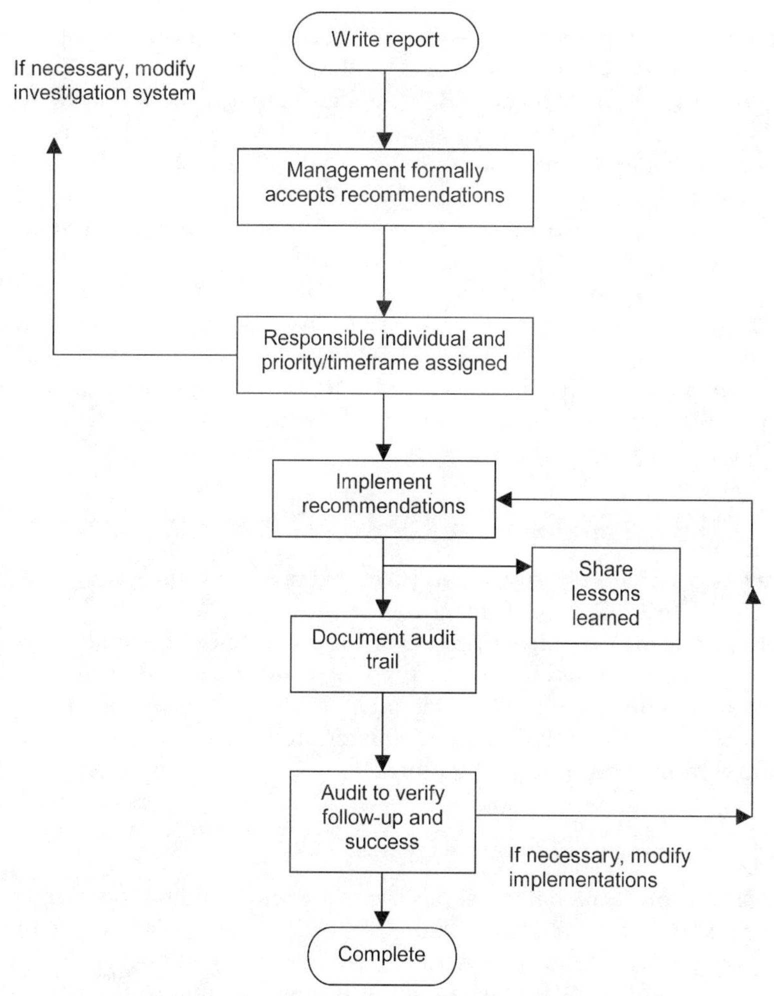

FIGURE 13-1. **Flowchart for implementation and follow-up.**

13.2. What Happens When There Is Inadequate Follow-up?

History unfortunately includes examples of repeat incidents that might have been prevented or mitigated if acceptable follow-up had been completed following a previous incident investigation.

13.2.1. Nuclear Plant Incident

There were several root causes for the Three Mile Island Nuclear Power Plant incident that occurred in March 1979.[2] Inadequate follow-up to previous similar incidents played a part in the occurrence. The actual initiating event for the incident was a malfunctioning pilot-operated pressure relief valve. It has been widely publicized that this valve stuck in the open position and allowed the steam system inventory to decrease, ultimately exposing the reactor core. The cause of the stuck valve has not received as much publicity. The valve malfunction was due to water contamination of the instrument air system. The instrument air system was contaminated by inadvertent introduction of water during an attempt to clear a blockage in a deionized water system using compressed air. The blockage was caused by ion exchange resin breakthrough from a water polisher bed. What is also not well publicized is that the instrument air contamination scenario had occurred twice before (in October 1977 and May 1978) in exactly the same manner, yet the follow-up resolution was inadequate.

13.2.2. Aircraft Incident

The Concorde aircraft tragedy in July 1999 involved a fuel tank puncture caused by tire failure.(3) During the investigation, it was disclosed that there had been at least five previous incidents of almost the same scenario. In one of the previous incidents, a tire failed and punctured the fuel tanks, yet there was no ignition. The incident investigation management systems used by the airlines failed to adequately follow-up and apply lessons learned from previous investigations.

13.2.3. Petrochemical Plant Incident

At a facility in Pasadena, Texas, a serious fire and explosion occurred on a compressor section involving failure of a check valve. (1) During the investigation by regulators, it was disclosed that a similar occurrence had recently taken place. The company was cited for failure to adequately apply lessons learned from previous incidents.

13.2.4. Challenger Space Shuttle Incident

The *Challenger* space shuttle disaster can also be viewed as an example of less than adequate incident investigation follow-up. (4) Failure of an O-ring seal in the solid rocket booster is one of the widely publicized causes of the *Challenger* space shuttle disaster. The presidential commission investigation report disclosed that O-ring failure had been previously identified as a serious problem. The potential severity of consequences resulted in a formal launch restraint being imposed six months before the January 1986 incident. O-ring failure was specifically identified as a problem and follow-up action had been initiated to resolve the problem. Unfortunately, the actions taken were inadequate. As pointed out by Diane Vaughan's study, *The Challenger Launch Decision*, there is a process in organizations that can be defined as "normalization of deviance."(5) Over time, deviations from specified practices are tolerated until, in some individual cases, the deviant practice becomes the norm. This is a trend that investigators look for in near-miss investigations.

13.2.5. Typical Plant Incidents

In many chemical processing industry facilities, repeat incidents with identical, or nearly identical, causes have been a common problem in the past. Common examples of statements in incident reports include the following. (Note that these are not examples of recommendations.)

- One of the causes of this incident was due to employee failure to follow established procedure.
- Failure to depressurize the hose before disconnecting caused the exposure.
- Overfill and overflow of storage tank despite presence of high-level alarm.
- Leak caused by pump seal system failure.
- The premature failure of bearings was due to water contamination of lubrication.
- Leak caused by external corrosion underneath insulation covering.
- Chemical release caused by inadequate equipment isolation associated with lockout/tagout activities.
- Process relief valve lifted due to improper pressure testing practices (added nitrogen too fast).

A large portion of these common events can be linked to less than adequate follow-up of recommendations from previous incident investigations. The earlier investigation teams may have properly identified underlying root causes, submitted suggested preventive actions, and attempted to share results; yet the repeat occurrences continue due to incomplete implementation of recommendations or ineffective sharing of lessons learned between potentially affected parties.

13.3. Management System Considerations for Follow-up

A formal management system should be in place to promptly and thoroughly address each recommendation. In the United States, Occupational Health and Safety Administration (US OSHA) and Environmental Protection Agency (US EPA) regulations specify that the employer should establish a system to promptly address and resolve the incident report findings and recommendations.[6] A system to follow-up and track each recommendation until completion or resolution must be in place. Chapter 2 addresses the overall system needs. Specific implementation and follow-up requirements that should be considered in the overall management system are:

- Formal acceptance of recommendations
- Assignment of a responsible individual
- Determination of action priority
- Implementation of actions
- Changes to the management system
- Audit trail
- Action tracking
- Sharing lessons learned
- Follow-up audit
- Incident trend analysis

13.3.1. Understanding Responsibilities

The incident investigation team has the responsibility to develop practical recommendations and submit them to management. It is then the responsibility of management to review the recommendations, ask for clarification or revisions, or approve them as written, implement them, and follow-up on implementation. When investigating incidents, regulatory agencies usually take special interest in previous recommendations. Lawyers give this issue significant attention. The assumption is that prudent and responsible managers should apply lessons learned from an incident promptly.

Some pertinent management responsibilities include:

1. Addressing the technology deficiencies involved in causing the incident
2. Examining possible risk reduction measures to (a) reduce potential consequences, (b) reduce probability of occurrence, (c) improve the initial detection, diagnosis, alert, and alarm systems, (d) improve emergency response, (e) eliminate the hazard
3. Ensuring that necessary resources are available (initially and continuing) for full implementation of the recommendation in a timely manner
4. Monitoring the status of the action items and periodically reviewing the target completion dates
5. Conducting a post-implementation audit to ensure that the intent of the recommendation was actually achieved

6. Reviewing the incident investigation management system for opportunities to improve the actual investigation procedures
7. Sharing the results of the investigation with all departments that could possibly benefit
8. Seeking out any similar hazards in the facility and company to capitalize on the lessons learned
9. Sharing the information and lessons learned with other companies and organizations.

Employees are affected by the recommendations. Their responsibilities include:

1. Using new or modified equipment properly
2. Abiding by procedural improvements
3. Giving feedback to management when something is not working as expected
4. Sharing their knowledge when they find a better or safer way to address the problems identified in the investigation

In summary, developing the recommendations is a responsibility of the incident investigation team. Implementing the recommendations is a management responsibility. The inclusion of the elements of the recommendation in daily work practice is the responsibility of each individual affected by the recommended action.

13.3.2. Formally Accepting Recommendations

The decision to accept, reject, or modify a recommendation is a management responsibility. These decisions and their resulting actions should be documented. The basis for the decision should be specified along with any new information or new options that were considered.

If management decides to reject, modify, or delay implementation of a recommendation, it is a good practice to document the reasons for not accepting or implementing the recommendation as submitted by the investigation team. A common and avoidable mistake is to provide inadequate documentation for this decision. An example of inadequate documentation would be a brief entry in the record to the effect that this recommendation is "not considered justifiable," with no further explanation. Examples of more adequate reasons for not accepting a recommendation may be the recommendation did not specifically address a root cause, or the recommendation is not technically feasible. In either case, an alternate recommendation should be presented.

13.3.3. Assigning a Responsible Individual

An individual rather than a department or division of the company should be named as responsible for each recommendation. A key responsibility is also determining the most appropriate action to address the recommendation. This person should be responsible for implementation, monitoring the status, resolving any problems, and finally verifying and documenting that the intended preventive action has been completed.

13.3.4. Determining Action Item Priority

Management is also responsible for determining priorities, setting initial target completion dates, and the allocation of resources (personnel, equipment, and finances) to complete the preventive action in a timely manner.

Each recommendation should have a suggested target completion date reflecting both urgency and practicality. Complex actions that require several steps or an extended time to complete should be assigned intermediate milestones to monitor progress of the action. In these circumstances, it may also be appropriate to consider additional temporary safety measures until the main actions have been completed.

As an aid to priority determination, it is often helpful to risk rank each recommendation. Several CCPS publications provide guidance on the use of a variety of risk ranking techniques. [7,8]

13.3.5. Implementing the Action Items

A variety of challenges can influence effective resolution of recommendations identified by the investigation team. In most cases, the investigation team conducts a preliminary examination of any possible adverse impacts of implementing the proposed recommendation. There have been instances of implementing a fix for one problem and creating a new problem that did not exist before if the management of change system is bypassed. The team should conduct a preliminary examination (management of change) of possible adverse impacts of implementing recommendations before the recommendation is published in the final report. When the recommendation reaches management for action, there is usually another examination of the cost, benefits, and potential consequences of implementing each recommendation.

Cost benefit analysis is not always easy or straightforward. One of the tasks is to accurately determine the risk if the recommendation is rejected. Cost benefit analysis is also used for comparing different actions (options) for addressing the recommendation. These determinations require examining a set of scenarios, each with different likelihood and consequences. Accurate identification/selection of this set can be a challenge. Layer of protection analysis (LOPA) is one tool that may be useful in this application. [9]

Another challenge to effective recommendation resolution appears when the action intended by the investigation team is not stated clearly or completely. This avoidable mistake can lead to misunderstandings on the part of management decision-makers. A common example of obscure wording is ineffective use of the terms "consider" or "review." If the investigation team believes that a particular system defect exists and should be corrected, then the team should make this finding very clear and recommend a specific measurable task. In other cases where the investigation team does not believe the recommended action is mandatory, this distinction should also be clearly stated. An example would be the recommendation of a best practice activity, which could be rejected by management without major consequence. This is one of the reasons that each recommendation statement should include comments on the consequences to be averted and the benefits associated with implementing the recommendation. Effective recommendations include phrases such as "in order to prevent *x*, implement *y*." It should be noted, however, that some companies have recommendation language protocols in place that may differ from this advice. Their systems may be based on potential or perceived legal liabilities. (See Chapter 10 for guidance on writing clear recommendations.)

Resolution of action items also has a legal dimension. In the US tort litigation system, there is an accepted concept that once management decision makers have knowledge of a hazard, there is an accompanying responsibility and duty to act to address that hazard. US OSHA enforcement representatives also apply the concept of "repeat willful," in cases where there has been a previously identified hazard and the risk has not been reasonably mitigated.

In addition to a management system for recommendation follow-up, there must also be a system for managing the changes created by the recommendations. More often than not, these two management systems are handled as separate activities. Management of change is a concept recognized by the CCPS as one of the twelve fundamental elements for successful process safety management. [7] An evaluation should be made to clearly understand the possible impact of all recommendations *before* implementing any action. Additionally, a rigorous management of change system helps achieve a thorough follow-up for action items.

As an example, consider an incident where an employee is exposed while sampling chlorine. One of the root causes is determined to be an inadequate standard operating procedure (SOP) for sampling. A management of change system can provide a mechanism to identify things that may be affected by changing the sampling procedure. Some of these impacts might be:

- The operator training program (training manual, qualification criteria, lesson plan agendas)
- Reference manuals in the quality control laboratory
- Specifications in purchasing and stores (warehouse) for ordering and stocking sample containers
- Respiratory protection program
- Maintenance Department preventive maintenance program instructions and records
- Engineering design standards for sampling stations.

One of the root causes of the *Challenger* disaster was inadequate management of change for an earlier recommendation that attempted to prevent O-ring seal failures.[4,10] The pressure testing procedure for the seal was changed in 1984 and actually resulted in an *increase* in the risk rather than the intended decease in risk. A more thorough analysis of the proposed recommendation might have identified and corrected this problem before the disaster. Additional information on case histories can be found in Chapter 15.

If a recommendation asks for a change in the process, the action must undergo a formal process hazard analysis (PHA) study, such as a HAZOP or other methodology, before implementation. This systematic and formal approach identifies and evaluates hazards associated with the proposed revisions. The study may uncover failure scenarios, adverse consequences, and obscure relationships that are not immediately apparent. The CCPS publication *Hazard Evaluation Procedures*[8] is an excellent guide to selection and proper application of PHA methodologies.

13.3.6. Documenting Recommendation Decisions—the Audit Trail

Once a recommendation has been accepted for implementation, a clear, auditable document trail should be established and maintained. It is the prevailing opinion of many regulatory agencies in the US that any changes in the originally accepted recommendation should be thoroughly documented. If a recommendation is modified in scope or time commitment, or is otherwise not implemented as originally planned, then the basis for this decision should be documented.

13.3.7. Tracking Action Items

A challenge to effective recommendation resolution is adequate tracking of action item progress and completion. Initial responsibilities and target dates may be properly assigned, and then in the course of normal business, competing priorities require reevaluation of realistic target completion dates. A high quality tracking system will address this common situation and prevent action items from becoming lost. A good system will automatically handle personnel changes as well. Some organizations require the personal participation of upper levels of management in tracking and documenting action item recommendations from process safety incident investigations.

Line management should establish a system to provide consistent information about incidents and compliance issues, together with associated follow-up actions. Progress should be reviewed regularly, and can be recorded on a punch-list-style report showing status, estimated completion date, and whether an item is open or closed. When action items are completed, they can be moved to a closed punch-list for future audit trail purposes.

This information will help management to target interventions where they are most needed to deliver performance improvement. It can be used to drive a significant boost to the management of corrective action in the organization, and maintain a "living" process safety assurance program designed to address and anticipate risk within the facility.

13.3.8. Revising the Incident Investigation Management System

Lessons may be learned during the investigation that result in improvements to the incident investigation management system. Each investigation is unique and can be used as an opportunity to improve the incident investigation system. A critique should be conducted and include everyone who participated on the investigation team. It should identify potential changes that would make the next investigation more successful as well as those items that were successful and do not need change. Evaluation of the planning, team training, team competencies and skill sets (both individual and general baseline capabilities), team supplies, coordination, and communication issues often will show areas for improvement. Finally, auditing is an integral step to verify that the actions have been correctly implemented and that the lessons have been retained within the management system.

Management should carefully review any changes that would enhance the quality of the next incident investigations. Proposed changes to the incident investigation system should be evaluated and executed using the management of change system. Adopted changes should be implemented and tracked as any other recommended action item. See Chapter 14 for additional guidance.

13.4. Sharing Lessons Learned

It makes good business sense to capitalize on newly discovered (sometimes rediscovered) information produced by the incident investigation. In production operations, there is a strong emphasis on sharing incremental knowledge discoveries among all potentially affected production units. In chemical process facilities, a small process improvement can result in significant economic benefit when properly shared and applied across all similar production units in the corporation. To a lesser extent, this mindset is applied to lessons learned from incident investigations, but there are additional barriers to overcome. One of these barriers is the potential legal liability exposure from disclosure of any safety-related deficiency. Admission of mistakes can result in increased frequency and size of damage awards to third parties. The expectations of regulators have also changed since the implementation of the process safety regulations, such as US OSHA, *Process Safety Management of Highly Hazardous Chemicals,*[6] US EPA, *Accidental Release Prevention Requirements: Risk Management Programs*[11] and DOT, *Pipeline Integrity Management in High Consequence Areas.*[12]

13.4.1. Performing the Follow-Up Audit

A follow-up audit should be conducted after an appropriate period following the implementation. The objectives of this audit are to verify that recommended actions remain in place, and are still working as intended. It also presents an opportunity to review the retention of the lessons learned and to identify any practices, knowledge or awareness items that are being lost. The audit should therefore consider whether the lessons learned were communicated as appropriate throughout the company and to others in the industry.

The audit may uncover that some recommended action was ineffective. Engineers, designers, or the person responsible for implementation may find a reason why the original recommendation did not work or was not effective.

Although not essential, it may be helpful to include a member of the original incident investigation team on the audit team to help assure that the final implemented actions address the original issues in an acceptable manner.

13.4.2. Internal Sharing

An internal communication plan can be formulated to determine how investigation information will be shared, with whom it will be shared, and to what level of detail. Not all groups have the same information needs, so it may be practical at times to prepare more than one information release. The personnel who work directly in the process unit that experienced the incident have a need to know what happened, why, and what direct specific changes are to be made. Indeed, US OSHA regulations specify that the report should be reviewed with all affected personnel whose tasks are relevant to the incident findings including any contract employees. [6] Personnel in adjacent process units may not be exposed to the same specific hazards, and therefore may need a more generic presentation of the lessons learned. Workers in other similar facilities within the same parent organization may be exposed to the same hazards, and yet will remain uninformed without proactive communication measures. Internal company intranet webpages can be devoted to lessons learned, or summaries of incidents. These internal webpages are a powerful tool for collective corporate memory.

Sharing of findings and lessons learned could take several practical forms. Some suggestions to achieve optimum effectiveness follow.

- Communicate lessons learned in a user-friendly method with enough information for the intended audience. Include what happened, why, and how it can be prevented.
- Include lessons learned in the organizational policy statements as appropriate.
- Include the knowledge from the lessons learned in training, orientation, process safety information, and written procedures for employees.
- Include the knowledge from the lessons learned in key performance indicators, expectations, and accountability for management.
- Ensure that the company's legal counsel understands and accepts the lessons learned.
- Ensure that training systems periodically refresh personnel's memory, address personnel turnover, and early retirements.
- Document lessons learned in the incident investigation management system.
- Audit implemented recommendations for effectiveness.
- Include the lessons learned in a quality assurance (for example, kaizen) component.

Lessons learned from one facility's incident often have applicability to other facilities within the same parent organization. A management system should be in place to ensure that the understanding of the lessons learned is not isolated to a single location. Another way to express these thoughts was presented in Trevor Kletz's paper "Organizations Have No Memory."[13] In practice, organizations may find it extremely difficult to maintain continuity regarding lessons learned from any one incident. This challenge remains regardless of the quality of the initial investigation. A well-designed and operated incident investigation system can minimize the experience of relearning the same lesson.

The American Chemistry Council identifies this challenge in its Responsible Care® Process Safety Code of Management Practices, Management Practice #5, "*Sharing of relevant safety knowledge and lessons learned from such incidents with industry, government and the community.*" addresses this issue. [14] The incident investigation system should set up a mechanism to communicate lessons learned to all appropriate company groups. This includes maintaining an incident log, and ensuring that incident reports become a part of the process safety information.

Some companies have found success with a periodic publication of incident abstracts. Each incident may appear as a one or two paragraph summary in a quarterly bulletin. This information is circulated widely within the corporation. If a site has a special interest in one particular incident, then full details are requested by direct contact between the two sites. The highly abbreviated summary has other uses, for instance, employee safety training meetings or bulletin board postings. Summaries of case histories can be found in Chapter 15.

13.4.3. External Sharing

Process safety is a dynamic field, and enhancing process safety knowledge is one of the twelve key elements recognized by CCPS for successfully managing process safety. [7]

Management commitment to using all available resources for enhancing process safety knowledge of all levels in the organization is a key difference in having a minimum, adequate, or outstanding process safety program.

As previously stated, the American Chemistry Council also recognizes this concept as an integral part of the code of Responsible Care®, Process Safety Code of Management Practices. [14] Companies should openly share nonproprietary results from internal process safety incident investigations and related research through the auspices of trade and professional associations, and other networking opportunities. Other organizations should strive to benefit from the lessons learned by others and keep abreast of latest developments and safety alerts.

In practice, external sharing of lessons learned from investigations is not an easy task. Those who wish to share details of serious incidents publicly encounter numerous challenges. Sometimes process safety incidents involve fatalities or adverse impacts to third parties that may ultimately end up in litigation. Inevitably, this can inhibit the open exchange of valuable information relating to immediate and root causes, and measures taken to prevent an occurrence, which should be of interest to operators of similar process facilities.

AIChE has sponsored public presentations of actual incidents with accompanying lessons learned. Some of these case histories are contained in the following AIChE publications.

- *Ammonia Safety Symposium*
- *Loss Prevention Symposium*
- *Chemical Engineering Progress*
- *Plant Operations*

The American Petroleum Institute also publishes a series of booklets on specific lessons learned. In the UK, the Institution of Chemical Engineers publishes *Loss Prevention Bulletins* that contain specific case histories.

More recently, CCPS has begun publishing the *Process Safety Beacon*. This brief newsletter offers incident lessons on a monthly basis. It is aimed at building awareness among plant operators and technicians of how process incidents can occur so that they can avoid these accidents in future. Each issue presents a real-life accident and describes the lessons learned as well as practical means to prevent a similar accident from occurring in other plants. This material is suitable for sharing at crew, toolbox or safety meetings, or through postings in break or control rooms. The *Process Safety Beacon* can be downloaded from the CCPS website: http://www.aiche.org/ccps/safetybeacon.htm. A sample of the *Process Safety Beacon* is located on the accompanying CD.

Industry databases containing details of process safety incidents are addressed in more detail in Chapter 11.

Valuable case history information can often be exchanged informally during conferences and seminars. This is especially true for those incidents in which litigation is not expected. Additional information on case histories can be found in Chapter 15. Examples of conferences and seminars are the annual AIChE Loss Prevention Symposium and the CCPS 2000 International Conference *Process Industry Incidents*. Members of the American Chemistry Council have endorsed a policy for sharing information with outside groups, and other industry associations encourage sharing.

Information sought and shared openly can be the most useful. Unrecognized latent properties of materials can and do receive publicity in technical journals. Information found in research is more easily communicated than that found through serious incident investigation. Previously unrecognized equipment failure modes are often proactively publicized in response to the manufacturers' and suppliers' obligation to warn and inform. Many probabilistic risk assessments done for or by government regulators are available as part of the public record. One example is the "Reactor Safety Study WASH-1400," published by the US Nuclear Regulatory Commission, which is frequently used by industry. [15]

13.5. Analyzing Incident Trends

Historical incident trends provide the information to:

- Allow precautions to be taken at the affected facility and other facilities
- Apply lessons learned in future design
- Help identify trends not apparent from single incidents

Because incidents have many causes, some causes may not be identified in the investigation of a single incident. For example, if an incident reoccurs on Saturdays, this may simply be coincidence, or it may be a symptom of deficiencies in management systems on weekend shifts. When analysis shows a pattern of incidents, management can take informed action. Without incident recording and analysis of the record, such patterns may go unnoticed and lessons to improve process safety management may go unlearned.

Common cause and trend identification analyses provide an opportunity to apply deductive hindsight to past incidents including near misses. This process allows a more productive allocation of resources and can maximize incident prevention and control. Establishing relevant categories for classification of incident causation and trend data is not always as easy. There will be hard choices between having too many small categories and too few broad, somewhat nonspecific, categories. Searchers should be able to clearly focus on commonalities. The categories should include specific system deficiencies or breakdowns such as design, training, process mechanical integrity, and equipment specifications. Other categories may be by specific hazard exposure. Two examples of predefined trees containing root cause categories, the Comprehensive List of Causes and the Root Cause Map™, are located on the accompanying CD.

Computer-aided trend identification offers potential benefits, but is dependent on the quality of the input information. Expert systems and artificial intelligence are tools being tested and may give improved insight into identifying common causes and trend analysis.

Endnotes

1. US EPA, *EPA/OSHA Joint Chemical Accident Investigation Report, Shell Chemical Company, Deer Park, Texas*. (US EPA document # 550-R-98-005)
2. Ford, D. F. *Three Mile Island, Thirty Minutes to Meltdown*. New York: Penguin Books, 1981.
3. Weir, A. "Concorde Crash Raises Questions without Answers." *Journal of System Safety*, System Safety Society. Second Quarter 2001.
4. Feyman, R. P. *What Do You Care What Other People Think*? New York: Norton, 1988.
5. Vaughan, D. *The Challenger Launch Decision*. Chicago: University of Chicago Press, 1996.
6. US OSHA. "Process Safety Management of Highly Hazardous Chemicals" *29 CFR 1910.119*. Washington, DC: Occupational Safety and Health Administration, 1992.
7. CCPS. *Guidelines for Technical Management of Chemical Process Safety*. New York: American Institute of Chemical Engineers, 1989.
8. CCPS. *Guidelines for Hazard Evaluation Procedures 2nd Edition*. New York: American Institute of Chemical Engineers, 1992.
9. CCPS. *Layer of Protection Analysis, Simplified Process Risk Assessment*. New York: American Institute of Chemical Engineers, 2001.
10. Winsor, D.A. "*Challenger*: A Case of Failure to Communicate." *Chemtech Magazine,* American Chemical Society, September 1989.
11. EPA. "Accidental Release Prevention Requirements: Risk Management Programs." *Clean Air Act* Section 112(r)(7). *40 CFR Part 68,* Washington, DC: Environmental Protection Agency, 1996.
12. DOT. "Pipeline Integrity Management in High Consequence Areas." *49 CFR 195.452*. Washington, DC: Department of Transportation, 2000.
13. Kletz T. A. "Organizations Have No Memory." Paper presented at AIChE Loss Prevention Symposium, April 1979.
14. ACC. "Resource Guide for Implementing the Process Safety Code of Management Practices." Washington, DC: American Chemical Council, 1999.
15. Rasmussen, N.C. *Reactor Safety Study WASH-1400*. U.S. Nuclear Regulatory Commission, 1975.

14

Continuous Improvement for the Incident Investigation System

A key element in the process safety incident investigation management system is a mechanism for ensuring that the process is actually providing the value expected and that there are systems in place for continuous improvement. The objective, of course, is to conduct an investigation that finds the root causes of the incident and then to communicate those findings so that that an identical incident, similar incidents, and incidents with associated root causes never occur again. In addition, the investigation must be conducted so that it meets certain regulations, rules, and defined expectations of the sponsoring organization. This process and the process of continuous improvement are most effectively implemented as an element of the management system.

Poorly managed investigation programs are usually limited at best to preventing a reoccurrence of an identical incident at the same location. These programs typically lack the infrastructure needed to assure compliance with regulations and expectations, and have little or no provisions for continuous improvement. On the other hand, a well-managed incident investigation program values the need to assure compliance and seeks out quality improvement opportunities. The difference in the two is exhibited in down stream performance—one allows a repeat of incidents and the other prevents similar incidents throughout the organization.

Continuous improvement is a recommended component of safety management systems as addressed and included in API Recommended Practice 9100 A&B for Safety Management Systems,[3] in the proposed UK standard for HSE Management Systems OHSAS 18001,[4] and in the ISO Quality Standards 9000 and 14000 series.[5,6]

Each incident and the ensuing investigation is unique and can be used as an opportunity to improve the investigative system. The investigation team members and others from management should conduct critiques like the examples in this chapter to assure that the current investigation is comprehensive, and to identify potential changes that would make the next investigation more successful. These critiques address regulatory compliance, investigation quality, recommendation quality and follow-up as well as potential optimization methods. They should evaluate each phase of the investigative process (for example, planning, team composition, approach, gathering, and preservation of evidence) and should recommend changes where appropriate as well as identify and capture the positive aspects of the investigation for future use.

When or if changes to the investigative process are recommended, they should be evaluated using the facility's management of change process to ensure a clear understanding of the benefits of the recommended change and the potential undesired consequences before implementation. Approved changes should be integrated into the incident investigation training system.

14.1. Regulatory Compliance Review

The US OSHA process safety management (PSM) regulation in 29CFR 1910.119 (m) clearly defines the requirements of investigations conducted in "covered" facilities. The US EPA RMP regulation in 40CFR Part 68.81 mirrors the US OSHA requirements. During reviews of the effectiveness of the investigation system, it may be helpful to confirm that the investigations address all necessary regulatory requirements. Section 68.42 of the US EPA risk management program (RMP) standard requires documentation of certain specific information that must be included in each incident that is included in the five year summary of incidents. Some of the required data includes:

- duration of the release,
- quantity of the release,
- notification of offsite responders, and
- changes to the process that resulted from the investigation.

Table 14-1 lists these requirements and provides a record of compliance for future analysis. Requiring completion of this record for each process incident investigated enhances the probability that all elements are covered. Auditing of incident reports against these requirements provides the forum for continuous improvement in meeting compliance requirements.

This table may also be incorporated into the PSM program assessment/audit protocol and used during periodic PSM program evaluations.[1]

TABLE 14-1
Requirement Compliance Checklist

Requirement Statement	Compliance?	
The Investigation Itself:	**Yes**	**No**
1. You must investigate each incident in a covered process that did or could reasonably have resulted in a catastrophic release of: –a highly hazardous chemical per US OSHA PSM or, –a regulated substance per US EPA RMP. 2. The investigation should start as soon as is reasonably possible, but MUST start within 48 hours following the incident. (This requires documentation of date and time at which the investigation began.) 3. The investigation team is to be composed of: –at least one person knowledgeable in the process involved, –a contract employee if the incident involved work of the contractor, –any other person with appropriate knowledge and experience that is required to thoroughly investigate and analyze the incident.		
The Report and Findings:	**Yes**	**No**
1. A report is required at the conclusion of the investigation and the report must include: –date of the incident –date the investigation began –a description of the incident –the factors that contributed to the incident –recommendations resulting from the investigation		
2. The report must be reviewed with all affected personnel whose jobs are relevant to the investigation findings, including contract employees where applicable.		
3. A system must be in place and utilized to promptly address the incident report findings and recommendations.		
4. The investigation report must be retained for five years.		

14.2. Investigation Quality Assessment

A thorough investigation contains several key elements. Even though a variety of tools and techniques may be used throughout the investigative process, the result must discover the underlying issues that caused or allowed the incident to occur in the first place. When this is done, appropriate corrective recommendations can be made.

Because organizations are dynamic and ever changing, a conscientious effort must be made to assure that investigations continue to hit this target. To do this it is necessary to periodically review and update the entire process and management system, the individual components, and relevance of findings. Although you cannot inspect quality into a product, you can gather enough data to make adjustments and corrections so that future products meet the needs more closely. For incident investigations and resultant reports, this can easily be done by simply listing the critical elements that should be addressed in an investigation and critiquing actual performance against those criteria. Table 14-2 is an example critique sheet.

14.3. Recommendations Review

To effectively use the findings of an investigation, appropriate recommendations must be drafted and eventually acted upon. Recommendations should accurately translate the investigation findings into realistic actions that when implemented resolve the issue that prompted them. (They should solve the problem!) They must clearly define what is to be done so that the implementer understands not only *what* to do, but *why*. A well-written recommendation will also identify the consequences that are being avoided or abated, and/or the likelihood of a reduction of consequences or risks. Periodic checks or audits of recommendations coming from incident investigations provide managers a better understanding of the location and nature of problems. Table 14-3 is an example of a recommendation review checklist.

14.4. Potential Optimization Options

14.4.1. Follow Up

Table 14-4 offers ways for optimizing the effectiveness of incident investigation follow-up. Not all options are appropriate for all investigation management systems or every investigation. The reader should determine which ones should be used and where.

14.4.2. Causal Category Analysis

Each company's management style and safety systems have strengths and weaknesses. These strengths and weakness tend to influence the types and severity of incidents that might occur. An analysis of incident investigation findings in terms of causal factors, immediate causes, contributing causes,

TABLE 14-2
Investigation Key Element Critique Checklist

Investigation Key Element Query	Yes	No
1. Are there written procedures or protocols for reporting and investigating process safety incidents?		
2. Has the investigation team leader been trained (qualified) to lead investigations and to use appropriate investigative tools?		
3. Does the investigation team leader have independence from the issue to be investigated to the point that there is no question as to that person's objectivity?		
4. Were the necessary skills available either on the investigation team or readily available to the team when needed?		
5. Have pertinent causes and discovery processes, including data gathered, been recorded and documented?		
6. Were the proper investigative techniques applied correctly?		
7. Did the investigation go beyond the immediate or obvious causes and discover contributing causes?		
8. Was evidence gathered and preserved properly, including a documented chain-of-custody?		
9. Did the investigation address all facets of all causes?		
10. Were the underlying root causes identified?		
11. Were the management system failure(s) identified?		
12. What other resources, techniques and/or tools could be used to make the next investigation better? Discuss below.		
13. Were critique forms completed for each investigation?		
14. Were there any legal issues from this latest investigation that were related to incident investigation reports or documentation that need to be resolved before the next major incident investigation?		
15. Do we need to change any internal communication practices?		
16. Do we need to change any team training or team procedures?		
17. Would we do anything differently to change the litigation exposure or litigation discovery phase items?		
Discussion:		

TABLE 14-3
Recommendations Review Checklist

Recommendations Review	Adequate?	
	Yes	**No**
1. Do the recommendations address the underlying or root causes?		
(a) Is there a recommendation that addresses each root cause?		
2. If there are contributing or enabling causes identified, are there corresponding recommendations if warranted?		
3. Do the recommendations clearly identify what is to be done and why?		
4. Is each recommendation feasible?		
5. Will the recommendation(s) actually reduce the risk by lowering either the probability of occurrence or lessening the consequences?		
6. Is a system in place for tracking each recommendation, including:		
(a) Assignment of an individual responsible for completion of each recommendation?		
(b) Target-for-completion dates for each recommendation?		
(c) Periodic status checks and reports?		
(d) Documentation of final resolution of each recommendation?		
7. Is a formal documented system in place that assures each recommendation is evaluated through the management of change program before being implemented?		
8. Is there a system in place that assures communication of pertinent facts regarding the incident, the recommendations, and status to affected employees and contractors?		
9. Have the details of the incident and the recommendations been communicated to other facilities in the organization that may have a need to know?		
10. Is a system in place that evaluates the usefulness and advisability of sharing this information with other companies in a similar business, per Responsible Care® Process Safety Code of Management Practices 4.3 and 5?		

TABLE 14-4
Example Follow-up Checklist

Follow-up Issues	Addressed?	
	Yes	**No**
1. Are the incident investigation follow-up expectations clearly stated in the incident investigation policy statement?		
2. Are the incident investigation follow-up expectations and requirements included in key performance indicators?		
3. Does the incident investigation management system include: –Strongly encouraging near miss reporting and investigation? –Requirements for formal periodic status reports of recommendations? –Requirements for documentation of a formal plan for sharing lessons learned? –Provisions for providing appropriate report information to various levels as needed? –Provisions for modifications of original recommendations?		
4. Are appropriate levels of upper management aware of and involved in monitoring the implementation or resolution of recommendations and resultant action plans?		
5. Have audit protocols been established that include examination of effective implementation of: –Investigation follow-up measures? –Recommendations?		
6. Are incident investigation follow-up expectations included in training and competency systems?		

enabling causes, and especially root causes, may identify broad areas or management systems that contribute to or play a part of more incidents than others may. The determination of these management system failures allows a more global approach to reduction of common cause weaknesses and prevention activities than addressing individual causes might. Table 14-5 is an example of one way to accumulate this data for analysis by using causal categories.

This approach is used for statistical analysis only AFTER the investigation is complete and the causal factors, including the root causes, have been determined. This approach is NOT appropriate for use as an investigative tool in the sense of finding causes of the incident. Rather, it should be used to define the broad categories into which a larger portion of incident investigation findings has occurred for a holistic approach to prevention.

TABLE 14-5
Example Categories for Incident Investigation Findings

Instructions: *Review each classification statement to determine if it is TRUE or FALSE for the incident investigation finding in question. Any statement that is answered with FALSE presents a causal path and an associated management system improvement opportunity.*

Category	Circle	Defining Statements
Design	T / F	The current design used the correct specifications and was built so that it was adequate for the intended service. (This includes design logic, hardware, installation accuracy, arrangement, and ergonomic factors.)
Process Controls	T / F	The control system(s) for the equipment or activity in question performed in accordance with the design logic, programming, or other instructions. (This deals with the actual control operation or execution. It would not include control logic that is in the "design" category.)
Administrative Procedures	 T / F T / F T / F T / F	The administrative procedures were: • available (you had them) • adequate (they were usable) • accurate (they were correct) • approved and enforced (you really meant it) These are the procedures covering broad organizational needs such as management of change, design and installation expectations (including avoiding headknockers and providing logical labeling), procurement (including approving substitutions and vendor equivalents), implementation (including defining training requirements and administrative support systems), safety (including specifying appropriate protective gear), environmental compliance, housekeeping standards, and emergency response.
Operation Procedures	 T / F T / F T / F T / F	The operational procedures were: • available (you had them) • adequate (they were usable) • accurate (they were correct) • approved and enforced (you really meant it)
Maintenance Procedures	 T / F T / F T / F T / F	The maintenance procedures were: • available (you had them) • adequate (they were usable) • accurate (they were correct) • approved and enforced (you really meant it) (The focus of this category is the actual maintenance tools, techniques, and standards for work that go beyond the traditional scope of normal inspection and preventive maintenance activities.)

Category	Circle	Defining Statements
Training	 T / F T / F	Training was: • available, timely (initially and in reviews) • adequate and verified to be effective to achieve functional and compliance requirements
Inspection and Preventive Maintenance	T / F	Inspection and preventive maintenance were in accordance with applicable procedures, manufacturer's or experience-based recommendations and governing standards, and were adequate for the service conditions.
Equipment and Materials	T / F	The equipment, parts, and materials as initially procured were as specified, were not defective, and met or exceeded the applicable specifications. (If you had received what you thought you were getting, the equipment or material involved would not be an issue.)
Personnel Fitness	T / F	Personnel were "fit for duty." (Includes physical/mental/emotional states and addresses preexisting physical conditions, substance abuse, and other related concerns.)
Human Actions	T / F	Personnel actions, activities, and decisions were in accordance with procedures, training, and expected workplace standards.
External	T / F	External items including weather and external third party actions/events were not creating out-of-design conditions.
Other	T / F	The incident has been satisfactorily classified in one or more of the above categories

By gathering these data from each incident investigation, a database is established that will, over time, indicate the broad categories or management systems in which incident investigation findings tend to accumulate. The company can then devise and implement a more holistic approach to prevention than the one developed by addressing individual root causes.

Endnotes

1. US OSHA. "Process Safety Management of Highly Hazardous Chemicals" *29 CFR 1910.119*. Washington, DC: Occupational Safety and Health Administration, 1992.
2. US EPA. "Accidental Release Prevention Requirements: Risk Management Programs." *Clean Air Act* Section 112(r)(7). *40 CFR Part 68,* Washington, DC: Environmental Protection Agency, 1996.

3. American Petroleum Institute (API). Recommended Practice API 9100, Model Environmental, Health & Safety (EHS) Management System and Guidance Document
4. Health and Safety Executive (HSE), OHSAS 18001 (proposed), *Management Systems Standard* .
5. International Organization for Standardization (ISO), ISO 9000 series, *Quality Management Systems*.
6. International Organization for Standardization (ISO), ISO 14000, *Environmental Management Systems*.

15

Lessons Learned

Progress, far from consisting of change, depends on retentiveness. Those who cannot remember the past are condemned to repeat it.

—*George Santayana*

The paramount importance of a good investigation is its influence on the behavior and performance of an organization following an incident. Although recommendations are usually aimed at correcting specific faults and weaknesses in management systems, the organization as a whole can recognize the need for system-wide change when an eye opening revelation presents itself. The lessons learned approach pursues themes that are likely to leave a lasting impression on management and workers and uses this knowledge to bring about change. Lessons learned from incidents can provide a clear warning of the potential dangers of ineffective implementation of process safety management. Learning from incidents is an effective means of ensuring continuous improvement in an organization.

15.1. Learning Lessons from Within Your Organization

Lessons learned from incidents can take on two forms:

1. A simple reminder or jogger that an operation is vulnerable or that an existing standard or practice is critical. The effectiveness of such a reminder depends on the background of the reader. For those with considerable technical and operations experience, "lessons learned" will drive a quest for more information. Ultimately, this

could lead to a critical examination of management system effectiveness on the home front.

2. New scientific or technical information relating to the cause of the incident or its subsequent effect. A first time failure in a specific type of equipment or a higher loss than previously experienced are examples of this second category. In fact, data from actual incidents are a primary source for the development of hazard models.

If the investigation team takes the time to explore the unique features of an incident and to truly understand how it might have occurred, the recommendations can be tailored to the organization and made more effective. Rationalizing that an incident simply occurred because of failure to follow standards and practices does not adequately address the underlying system failures. A thorough investigation should attempt to determine the underlying cause(s) of the occurrence and share the findings across the organization in order to reduce the chances for a repeat occurrence. Recommendations should be structured in such a way that they teach and promote a better way of doing things. Management must play a role in ensuring that the proper lessons are communicated across organizational boundaries. This includes different levels of personnel and different business units or plant sites. In this way, the organization will be better prepared to deal with problems and repeat incidents may be avoided.

15.2. Learning Lessons from Others

Another category of lessons learned includes those at other sites or companies. Many of these may have captured media attention when they occurred. If the conclusions of an investigation can reference similar losses elsewhere, the recipients of the recommendations may better appreciate the significance of the loss and the importance of pursuing follow-up. One way of doing this is to include a discussion of relevant external incidents in the findings section of the investigation report. The discussion should highlight the differences and similarities between related incidents and the incident under investigation. Formal investigation reports should be written in such a way that they might educate and influence persons remote from the accident scene including future generations of managers and workers. These should be circulated to other sites within the same organization. A factual and logical narrative can help portray how an incident unfolded within an organization.

The application of lessons extracted from previous or distant incidents can be used in a preventive capacity. Safety managers should scan the media as well as published accident databases and periodically raise an awareness of incidents that have occurred elsewhere. One approach to doing this is to issue safety alerts and incident bulletins pertaining to technology or safety practices in use at a given facility.

The opportunity to learn from experience applies to the incident investigation process as well as to incidents themselves. How previous investigations describe losses, analyze causes, present recommendations, and influence change provides some insight into how successful a subsequent investigation might be. Just as the legal profession relies heavily on case precedents, so too, incident investigation can benefit from the knowledge of what has worked well previously. To that end, it is recommended that an investigator become acquainted with other successful investigation reports involving similar technology and equipment.

Learning from incident case studies requires the ability to step back and examine the similarities and differences between a current situation and an incident that was experienced elsewhere. One should ask, "Why can't the incident that happened at company *x* happen here?" Be alert for of "action overload," that is, incident actions competing with existing actions from audits or other sources. A management team may be needed to prioritize multiple actions and their close-out. Learning from incident case studies also requires a willingness to change and adopt new and improved practices. Table 15-1 lists questions that can be applied to conduct a critical analysis of accident case studies.

15.3. Cross-Industry Lessons

There is a tendency in many organizations to recognize only those incidents that have occurred in similar operating environments or in similar processes within the same industry sector. For example, at this time the chemical industry does not have a common platform to exchange incident information with the oil and gas sector, pulp and paper or other industries. Given that the type of hardware and the processes used may be very similar in these industries, there is a significant need and opportunity to share lessons across geographical or industry boundaries. Case in point, the *Challenger* disaster described below, emphasizes the importance of tight seal integrity in critical service and organizational effectiveness in addressing high-level concerns. While some may discount the Challenger as a non-process related incident, it was nonetheless a loss of containment occurrence of catastrophic proportions. The lessons from the Challenger are as appropriate today as they were in 1986 and they apply to *all* process industries.

TABLE 15-1
Questions for Critical Analysis of Accident Case Studies

1. What are the synergies or similarities between the reference case and your own operation?
2. Are there similar chemicals used at your facility?
3. Do you have similar processes or equipment at your facility?
4. How does your site layout and infrastructure compare to that of the reference case?
5. How does your operation or organization compare to that of the reference case?
6. Are there any trends or patterns in your own operations that reflect those in the reference case?
7. Could a similar incident occur at your facility? Why or why not?
8. What are the potential consequences of this type of incident occurring at your facility? What is the most probable outcome? Why?
9. Have the immediate causes of this incident ever contributed to a loss at your facility? What were the consequences?
10. How have incidents such as these typically been dealt with at your facility?
11. How would an investigation at your facility have likely dealt with this type of incident? Would it have drawn the same conclusions?
12. How do the PSM systems at your facility compare to those described in the case study?
13. Were there effective process safety initiatives or other positive factors that limited the consequences in the reference case? Does your facility or operation provide such benefits or opportunities?
14. How effectively would your PSM systems have prevented or limited such an incident at your facility?
15. Is there anything you could do to reduce the likelihood of a similar incident at your facility?
16. Is there anything described in the reference case that your facility should eliminate or avoid?
17. What direct learning can you apply to your own PSM systems or operation as a result of this incident?
18. Is there anything you should implement at your own facility as a result of what you learned from this case?
19. Is there anything from this incident that could influence future decisions and direction within your own organization?
20. Is there someone else who would benefit from having this information? If so, who are they and how can you get it to them?

15.4. Trends and Statistics

Incident trends and statistics carry a powerful message. Incident statistics can forewarn of possible problems at a user site. Analysis of causal factors extracted from a statistical base of several incidents can help to establish process safety priorities and drive follow-up commitment. Recurring trends and statistics often drive codes and regulations. If an incident at a facility matches a common industry pattern in terms of common failure mechanisms, process hazards, equipment details, or associated consequences, there might be an opportunity to match and compare the logic leading up to the incidents. The successful actions taken by others to deal with historical occurrences might provide a clue to developing effective follow-up at a user site.

15.5. Management Application

A high level of technical competence is required to oversee the operation of a large chemical plant. Senior management must understand the full consequences of improper decisions or failure to support process safety initiatives at the field level. While experience is a good teacher, few senior executives have had sufficient exposure in an operating environment to understand all the things that can possibly go wrong and contribute to major losses. In addition, it should be recognized that major losses are so rare—even in large corporations—that it is necessary to look outside to gain sufficient knowledge. Well-documented incident case studies can help fill that void. By comparing the incident information with proactive evidence from deep audits, one may determine if an incident is symptomatic of a wider problem.

If incident case studies are periodically communicated to executives, they can help to reenforce the importance of doing things correctly. This typically translates into critical tasks that must be performed by skilled workers and accountability for those activities at an executive stewardship level. Such a commitment will carry over into incident investigation follow-up.

15.6. Case Studies

Case studies provide a reality check for demonstrating where an ineffective operation could end up. The condensed case studies below provide examples of physical failures, incident mechanics, wide-ranging consequences, and management system failures. These are intended primarily to illustrate the range of issues likely to be encountered in a formal incident investigation. Direct references for these case studies have not been provided; the information included herein has been extracted from a number of publicly available sources. There may be discrepancies or inconsistencies between these summaries and the official investigation reports released within the corporations involved. It is not the intent of these summaries to provide technical accuracy.

Some of the cases below reflect situations that might be encountered in a typical process plant. Others are less obvious and may only appear remotely connected to an operating plant. The Concorde crash and the Challenger disaster are examples of this latter category. These clearly demonstrate loss of containment scenarios that could be encountered in any operating environment.

15.6.1. Esso Longford Gas Plant Explosion

On September 25, 1998, a catastrophic explosion and fire destroyed one of three large gas plants and disrupted the energy supply to the entire state of Victoria, Australia. Two workers were killed, eight were seriously injured, and property damage exceeded $75 million (US). A total gas outage lasting more than two weeks shut down the entire manufacturing industry in the state of Victoria and business losses exceeded $800 million. Gas was partly restored within three weeks but strict rationing was carried out for at least one year pending repairs. The common control room was destroyed making it impossible to operate the other two plants reliably. This control room was located close to the plant that sustained the incident.

The gas plant that was destroyed had been initially commissioned in 1969. It had not undergone a mechanical inspection for six years. Shutdowns were initially scheduled at three-year intervals, but this interval was extended to five years. The plant had not been retrofitted with modern instrumentation. Pen recorders were prevalent in the control room and operating problems had been encountered with several control valves. Three months before the incident, ice plugging in lines caused several process upsets that were not fully investigated. A defective bypass valve had leaked gas to the atmosphere and had been awaiting maintenance up until one day before the accident.

On the day of the incident, a pump that supplied hot lean oil to a large heat exchanger on the absorber column tripped. A chart recorder that should have indicated the upset was out of ink. The temperature in the exchanger dropped because of cold condensate flowing through the tubes. After several hours, two supervisors discovered the problem and restored hot lean oil flow to the exchanger to warm it up. This act was carried out without communicating with the control room operator. The extremely cold exchanger ruptured and released a large vapor cloud, which subsequently ignited. The failure was determined to be a brittle fracture, the result of thermal shocking an extremely cold exchanger. The explosion and subsequent fire destroyed the plant. Both supervisory personnel who initiated the procedure were killed. The fire was not fully extinguished until two days later.

The system that experienced the failure had been modified within the previous decade. Instrumentation and control changes made it difficult for field crews to diagnose and troubleshoot operating problems. Process safety information had not been properly documented and there was little evidence that the changes had been formally evaluated. A few days before the incident, a tag was installed on a faulty letdown valve instructing that a bypass must not be used. No explanation was given for this instruction.

The loss of a critical flow in a process should have initiated emergency action. It could have been avoided through design modifications had the concern been highlighted. No formal HAZOP had been conducted on this plant despite recommendations to do so by external safety inspectors. There were no emergency procedures for dealing with loss of flow nor was any staff trained to deal with such a situation. Interviews with personnel after the accident revealed that no one on the plant site was familiar with the hazards of cold temperature embrittlement. Operator training was carried out in an open book fashion with trainees being prompted to provide the right answers to skill testing questions. There were no criteria for passing or failing the operator skills tests.

There were no formal operating procedures for dealing with the upset conditions that preceded the pump trip. The investigation revealed that fundamental operating practices were violated. Included were failures to monitor plant conditions, respond appropriately to alarms, report process upsets to supervisors, and undertake appropriate checks before making operating adjustments.

Since 1991, significant staff cuts had taken place. These were particularly noticeable among the supervisory ranks with a 75% reduction in supervisors over the previous three years. The engineering and technical staff had been relocated to a central office in the city of Melbourne 100 kilometers from the plant. The resident plant staff was not experienced and knowledgeable in dealing with upset conditions. This lack of skill and experience contributed significantly to the incident.

Analysis of Esso Longford as well as analysis in the UK Health and Safety Executive (HSE) investigation report into petrochemical complex major incidents all show that common underlying causes are often repeated. The Longford incident clearly illustrates the multiple root cause concept. A number of PSM system failures occurred either in sequence or collectively. These involved Process Safety Information, Process Risk Management, Operating Procedures and Practices, Training and Performance, Management of Change, Emergency Preparedness and Response and PSM audits. Serious deficiencies in PSM elements appeared to have spilled over into other elements. This has been a common finding in many large-scale incidents elsewhere.

Although the Esso "Operations Integrity Management System" nominally meets all the requirements of process safety management, the Longford plant failed to enforce or comply with its own corporate requirements. Notably, supervisors violated the rules they were expected to enforce by operating field hardware and failing to communicate with skilled operators. Staff cuts had reduced the number of experienced supervisors and technical personnel to an unacceptably low level. Management did not conduct formal audits or facility inspections nor did it arrange for process hazard analyses on the plant that was destroyed. Substandard conditions that contributed to the incident had preexisted and were tolerated. Despite an enviable safety record, several minor process upsets had not been analyzed or investigated. In short, the effectiveness of the safety management system at the Esso plant was seriously compromised.

15.6.2 Union Carbide Bhopal Toxic Gas Release

On December 3, 1984, a toxic gas release from a pesticide plant in India killed nearly 3000 people and injured at least 100,000 others. The chemical that leaked was methyl isocyanate, a chemical intermediate that was supposed to be stored in a cooled bunker near the plant's outer boundary. The vapor is highly toxic and causes cellular asphyxiation and rapid death. Despite engineering and procedural provisions to prevent its release, a total system breakdown resulted in the release of 40 tons of the deadly material into the densely populated community of Bhopal. Because of this incident, the plant was dismantled and ultimately the parent corporation, Union Carbide, was forced to make a number of organizational changes. The occurrence is considered by many to have been the most tragic chemical accident in history.

Methyl isocyanate (MIC) is a colorless liquid that must be stored in a cooled enclosure before it is subsequently used in the manufacture of carbamate, a common insecticide. MIC liquid is highly reactive in the presence of water and iron oxide, and it generates heat. In sufficient quantities, this heat may generate vapor, which, as explained previously, is highly toxic. Three adjacent bunkers were used for MIC storage. These were mounted in a berm and a refrigeration coil was used to ensure that the temperature did not exceed 5°C. A vent gas scrubber was used to prevent vapor escape, and despite a low operating pressure, a closed relief and blowdown system was provided. There were several redundancies in the design, but ultimately these required that procedures be followed.

For several months before the accident, conditions at the plant had been deteriorating. Procedures were not carefully followed and several mechanical features were either shutdown or compromised. Examples include the refrigeration circuit that was depleted of coolant and the vent gas scrubber that was out of service. The temperature indicator on one tank was defective. The temperature in one of the tanks had been allowed to exceed the maximum limit by as much as 15°C with no corrective action.

On the night of the accident, operators heard a screeching noise from the relief valve on one of the tanks. Unfortunately, the closed blowdown system had been taken out of service for maintenance. It was later established that while operators were on their shift change or on a break, someone disconnected a pressure gauge from the cover plate on one of the tanks and attached a water hose. A quantity of water estimated between 450 and 900 kg entered the tank and caused a severe upset and release of MIC vapor. With no means of notifying the public and evacuating the community, thousands were exposed to the vapor cloud, resulting in the deaths and injuries.

There are competing theories as to what caused the Bhopal incident. While the entry of water into the tank was ultimately established as the physical cause, either human error or sabotage may have played a role. In either case, local plant management had set the stage for the disaster by allowing hazardous conditions to prevail and by not taking appropriate corrective action. Following select interviews, it was determined that many operators also knew that things were not right but they were afraid to speak out. In this regard, many shared responsibility. The Bhopal tragedy clearly demonstrates the need for a holistic process safety management framework. This accident demonstrates the need for trust if an operation is to be safe and reliable.

The investigation into the Bhopal tragedy is itself a case study. The extensive technical work required to support the Bhopal incident investigation went on for many months after the occurrence. The investigation serves as an example of the rigor and determination required for closure in difficult circumstances. Public outrage and intervention by the government of India prevented a proper investigation from being initially carried out. The authorities were intent on establishing liability and collecting damages from the parent company, Union Carbide. Although emergency relief funds were quickly released, these did not reach the survivors and it is believed that the funds were diverted to government officials. Nonetheless, diligent efforts by the parent company resulted in a detailed forensic analysis being conducted to establish the physical causes of the accident. The sequence of events was reconstructed several times and challenged to get a precise fit. This was highly valuable because it established that conditions normally not tolerated in other Union Carbide plants had preexisted prior to the incident. Ultimately, a breakdown in management systems had played a more significant role than a single physical act.

15.6.3. NASA Challenger Space Shuttle Disaster

On January 28, 1986, the space shuttle *Challenger* exploded shortly after lift off destroying the craft, killing its crew of six, and causing a severe setback to the US space program. The accident was the result of a rubber O-ring seal failure between adjacent sections of the solid fuel rocket boosters. Unprecedented cold weather on the day of the launch made the rubber brittle, which, combined with the faulty design of the joint, allowed hot combustion gases from the burning rocket to escape. The flames and hot gases burned through the metal supports holding the rocket in position. When the rocket assembly released, it ruptured the side of the external fuel tanks allowing liquid hydrogen and oxygen to mix prematurely and explode.

A presidential commission was appointed to formally investigate the *Challenger* accident. More than 6,000 people were involved in the commission's four-month investigation and some 15,000-transcript pages were recorded during public and private hearings. It became apparent that there was a well-documented history of problems with the rocket booster design including the integrity of the rubber O-ring joints. When O-ring damage was initially observed after the second shuttle flight, the NASA team simply made changes to the assembly process (but not the design) and continued with future flights. The growing momentum to keep the shuttle flying was adversely affecting the team's reaction to the O-ring problem. NASA middle managers repeatedly violated safety rules requiring the prompt resolution of technical problems. Over time, the managers normalized the deviation, so that it became acceptable and nondeviant to them. Given the success of previous missions with known problems, middle management came to accept the risk and failed to communicate their concerns to top decision makers. While it may have been painful to accept defeat in terms of project schedule, such pain is insignificant when compared to destruction of the mission.

The NASA organization was highly complex and several parts of the project were contracted out to private consultants whose livelihood depended on the success of the project. Production objectives emphasized the importance of the launch date for each mission. In light of the safety concerns and potential setbacks, work continued on the project. Had the team collectively recommended the suspension of further work until all the technical problems were solved, the accident could have been avoided. The responsibility for making such unpopular decisions in large organizations geared for success is often not clear. The final commission report included recommendations affecting hardware design as well as NASA work procedures and communication protocols. An Office of Safety has since been established to oversee safety and quality related issues for all NASA operations. This office ensures that the highest levels of NASA's management team are aware of safety. Overall, more people have been assigned to safety and mechanical integrity programs, improved communications have been initiated, and the review system for compliance to new procedures is rigorous and well defined.

The *Challenger* Space Shuttle disaster can also be viewed as an example of less than adequate incident investigation follow-up. Failure of an O-ring seal in the solid rocket booster is one of the widely publicized causes of the *Challenger* Space Shuttle disaster. The Presidential Commission Investigation Report disclosed that O-ring failure had been previously identified as a serious problem. The potential severity of consequences resulted in a formal launch restraint being imposed six months before the January 1986 incident. O-ring failure was specifically identified as a problem and follow-up action had been initiated to resolve the problem. Unfortunately, the actions taken were inadequate. There is a process in organizations that can be defined as "normalization of deviance." Over time, deviations from specified practices are tolerated until, in some individual cases, the deviant practice becomes the norm. This is a trend that investigators look for in near-miss investigations.

While the *Challenger* disaster was not a process incident in the strictest sense, the nature of the failure was similar to many piping system failures that typically occur in the process industries. More importantly, organizational failure was a fundamental cause of the incident. This case study serves as a classic example of the type of loss that can occur in a large complex organization if management systems are not effective.

15.6.4. Tosco Avon Oil Refinery Fire

On February 23, 1999, a flash fire occurred at the Tosco Avon, California refinery killing four workers and critically injuring one other. The accident occurred while a critical piping repair job was underway on the crude fractionation unit. A small naphtha leak had been reported on February 10, but almost two weeks had lapsed while workers attempted to isolate the circuit and physically remove a corroded section of line. The unit remained in production while this work was taking place.

The Tosco incident illustrates the importance of safe work practices and, in particular, the need to provide strong principled leadership in the field. Over the past few years, the quality of raw feedstock had varied and problems had been reflected in the operation of the chemical desalter. Consequently, a marked deterioration in quality was noted in the crude unit operation. Before 1999, a number of serious incidents had occurred at the refinery, some of which involved fatalities. The refinery had also undergone several recent personnel changes.

After attempts to properly isolate the circuit failed, a plan was devised to drain excess naphtha to a collection drum and cold cut the line at several locations to facilitate removal of the line. Ultimately, the loss of liquid head as the naphtha drained resulted in a pressure surge through a partly blocked line, and caused several gallons of naphtha to spray onto the workers who were trapped on elevated scaffolding. The naphtha contacted hot surfaces on the crude column and ignited immediately. Three of the workers who were killed were contractors. One employee jumped from an elevated platform and was seriously injured. The fire burned for only 20 minutes causing minor property damage. The refinery was shut down for an extended period pending a formal investigation and was sold to another oil company in 2000. Regulatory fines were extensive.

The initial incident involved a pinhole leak on a 6-in. line located 112 feet above grade. The unit might have been shut down at this point, but a decision was made to keep it on line. Efforts to isolate the circuit by closing four block valves failed. The line remained hot and the level in the downstream stripper vessel continued to rise. Several attempts were made to depressure the circuit and free drain all the residual naphtha. At least 15 work orders were issued over a 13-day period and most of these failed to meet the company's own work standards.

After a blind flange had been installed at the main fractionator, the work progressed quickly. Believing the circuit to be free from additional hydrocarbon, a maintenance supervisor instructed the line to be cut in several places. Some process operators warned that the downstream valves at the stripper might not hold and that the unit should be shutdown. Nonetheless, the job proceeded and one of the valves passed a significant quantity of naphtha. The mainteannce supervisor was present at the unit and met with workers throughout the morning. He was not present at the unit after lunch, when the incident occurred. On the morning of the incident, there were no operations supervisory personnel present.

Both the failure of the isolation valves to hold and the difficulty in draining the system were due to solid build up in the circuit. In the previous two-year period, the refinery had changed its crude feedstock resulting in operating difficulties on the unit desalter. Carryover of salts led to accelerated corrosion in the crude column and associated piping circuits.

Corrosion products plugged the lines making it difficult to operate the stripper. Ultimately, the stripper operation was discontinued while the crude column remained in operation. This should have signaled the need for a major turnaround. Management did not have the perspective to analyze and probe the reasons behind these operating problems.

Auditing and performance measurement of precursors is an essential part of preventing future major incidents. Failure to provide appropriate leadership in the field and to apply safe work practices can be cited as the key management system failures. Problems were apparent long before this tragic incident. Inconsistencies in management and inability to recognize and address fundamental technical problems set the stage for a high-level process incident. Government investigators were able to demonstrate major deficiencies in many elements of process safety management.

15.6.5. Shell Deer Park Olefins Plant Explosion

On June 22, 1997, an explosion and fire destroyed an olefin plant at the Shell Deer Park, Texas chemical complex. Thirty workers received minor injuries and extensive damage resulted to both the plant and adjacent community. The incident was caused by the sudden failure of a pneumatic check valve in a 300-psig gas compressor discharge line that led to a large release of propylene vapor. A few minutes after the release, the gas ignited causing a vapor cloud explosion and flash fire.

The damage from the explosion was so widespread that several dozen possible release sources were identified. This posed a challenge for the investigation team making it necessary to systematically analyze each hole or breech before accepting or discounting it as the initiating failure. Ultimately, it was determined that a shaft had blown out of a large valve providing an open passageway for gas to escape from a closed system. The nature of the observed failure was such that fire or explosion could not have caused it.

The valve that failed was a 36-in. pneumatic check valve equipped with an external damping mechanism to minimize mechanical shock associated with rapid closure. The damping mechanism weighed 200 pounds and was attached to the internal shaft of the valve through a 3½-in. connection welded onto the side of the valve body. The valve shaft was the only obstruction to gas escaping. To facilitate assembly and maintenance, the valve was constructed in pieces. The shaft was attached to the disk assembly by a mechanical key and secured in place by a dowel pin. Over time, an unbalanced axial thrust developed on the shaft, causing the pin to fail, and forcing the key to drop out. The shaft and external damping mechanism were propelled a distance of 40 feet and struck an adjacent part of the plant. It is believed that cyclic loads and unsteady operating conditions may have contributed to the failure.

Before the incident, the facility had experienced a power failure that led to the shutdown of an olefins unit. For several hours, a restart had been attempted and the compressor had tripped on high vibration at low speed. This caused the check valve to close, which placed additional stresses on the valve shaft. Failure to execute a smooth planned start-up was a key contributor to the timing of this incident. However, given the design defects inherent in the valve, an ultimate failure was inevitable.

The investigation revealed that the valve design did not recognize or compensate for the forces experienced in normal operation at 300 psig. Alternatively, the design did not specify the correct hardware for the service conditions. Use of nonstandard hardware should be avoided until such time that its operation is well understood and documented. There are certain hazards and risks associated with the use of specialty hardware and these are magnified when large size equipment is used.

Apart from the technical details of this incident, there were some organizational issues worthy of note. Shell operates refineries and chemical plants in many locations around the world. Before the incident at Deer Park, a similar failure had occurred in a pneumatic valve manufactured by the same supplier at another Shell refinery in the Middle East. A company-wide bulletin had been issued warning of the potential for failure with this type of valve, but no follow-up action appeared to have been taken.

Given the large volume of gas that escaped in the incident, the potential for loss was much higher. The considerable time interval between the initiating failure and the blast provided opportunity for some personnel to exit the area. Although damage was widespread, many steel frame buildings on the site remained standing and did not collapse. Although they had to be demolished and replaced, they clearly demonstrate the importance of building siting and integrity reviews in protecting personnel from process hazards. Some of these factors may be coincidental. Nonetheless, several did affect the outcome of the loss and may have contributed to its occurrence.

15.6.6. Texas Utilities Concrete Stack Collapse

On November 14, 1993, a 600-foot concrete stack collapsed onto a coal-fired electric power plant killing one person and injuring four others. The incident occurred at a Texas Utilities plant near Mt. Pleasant, Texas while the facility was shut down for maintenance. Extensive damage occurred within the plant resulting in an extended outage. Destroyed in the incident were a 750 MW generator, a steam boiler, compressor, two scrubbers, and a precipitator. There were 60 employees in the plant at the time of the incident.

The stack was being cleaned internally to remove a buildup of fly ash and lime from the inner walls. This had resulted from the earlier addition of a lime-scrubbing unit to reduce environmental emissions. The maintenance crew was attempting to dislodge a large chunk of material from the wall using firewater when it fell to the base of the stack 80 feet below. The force of the falling material destroyed the base of the inner stack and caused the entire structure to collapse. Wind gusts had been experienced before the failure and may have been a contributing factor.

This tragedy emphasizes the importance of safe work practices especially for situations that have not previously been encountered. A contingency plan should have been drawn up to address the possibility of the solid mass releasing from the wall. Complacency, inexperience, and desire to get the job done may have overshadowed the need to develop safe work procedures.

The design of the stack was inadequate for the service especially since the scrubbing process introduced solid material into the stack. A good structural or mechanical design does not necessarily ensure suitability for process conditions. The stack should have been reinforced and should have been equipped with some means of minimizing deposits within the inner liner.

The stack was constructed of an inner liner of brick shrouded by a concrete outer shell constructed of reinforced steel and concrete. The brick and mortar liner had no reinforcing steel to support the assembly from side loading. The concrete shroud was primarily for protection against wind. A space of a few feet existed between the wall of the brick liner and the outer concrete shroud. The brick wall was about 3 to 4 feet thick at the bottom, and the wall thickness lessened with height. The concrete portion was of a similar design.

An external contractor was assigned to carry out the cleaning operation. This company was highly specialized in the design, construction, and maintenance of industrial chimneys and stacks. A team of eight workers had been involved in the cleaning operation. Work was carried out from a two-point suspended scaffold with the use of a high-pressure water washer, picks, shovels, and air chisels to remove the material clinging to the stack walls.

A large buildup of fly ash and other material had accumulated on the stack wall opposite the breech where the duct entered the stack. It was formed by solid particulate material in the flue gases impinging on and sticking to the stack wall due to the momentum of the particles in the fast moving flue gas. This buildup was estimated to be up to 15 feet thick at the center and 30 to 40 feet in diameter.

The stack cleaning had progressed from the top down to the point where this large buildup near the mouth of the breech was encountered. The high-pressure water wash was found to be somewhat ineffective and it was replaced with a firewater hose and nozzle. The hose was operated by a worker on a scaffold that had been erected at the breech entry into the stack. Water was directed onto the large mass in an attempt to wash as much off as possible before using picks and shovels. The entire mass of the buildup suddenly fell in one piece, according to two witnesses. A few seconds later, the entire stack collapsed.

Two workers were blown several feet into the breech, apparently by wind from the falling brick and debris. The individual using the water hose was covered by debris and died. He had been secured with a safety belt and lanyard. One of the surviving employees had just disconnected his lanyard when he saw the mass fall.

A technical analysis indicated that the falling mass released a significant amount of energy. This energy was absorbed when the mass hit the bottom, partly by the base of the stack and by the sides. The unsupported brick walls could not sustain this force and were probably blown out at the bottom by the force. The collapse of the brick stack imposed outward force on the lower walls of the outer concrete shroud and it failed and collapsed.

The buildup of residue in stacks, breeches, and manifolds is believed to be a problem only in generating units having scrubbers. The scrubbing process adds materials to the combustion gases that can settle on the walls to form residues if the materials are not adequately removed from the flue gases. This incident could have been avoided with proper job planning, procedures, and management of change (service application of stack).

At first glance, this incident does not appear to be process safety related since there was no direct loss of containment; however, many factors make it important to review. The material involved in the incident was an undesirable byproduct of an engineered process. When unknown material accumulates in an undesirable location, it often creates a physical hazard. This hazard may directly trigger a loss when it is least expected. Accumulated solids in closed systems may contribute to corrosion, blockage, and physical damage. This accident case study should also be examined in the context of an enclosed system.

In addition to the factual findings of the investigation, this incident helps to illustrate the importance of process safety information. When unusual or unprecedented situations are encountered, every effort should be taken to understand the chemical properties and behavior of chemical substances before attempting their removal. The adhesive properties of the fly-ash/lime mixture were obviously affected by water and this led to a large release and structural failure. Had the development of formal procedures been attempted, the knowledge gap might have been identified and a different strategy undertaken. If the operation had been closely monitored, the mechanism contributing to the buildup of material might have been better understood resulting in process engineering modifications.

15.6.7. Three Mile Island Nuclear Accident

On March 28, 1979, a loss of containment incident occurred at the Three Mile Island nuclear power plant near Harrisburg, Pennsylvania. An overheated reactor released radioactive steam and water to the atmosphere resulting in a mass evacuation of the surrounding community. Although no direct injuries were attributed to the incident, environmental effects were later observed and public outcry resulted in a slowdown in the growth of the nuclear power industry.

The process at Three Mile Island involved nuclear fission and subsequent reactor cooling using circulating water. The primary water was kept under pressure to prevent boiling. Heat was transferred to a secondary water system that supplied power to a steam generator. Upon completion of this step, steam condensate was recovered and recycled. All radioactive materials, including primary water, were enclosed in a lined concrete containment building to prevent their escape to the atmosphere.

One significant event that contributed to the incident was a malfunctioning pilot-operated pressure relief valve. This valve had stuck in the open position and allowed steam system inventory to decrease, which ultimately resulted in loss of cooling and an exposed reactor core. The valve malfunction was due to water contamination of the instrument air system. The instrument air system was contaminated by inadvertent introduction of water during an attempt to clear a blockage in a deionized water system using compressed air. The blockage was caused by an ion exchange resin breakthrough from a condensate polisher unit. An identical situation had occurred twice previously, but it had not been studied or effectively resolved.

Problems within the polisher unit caused operators to respond by attempting to unblock a choked condition using instrument air. The air was at a lower pressure than the condensate and this caused water to enter the air system. This was not a standard procedure and the commercially supplied polisher unit was not built to standards consistent with the plant. Water in the instrument air system caused several instruments to fail and ultimately initiated a turbine trip. This interrupted heat removal from the radioactive core. The heat generation within the reactor was halted automatically within a few minutes by dropping metal rods to absorb neutrons within the core.

Radioactive decay contributed to further heat buildup. This caused water in the primary circuit to boil. The pilot-operated relief valve on the primary circuit lifted and the supply pumps engaged to replace the water. Unfortunately, the valve stuck in the open position. An error in the signal circuit showed that the valve was closed. Several other symptoms suggested the valve was open but the operators chose to believe the indicator light that signaled the valve was closed. Complying with standard procedures and assuming that all instruments were functioning properly, the operators shut down the water supply pumps. The drop in cooling system inventory exposed the top of the reactor core to steam. The protective zirconium alloy cans at the top of the core reacted with the steam to produce hydrogen gas. As the system pressure built, contaminated steam was released from the relief valve, accumulated in the containment building and ultimately was released to the atmosphere.

This accident highlights the importance of training and knowledge to deal with extraordinary upset conditions. Critical instruments should be provided with an alternate means of validating their position or status. The use of human factor evaluations prevalent today would certainly have enabled the operation to proceed reliably under upset conditions. In a large critical plant operation, it is important to apply recognized technology and high quality hardware throughout. In an upset situation, process operators can ill afford to spend time troubleshooting equipment that is supplied by others. Unsafe work practices allowed the introduction of incompatible fluids into critical process circuits.

15.6.8. Concorde Air Crash

On July 25, 2000, an Air France Concorde flight bound for New York crashed shortly after takeoff from Charles de Gaulle Airport in Paris. All 109 passengers and crew aboard the aircraft were killed. The plane struck a nearby hotel destroying it and killing an additional four people. All remaining twelve Concorde aircraft in the fleets of Air France and British Airways were immediately grounded and service was not restored for at least one year pending an investigation into the cause of the tragedy.

The Concorde accident involved a physical puncture of one of the fuel tanks from parts of a tire believed to have been damaged during takeoff. The entire incident occurred over a period of slightly more than one minute. During the investigation, it was disclosed that there had been at least five previous incidents involving debris on the runway, but none resulted in catastrophe. In one of those incidents, a tire failed and punctured the fuel tanks, but there was no ignition. The incident investigation system used by British Airways and Air France failed to adequately follow-up and apply lessons learned from previous investigations. Furthermore, internal communications between Air France and British Airways were ineffective at sharing of minor and near miss incidents, particularly those involving fuel tanks, tires, and runway hazards.

The flight had departed on schedule in the early afternoon of July 25. Heading eastbound, the crew encountered difficulty retracting the left landing gear shortly after takeoff. Ground crews observed fire on the rear of one of the wings and notified the crew. An attempt was made to carry out an emergency landing but there was insufficient time. The fire quickly spread to an adjacent engine causing the plane to crash.

The investigation into the crash determined that the right tire on the left landing gear had disintegrated shortly after takeoff. At high speed, parts of the tire punctured the inner left fuel tank. The breech was very severe causing most of the fuel to leak out and ignite. This is believed to have caused the next fuel tank to rupture and burn. It was later determined that a 40-cm-long piece of metal left on the runway damaged the tire at takeoff. There was evidence to connect this piece of metal with a plane that had departed earlier for the United States. This type of forensic analysis can establish the physical cause of an accident, but does little to help understand how such an accident occurred and how it might be prevented.

The investigation focused its attention on the design of the Concorde. To achieve high speed it was necessary to tightly position the fuel tanks adjacent to one another. This made the fuel system more vulnerable to a mechanical failure. Had there been a protective barrier between the two tanks on the left wing, the flight might have been salvaged. In the months following the crash, both airlines thoroughly inspected their Concorde fleets. In at least seven planes, severe cracking was observed on the wings. Had the crash not occurred, perhaps another failure mode would have contributed to a disaster in flight.

The Concorde accident highlighted the importance of physical inspection and maintenance. This is especially critical for equipment that has been in service for thirty years. Given such a long time span, there is also a need to review the original design against conditions that have been experienced in service. Had the wings and fuel tanks been examined for prior damage, and had previous incidents been investigated and acted on, the loss might have been avoided.

15.7. Sharing Lessons Learned

The effectiveness of communicating important lessons across organizational boundaries is critical to the success of any investigation. A management system should be in place to ensure that the understanding of the lessons learned is not isolated to a single location. The challenge of maintaining perspective and continuity of focus remains regardless of the quality of the initial investigation. A well-designed and operated process safety information system can minimize the experience of relearning the same lesson. The Responsible Care protocol which is practiced in forty-seven countries (International Council of Chemical Associations) has a code of practice that requires member companies to share specific safety incident lessons both internally and externally.

Some companies have found success with a periodic publication of incident abstracts. Each incident may appear as a one- or two-paragraph summary in a quarterly bulletin. This information should be circulated widely within the corporation. If a site has a special interest in one particular incident, then full details can be requested by direct contact between the two sites. The highly abbreviated summary has other uses; for instance, as material for employee safety training meetings or bulletin board postings.

For several years, the American Institute of Chemical Engineers (AIChE) has sponsored public presentations about incidents with accompanying lessons learned. Some of these case histories are contained in AIChE publications or have been presented at the Annual Loss Prevention Symposium. The American Petroleum Institute also publishes a series of booklets on specific lessons learned. In the United Kingdom, the Institution of Chemical Engineers publishes Loss Prevention Bulletins that contain specific case histories. The Chemical Safety and Hazard Investigation Board (CSB) website summarizes notable process safety incidents that occur around the world. A reference is usually given to the host document or media source. The CSB website contains full accident investigation reports of significant incidents it has investigated. These case studies are particularly useful since they serve as a template for setting up an effective investigation report.

The CSB is an independent, scientific investigatory agency created by the Clean Air Act Amendments of 1990. It began operations in January 1998. As the Senate legislative history states: "The principal role of the new chemical safety board is to investigate accidents to determine the conditions and circumstances which led up to the occurrence and to identify the cause or causes so that similar occurrences might be prevented." Although the Board was created to function independently, it also collaborates in important ways with US EPA, US OSHA, and other agencies. The CSB mission and additional information can be found at its website at www.chemsafety.gov.

CCPS sponsor companies may subscribe to a confidential database of accident case studies and causal data. Any operating company prepared to submit and share incident data may subscribe to this Process Safety Incident Database (PSID). Additional information on databases is available in Table 11-3 on page 284.

CCPS publishes incident lessons in a monthly one-page "Process Safety Beacon" aimed at building awareness among plant operators and technicians. Each issue presents a real-life incident and describes the lessons learned as well as practical means to prevent a similar incident from occurring in other plants. This material is suitable for sharing at crew, toolbox, or safety meetings, or through postings in break or control rooms.

References

The following references provide information on techniques for sharing of lessons learned as well as additional information on some of the case studies that were highlighted in this chapter. Other significant incidents are also cross-referenced in some of these documents. It is important to recognize that some of these publications have been written from a non-technical perspective and may not have recognized the concepts or methods covered in this text.

American Chemistry Council. *Responsible Care®, A Resource Guide for the Process Safety Code of Management Practices*. Washington, DC: 1990.

Perrow, C. *Normal Accidents: Living with High Risk Technologies*, Princeton University Press 1999.

Chiles, J R. *Inviting Disasters, Lessons from the Edge of Technology,* Harper Business, 2001.

Clancy, M. S. and Kelly, B. D. "Use a Comprehensive Database to Better Manage Process Safety," *Chemical Engineering Progress*, August 2001.

US EPA/OSHA Joint Chemical Accident Investigation Report, *Shell Chemical Company,* Deer Park, Texas. EPA-R-98-005.

Hazard Prevention Magazine, System Safety Society, Second quarter 2001.

Hopkins, A. *Lessons from Longford: The Esso Gas Plant Explosion*. Australia: CCH Books, 2000.

Kletz, T. *Learning from Accidents*. Oxford, UK: Butterworth-Heinemann, 2001.

Lees, F, P., *Loss Prevention in the Process Industries*, Volume 3. Oxford, UK: Butterworth-Heinemann, 1996.

Process Industry Incidents, CCPS International Conference and Workshop, Proceedings, October 3–6, 2000. New York: American Institute of Chemical Engineers.

U.S. Chemical Safety and Hazard Investigation Board. *Investigation Report, Tosco Avon Refinery Fire Incident,* 99-014-1-CA.

Vaughan, D. *The Challenger Launch Decision*. Chicago: University of Chicago Press, 1996.

Appendix A

Relevant Organizations

ACGIH
American Conference of Governmental Industrial Hygienists
www.acgih.org
1330 Kemper Meadow Drive
Cincinnati, Ohio 45240
(513) 742-2020

ACC
American Chemistry Council
www.americanchemistry.org
1300 Wilson Blvd.
Arlington, VA 22209
(703) 741-5000

ACS
American Chemical Society
www.acs.org
1155 Sixteenth Street, NW
Washington DC 20036
(800) 227-5558 (US only)
(202) 872-4600 (outside the US)

AIChE
American Institute of Chemical Engineers
www.aiche.org
3 Park Ave
New York, NY 10016-5991
(800) 242-4363

AIHA
American Industrial Hygiene Association
www.aiha.org
2700 Prosperity Ave. Suite 250
Fairfax, VA 22031
(703) 849-8888

ANSI
American National Standards Institute
www.ansi.org
1819 L Street, NW
Suite 600
Washington DC 20036
(212) 642 4900

API
American Petroleum Institute
www.api.org
1220 L Street, NW
Washington DC 20005-4070
(202) 682-8000

ASME
American Society of Mechanical Engineers International
www.asme.org
3 Park Avenue
New York, NY 10016-5990
(800) 843-2763

ASSE
American Society of Safety Engineers
www.asse.org
1800 East Oakton Street
Des Plaines, Illinois 60018-2187
(847) 699-2929

ASTM
American Society for Testing and Materials
www.astm.org
100 Barr Harbor Drive
West Conshohocken, PA 19428-2959
(610) 832-9585

CCPS
Center for Chemical Process Safety
www.aiche.org/ccps/
American Institute of Chemical Engineers
3 Park Avenue
New York, NY 10016-5991
(212) 591-7319

CSB
United States Chemical Safety and Hazards Investigation Board
www.chemsafety.gov
2175 K Street NW Suite 400
Washington DC 20037-1809
(202) 261-7600

EPA
United States Environmental Protection Agency
www.epa.gov
Ariel Rios Building
1200 Pennsylvania Avenue, NW
Washington DC 20460
(202) 260-2090

EPSC
European Process Safety Centre
165-189 Railway Terrace
Rugby CV21 3HQ UK
+44 (0)1788 534409

FEMA
Federal Emergency Management Agency
http://www.fema.gov
500 C Street SW
Washington DC 20472
(202) 566–1600

HFES
Human Factors and Ergonomics Society
www.hfes.org
PO Box 1369
Santa Monica, CA 90406-1369
(310) 394-1811

HSE
Health and Safety Executive (UK)
www.hse.gov.uk/
Caerphilly Park
Caerphilly CF83 3GG
United Kingdom
+44 (0) 8701 545500

IChemE
Institution of Chemical Engineers
www.icheme.org
Davis Building
165-189 Railway Terrace
Rugby CV21 3HQ United Kingdom
+44 (0) 1788 578214

NFPA
National Fire Protection Association
www.nfpa.org
1 Batterymarch Park
Quincy, MA 02269-9101
(617) 770–3000

NSC
National Safety Council
www.nsc.org
1121 Spring Lake Drive
Itasca, IL 60143-3201
(630) 285-1121

NTSB
National Transportation Safety Board
www.ntsb.gov
490 L'Enfant Plaza East SW
Washington DC 20594
(202) 314–6000

OSHA
United States Occupational Safety and Health Administration
www.osha.gov
US Department of Labor
Washington, DC 20210
(800) 488-7087

SOCMA
Synthetic Organic Chemical Manufacturers Association
www.socma.com
1850 M Street NW Suite 700
Washington DC 20036
(202) 721-4100

SSS
System Safety Society
http://www.system-safety.org
PO Box 70
Unionville, VA 22567-0070
(540) 854-8630

Appendix B

Professional Assistance Directory

Visit www.aiche.org for the current directory.

ABS Group Inc.
www.abs-group.com
10301 Technology Drive
Knoxville, TN 37932-3341
Telephone: (865) 966-5232
Fax: 865-966-5287

AcuTech Consulting Group
www.acutech-consulting.com
1948 Sutter Street
San Francisco, CA 94115
Telephone: (415) 923-9226
Fax: 415-923-9274

Apollo Associated Services, Ltd.
PO Box 1228
Friendswood, TX 77549-1228
Telephone: (989) 835-3402
Fax: 281-218-0187

Baker Engineering and Risk Consultants
www.wbeng.com
3330 Oakwell Court
Suite 100
San Antonio, TX 78218
Telephone: (210) 824-5960
Fax: 210-824-5964

Battelle
www.battelle.org
505 King Avenue
Columbus, OH 43201-4726
Telephone: (614) 424-7499
Fax: 614-424-3404

DNV
www.dnvda.com
16340 Park Ten Place
Suite 100
Houston, TX 77084
Telephone: (281) 721-6000
Fax: 281-721-6900

DuPont Company
www.dupont.com/safety
SHE Excellence Center
Nemours Building N2526-3
Wilmington, DE 19898
Telephone: (302) 774-9558
Fax: 302-774-3140

FM Global Research
www.fmglobal.com
1151 Boston-Providence Turnpike
PO Box 9102
Norwood, MA 02026
Telephone: (781) 255-4986
Fax: 781-255-4024

ioMosaic Corporation
www.iomosaic.com
93 Stiles Road
Salem, NH 03079
Telephone: (603) 893-7009
Fax: 603-251-8384

Primatech, Inc.
www.primatech.com
445 Hutchinson Avenue
Suite 200
Columbus, OH 43235
Telephone: (614) 841-9800
Fax: 614-841-9805

Risk, Reliability and Safety Engineering
www.rrseng.com
2525 South Shore Boulevard
Suite 206
League City, TX 77573
Telephone: (281) 334-4220
Fax: 281-334-5809

System Improvements, Inc.
www.taproot.com
238 S. Peters Road
Suite 301
Knoxville, TN 37923
Telephone: (865) 539-2139
Fax: 865-539-4335

TNO Initiative for Industrial Safety
www.tno.nl/homepage_nl.html
PO Box 6006, 2600 JA DELFT
2280AA
Telephone: +31-15-269-67-48
Fax: +31-15-262-73-19

US DOE
www.energy.gov
19901 Germantown Road, EH-52
Germantown, MD 20874
Telephone: (301) 903-6061
Fax: 301-903-7773

Appendix C

Photography Guidelines for Maximum Results

The following guidelines provide a list to review before taking photographs.

1. Log and document every photograph. Trying to reconstruct or remember what the photograph is of and why it was taken after a period has elapsed is difficult.
2. Promptness is critical in order to minimize disturbing the data; however, in no case should emergency medical treatment or emergency response activities be delayed by any photographic activity.
3. Begin with overall views of the general area from multiple directions. This will help show perspective of distance and relative locations of items of interest.
4. Taking multiple exposures is a technique of professional commercial photographers. Most of the exposures actually shot by the professionals are never actually used. It would be unrealistic to expect nonprofessional photographers (such as incident investigation team members) to outperform professionals in this aspect. It is therefore wise to take multiple exposures of essential shots from several angles at various settings.
5. Every shot should include an item of measurable scale as a size reference. It is common to include a ruler/scale or some other object of known size in any close-up view (3 feet or less, 1 meter or less). Tape measures can also be used to show the size of objects and the distance of objects from each other. The orientation of the tape measure can also be used to show the orientation of the photograph.
6. Flash units and motor drives are potential ignition sources. In many cases, a gas test and hot work permit will be required before using a flash unit, motor drive, or video camera. ***Each*** specific use of flash devices may require authorization. Infrared gas detectors may also be set off by flash units. When using a flash even with permission, it is a good practice to take the time to warn and alert all personnel who could see the flash (or a reflection of the flash). This will pre-

vent startled response actions, and could prevent an injury (due to fall or other response). It is also a sign of courtesy and respect.

7. Consider the location of the sun and the accompanying glare, reflections, and shadows generated during outside shots. Sometimes a specially timed series of photographs is taken to document the approximate lighting conditions at the time of the incident. If more than several weeks have passed, the relative position of the sun may have changed enough to make a significant difference.
8. One disadvantage of an autofocus camera is that the camera does not always focus on the desired object. If the object of interest to the photographer is in the background and another object is in the foreground, then the camera may select and focus on the closer object. A familiar example is the out-of-focus picture where the camera has focused on some background object in the gap between two people. Most autofocus cameras are now equipped with a selectable feature to overcome this limitation.
9. A common avoidable mistake is to expect the camera to duplicate the human eye in low light conditions such as dusk or heavy shade. The performance specifications for normally available 35-mm films and digital cameras represent a compromise of several factors. These include lighting conditions, technical quality, and image resolution. Special low light level and extended wide range speed films are available for 35-mm cameras at additional costs. The camera/film systems are designed to perform in a specific envelope. Operating near or beyond the edge of these specifications will produce correspondingly lower performance.
10. Camera battery life can be unpredictable! A fresh and complete spare set is a necessity rather than a luxury. In many modern 35-mm camera systems, functioning batteries are required for a simple task such as loading film. If the camera is part of a seldom used supply kit, special attention is needed to ensure fresh primary and spare batteries are available and that the film has not passed its expiration date.
11. Some type of portable background is often desirable when shooting data in the field. A light colored pastel cloth will usually give better results than black or white.
12. When documenting a witness statement, the photograph should be taken from as close as possible to the actual viewpoint used by the witness.
13. Film (exposed or not), will be adversely affected by heat and moisture. The temperatures inside vehicles can easily exceed 100°F (40°C). Some protective lens coatings will melt and distort from the excess temperatures found inside closed vehicles. A common prac-

tice is to store film in a refrigerator. Be sure to rotate your stock of unused film to avoid the use of expired film.

14. The possible damage to underdeveloped film caused by airport security X-ray machines has been extensively debated. The Investigative Photography guide published by the Society of Fire Protection Engineers clearly states that all unexposed film is affected by these X-rays.[1] Each time the film goes through the machine, a light fog is deposited on the film. This can be significant for multiple passes. There is no easy solution, yet the degradation caused by a single pass through the machine is tolerable in almost all cases. X-rays affect higher speed films more than lower speed films.
15. Backlighting can cause major problems, especially when using an automatic or semiautomatic exposure control camera. Backlighting is the condition where the subject of interest (in the foreground) is in relative darkness caused by a brighter background. The camera will sense the bright background and thus produce a photograph in which the desired object in the foreground appears to be in a shadow. Examples of this occur often when shooting in an upward direction in an attempt to capture some detail on an overhead pipe rack. Older (manual) cameras and many of the newer (automatic and semiautomatic) models have a feature that can be activated to help this situation.
16. A common mistake is to expect the camera to do the thinking for the investigator.[1] Some investigators have used the approach of taking a general barrage of pictures in the hope that somewhere in the large pile will be a "gold nugget" with the key to the investigation. Each shot should have an intended purpose. Planned shots have better results than random shots do. Test plans (discussed above) and the analysis techniques in this guideline will help to guide the investigation team to take the appropriate photos.
17. Professional photographers anticipate the instantaneous temporary shadow created by the flash itself and use various diffusers and backgrounds. These devices are also helpful for close-up shots of items. Special flash units that fit around the lens of some 35-mm SLR cameras are available to eliminate these shadows.

1. Berrin, E. *Investigative Photography*. Report 83-1. Society of Fire Protection Engineers. Boston, MA: Society of Fire Protection Engineers, 1982.

Appendix D

Example Case Study—Fictitious NDF Company Incident

The following case study describes the investigation work process for a hypothetical occurrence using a logic tree based multiple root-cause systems approach. An example incident investigation report follows the work process description. The example is intended for instructive purposes only; descriptions of process equipment and conditions are not intended to reflect actual operating conditions.

The Work Process

At the NDF Company in Georgetown, South Carolina, a major fire occurred in the catalyst preparation area on August 1, 2001. The fire originated at Kettle No. 3 at 11:10 A.M. An explosion of catalyst storage tank No. 2 followed at 11:20 A.M. Final extinguishment of the fire was accomplished by the local fire department and plant fire brigade at 12:10 P.M. One fatality and five personnel injuries resulted from this event.

On resolution of this incident, the catalyst preparation area was secured against unauthorized entry, and plant management assembled for a meeting to discuss immediate actions. They decided to call in a corporate risk analyst to lead the team. With his help, by teleconference, management selected the following incident investigation team:

- Corporate Safety and Risk Analyst, Team leader
- Process Engineering Supervisor
- Safety Supervisor (trained and expert in the multiple-cause systems-oriented incident investigation methodology)
- Catalyst Production Supervisor
- Outside Operator
- Polyethylene Process Unit No. 1 Foreman
- Maintenance Foreman

- Representatives from OSHA, the local fire department, and the property insurance carrier were also invited to participate in the incident investigation.
- Corporate Legal Representative

The selected team initially established a specific plan of investigation procedures for this occurrence. This strategy session listed priorities and necessary actions to ensure that all required information was obtained in a prompt manner. Needless delays in evidence collection were avoided by the use of this plan which helped to accelerate the rebuilding/restarting of the catalyst preparation area.

The investigation team visited the scene of this incident before the physical evidence could be disturbed. The maintenance foreman was given the duty of taking photographs of the damaged area with a simple 35 mm camera. He was careful to obtain overall views of the scene and individual equipment and logged where each photo was taken. All team members were provided with a field investigative kit and appropriate safety protective gear. Important evidence was gathered, preserved, and identified using a written log and tagging system. A plot plan was posted and the location of each physical piece of evidence was noted on the plan along with the tag number.

On completion of this task, preparatory work was performed by the team members for preliminary witness interviews. Emphasis before the actual interviews was placed on the downplay of blame and the need for confidentiality. One team member, the safety supervisor, was chosen to meet with the witnesses. A conference room in the Administration Building was allocated for this project. The setting was arranged informally to allow the person involved to feel at ease. After considerable debate within the team, a conclusion was reached to not use a tape recorder during the witness interviews. The interview process was started early the evening of incident and was continued throughout the next two days. At the end of each day, the investigation team met to discuss the information obtained from the interviews and other activities.

The catalyst preparation area supervisor, on-duty control room operator for the catalyst operation, and maintenance superintendent were key sources of information. Their written records and logs were examined in detail. Other personnel that were interviewed included two outside operators, fire brigade members, and associated maintenance employees. During these conversations, special attention was paid to nonverbal signals. The interview process generated several unanswered questions about operational and maintenance procedures that required further study.

Second interviews, further evidence collection and examination, and thorough evaluations of operational and maintenance records were conducted to try to find explanations for the questions created by the preliminary witness interviews. Due to a high pressure alarm occurring at Kettle No. 3 in the catalyst preparation area prior to the fire, an analysis of the software and hardware for the control panel that oversees this process was deemed essential for this study.

In the incident investigation team's daily meetings, they began a chronicle of the key times and events preceding and during the incident, using flip charts and sticky notes for easy modification as new information became available.

The team conducted a series of fact finding and evidence analyzing meetings. During each of these meetings, specific action item assignments were made in order to further understand the events, systems functions, systems interrelationships, and failure modes. The team used the MCSOII or Multiple Cause Systems-Oriented Incident Investigation process for their investigation.

After the team completed the preliminary sequence of events, they began to develop logic trees to describe the events. As the top event, they choose the last injuries in time, those to the four fire brigade members. They asked, "Why did these injuries occur?" Two events are required and are sufficient: the explosion at the storage tanks AND the presence of the fire brigade members. The team added these events to the tree and continued to ask "Why?" until system-level root causes were determined. (All logic trees are included in the investigation report.) To reduce the complexity of the trees, the team chose to treat the operator fatality, the contractor injury and the injuries to the fire brigade as separate trees.

Several times, the team created fact-hypothesis matrices where needed to determine which branch of the tree contributed to the incident. One of the fact/hypothesis matrices is shown in Chapter 9.

On completion of the investigative work, the team convened to discuss its findings. During the discussions, important recommendations for corrective actions were developed. Special attention was allotted toward determining the potential effects of these suggested alterations on the efficiency of the plant operations. After long deliberations, responsibilities and desired completion dates were designated for each recommendation.

The team presented its findings to the plant management and to the corporate safety department orally and handed out lists of the causes, the trees, the recommendations, and the criteria for restart. Management accepted the oral report and appointed the operations manager to be responsible for seeing that the action points were completed.

Over the next two weeks, the investigation team compiled and published a detailed report. The team leader appointed one member to edit the report; the editor used the criterion that the report would be understandable to a new operations or engineering person two years in the future. It was assumed that the new person would be experienced in polyethylene technology and NDF culture, but would not have any knowledge of the incident.

The investigation team members were consulted frequently during the design and installation of the repairs for restart. Several team members participated in the recommended process hazards analysis and the pre-startup safety reviews.

To reduce risk in the industry as a whole, NDF endeavored to share the lessons learned from the incident with others in the same or similar industries.

In October 2001, the site manager gave an oral summary of the incident to the local manufacturers' association.

In December 2001, the NDF representative on the coproducers' safety committee informally discussed the causes and corrective actions with the other coproducers.

In July 2002, the safety supervisor presented an overview of the incident, causes, and corrective actions to a safety meeting sponsored by the regional chemical industry council.

In March 2003, the process engineering supervisor gave a paper on the incident at a Loss Prevention Symposium of the American Institute of Chemical Engineers. The paper was published later in *Plant Operations Progress.*

Incident Investigation Report

Executive Summary

A major fire and explosion occurred August 1, 2001 at the NDF Company polyethylene manufacturing facility in Georgetown, South Carolina, resulting in one fatality, four personnel injuries and extensive damage. The fire originated in the catalyst area when a vessel was overfilled and the exit piping ruptured releasing isopentane, a flammable material, and aluminum alkyl, a pyrophoric material.

The first fireball, at approximately 11:10 A.M., caused an operator fatality and a contractor injury. Emergency response was impaired because the fire water pumps were down. The fire spread to the catalyst storage tanks. A subsequent explosion of an adjacent catalyst storage tank resulted in the injury of four firefighters. Extinguishment of the fire was accomplished by the local fire department and plant fire brigade at 12:10 P.M.

The root causes of the incident relate to several process safety management areas:

- mechanical integrity,

- contractors,
- emergency planning and response,
- process hazards analysis, and
- management of change.

Background

In 1989, the NDF Company opened a facility in Georgetown, South Carolina to produce low density polyethylene. Manufacturing of the polyethylene is done in two 50-ton reactors that are encased individually within their own 8-story-high process unit. The main raw materials for the manufacturing operations include ethylene, hexane, and butene. The polymerization is completed in the presence of a catalyst. The base chemicals for the catalyst are aluminum alkyl and isopentane. The reactor and catalyst preparation areas are on a distributed control system (DCS). A simplified process flow diagram is attached.

In the catalyst preparation area where the fire occurred, aluminum alkyl and isopentane are mixed in a batch blending operation in three 8000-gallon kettles. The flow rates of components are regulated by an operator at the control room. Temperature, pressure, and liquid level within the kettles are monitored by the control room operator. The formulated catalyst is stored in four 12,000-gallon vertical storage tanks within this process unit. Aluminum alkyl is a pyrophoric material and isopentane is extremely flammable. Each vessel was insulated and equipped with a relief valve sized for external fire.

The isopentane for the catalyst preparation unit is stored as a liquid in a 60-ton horizontal (bullet) storage tank. The aluminum alkyls and other required chemicals for this process are received in small truck trailers and kept beneath a metal canopy.

The catalyst preparation area is positioned between the two polyethylene production units with 60 feet separating each one. The aluminum alkyls storage canopy and isopentane horizontal storage tank are located at a remote area at an approximate distance of 250 feet away from the production and utility areas. The isopentane is transported to the catalyst preparation area through a 3-inch pipeline. A remote actuated isolation valve on this supply line that fails closed is located at the isopentane storage tank. This control valve and an associated isopentane feed pump are managed by the operator in the control room.

The catalyst preparation area is protected by an automatic water-spray sprinkler system that is actuated by associated heat detectors. Fixed fire water monitors surround this process area. The water for these fire protection systems is supplied through 8-inch underground water mains by three (two diesel and one electric) horizontal, centrifugal, 2500 gpm rated, 125 psi automatic fire pumps that take suction from a 750,000 gallon above-ground storage tank. The electric fire pump's power source is from an independent electrical feed. The water supply for this facility was designed to meet the highest water demand within the facility when one fire pump is out of service.

Sequence of Events and Description of the Incident

On August 1, 2001 at 10:30 A.M., a control room operator remotely started the feeds to Kettle No. 3 in the catalyst preparation area. The normal procedure was to fill the kettle to approximately 80%, but Kettle No. 3 was apparently completely filled this time. The level indicator showed a high level, but the alarm did not sound. (The alarm was later found to be inhibited.) A high-pressure alarm for this vessel was acknowledged at 11:03 A.M. by the control room operator. At 11:00 A.M., a severe thunderstorm had started and within 5 minutes caused a power outage throughout the immediate vicinity. The ambient temperature was about 83°F and winds were from the northwest at about 3 mph.

With an available diesel emergency generator supplying power to critical pumps, the control room operators initiated shutdown procedures for the two reactor areas. An uninterruptible power supply (UPS) kept power to the DCS screens and instruments; however, the DCS system was designed close all catalyst preparation and reactor feed valves on loss of power. Outside operators were sent to manually block in reactor feeds.

At 11:09 A.M., a high-LEL detector in the catalyst preparation area sounded on the DCS. The lead outside operator was contacted by radio communications to investigate the problem. He said he was just leaving the Reactor No. 1 area and would go right to the catalyst preparation area. The thunderstorm had passed overhead and the rain was diminishing. At about 11:10 A.M., a "whooshing" noise (assumed to be the fireball) was heard by many and the heat detector for the automatic water-spray sprinkler coverage in this area alarmed in the control room. The lead outside operator did not respond when called on the radio.

The plant fire brigade and the local volunteer fire department were notified by the supervisor of the catalyst preparation area by 11:12 A.M. On their arrival to the scene of the fire at 11:15 A.M., the plant fire brigade saw the lead outside operator down about 40 feet from the fire, in between the catalyst preparation area and reactor building No. 1. They also found a seriously burned unknown person about 120 feet from the fire, near the finishing building. (This person was eventually determined to be a service contractor who entered the premises at 10:30 A.M. to calibrate equipment in the instrument house for Reactor No. 1.)

The fire engulfed most of the catalyst preparation area. Also, the automatic deluge sprinkler coverage for this area had actuated, but no water was available. The fire brigade tried to activate a fixed monitor, but again got no water flow. With the limited water supply from the plant fire engine available as a shield, the fire brigade members felt they could reach the lead outside operator.

The commander of the plant fire brigade sent a team member to the fire pump house. The electric fire pump was inoperable due to the power outage. One diesel fire pump was known to be impaired due to mechanical problems and other diesel fire pump had failed to start because its batteries were dead. Several maintenance personnel were sent immediately to repair this diesel fire pump.

Another explosion occurred at 11:20 A.M. as the fire spread to the formulated catalyst vertical storage tanks. Hot metal fragments from this blast severely injured four fire brigade members involved in the rescue attempt of the lead operator. They were about 60 feet away from the fire at the time of the explosion.

The local fire department arrived just after the explosion at 11:22 A.M. With the limited water supply on two of the fire trucks and the utilization of another fire truck to pump water directly from a nearby cooling water tower basin, the firemen were able to slow the fire spread. By 11:30 A.M., the Maintenance Department was able to transfer the set of batteries from the impaired diesel fire pump to the other diesel fire pump. On completion of this task, this diesel fire pump was started. The automatic deluge sprinkler protection was severely damaged by the fire/explosions and had to be valved into the off position. Three fixed monitors were turned onto full flow and directed at the fire. Also, the firemen and fire brigade used two hose streams off nearby fire hydrants for fire fighting purposes. At 11:58 A.M., the fire was under control. Final fire extinguishment was accomplished by 12:10 P.M.

The lead operator died the next day due to lung damage attributed to inhaling the hot gasses. Five other people were seriously injured. The catalyst preparation area received extensive property damage. The production operations at this facility are estimated to be suspended for 2 months until this area including associated pipelines can be rebuilt.

Cause Analysis

The team developed logic trees to describe the events. To reduce the complexity of the trees, the team chose to treat the operator fatality, the contractor injury and the injuries to the fire brigade as a separate tree. Since explosion at the catalyst storage tanks resulted from the spread of the fire from Kettle No. 3, the trees are interconnected. All the logic trees are attached and a key to the Trees and subtrees is shown below.

Tree	Tree Title
A	Operator Fatality in Kettle No. 3 Fire (Branch L is on Tree A)
B	Pool Fire at Kettle No. 3 (Branch C is on Tree B)
C	Kettle No. 3 Exit Piping Cracked
D	Kettle No. 3 System Pressure Reached 120 psig (Branch F is on Tree D)
E	Kettle No. 3 Exit Piping Failed at 120 psig (Below Design Pressure)
F	Power Failure Occurred
G	Contractor Injury in Kettle No. 3 Fire
H	Contractor Did Not He Should Leave Area
I	Four Fire Brigade Members Injured By Metal Fragments
J	Fire Brigade Members ~60 feet from Fire When Explosion Occurred
K	Extended Pool Fire Under Catalyst Storage Tanks
L	Operator Near Kettle No. 3 During Flash

Some events were personnel or equipment doing what they were supposed to be doing at that time or have a very high likelihood of occurring; these events are depicted as a "house" symbol. For some events at the bottom of the tree, the team did not have enough information in their possession to answer the "Why?" question. Such events are shown as a diamond, indicating a team decision to stop the tree at that point. For many of the diamonds, the team recommended further study or investigation by other groups.

The incident investigation team concluded that the fire occurred due to failure of the Kettle No. 3 exit piping in the catalyst preparation area. The failure released isopentane, a flammable material, and aluminum alkyl, a pyrophoric material, from the vessel. Moist air and water in the curb initiated an ignition of the contents. The atmospheric temperature was just above the flash point for isopentane, resulting in flashing vapor and some auto-refrigeration of the liquid. This resulted in a jet fire from the release point combined with a pool fire which spread throughout the dike and under the catalyst storage tanks. Because there was no fire water available, the fire could not be fought or the adjacent tanks cooled. The fire brigade was about 60 feet from the catalyst preparation area, attempting to rescue a victim, when an explosion occurred. The explosion occurred when catalyst storage tank No. 2 failed. Even without water to fight the fire, the storage tank failed more quickly than would be expected for a tank with insulation in good repair and a relief valve sized for the fire case.

The root causes of the incident relate to several process safety management areas:

- Mechanical integrity
- Contractors
- Emergency planning and response
- Process hazards analysis
- Management of change.

Findings and Recommendations

The root causes are:

The carbon steel piping in the catalyst preparation area and the isopentane feed lines to the area was weakened by external corrosion. The lines were Schedule 40, carbon steel lines which are suitable for this service. However, the lines were 12 years old. Physical evidence indicates that the failure most likely occurred at an elbow in the Kettle No. 3 exit piping. Pressure data from the system indicates the failure occurred when the system pressure was 120 psig, which is below the pressure rating for the vessel. Inspections of remaining parts of the catalyst mix and isopentane feed lines revealed deterioration of insulation and missing parts of the external shield (designed to prevent water from getting into the insulation). Corrosion under insulation especially in a heat affected zone is consistent with a failure in the kettle exit piping. *(Mechanical integrity)*

The existing mechanical integrity program did not seem to cover the catalyst preparation area. While records were found for inspections of the Reactor systems and the isopentane storage area, no inspection records were found for the catalyst preparation area. Interviews suggest these inspections were delayed by the 2000 budget crunch. *(Mechanical integrity)*

The No. 1 diesel fire water pump was down because it overheated during an outside agency annual performance test on July 15, 2001. The pump probably had problems prior to the test, but overheating may not be detected in monthly maintenance tests because the 5 minute run time may not sufficient to find the overheating. *(Mechanical integrity)*

The No. 2 diesel fire water pump was down because its batteries were dead. The dead batteries were detected and recharged in the 7/3/2001 monthly check, but they were not replaced or rechecked. *(Mechanical integrity)*

Interviews suggest that the fire water pumps had not been repaired due to a mechanical department perception that because of the budget crunch, the expensive repairs needed to be delayed until the first of the year. It is interesting to note that although several people knew that one fire water pump was impaired, no one person in the department knew that both pumps were impaired. In interviews, several upper management representatives stated that fire water pump repairs would be critical and would be completed immediately, so there is a mismatch between the employee and management perspectives on the severity of the budget constraints.

The catalyst storage tank failed earlier than expected if the fireproofing insulation was in good condition and the relief valve was adequate for the fire case. Witnesses indicate that several sections of the insulation had either fallen off or been removed from the tank 2–3 months ago. The insulation had not been repaired. *(Mechanical integrity)*

A check of the catalyst storage tank relief valve sizing calculations indicates the valve was large enough for the fire case assuming the tank had fireproofing insulation, but it was undersized for an un-insulated vessel. The original relief valve design calculations could not be found. The relief valve may also have been compromised by improper maintenance or pluggage. The last relief valve preventative maintenance and pop test occurred in 1996. No records were found for pre-1996 pop tests. *(Mechanical integrity)*

Although, the system failed below its design pressure, the overfilling of Kettle No. 3 caused a higher than normal pressure in the system. Also, the isopentane pump was deadheaded which could have resulted in the isopentane pump overheating, a potentially serious event by itself. There were several causes for the Kettle No. 3 system being filled completely:

- The control room operator did not stop filling Kettle No. 3 at the normal level of 85%. *(An operator error, but one that would be expected to occur over the normal life of a process)*
- The Kettle No. 3 high-level alarm was inhibited, so it did not annunciate or log to the DCS alarm journal. *(Management of change)*
- The operators inhibited the alarm because it was set below the normal batch level. The batch size had been changed from a 70% level to an 85% level and the alarm was still set for 80%. *(Management of change)*
- There was no redundant back-up protection (second level or monitoring of the pump) to shutdown the pump in case it was blocked in. *(Process hazards analysis)*
- *Note:* The isopentane feed valve is designed to fail closed on a power failure to prevent reverse flow from the kettles to the raw material storage tanks. This is the appropriate failure position for this valve.

No fire brigade member reported to the fire pump house when the fire alarm sounded. Interviews suggest personnel were confused about whose responsibility it was to go to the fire pump house. This may be training or drill issue. *(Emergency planning and response)*

The fire brigade approached the catalyst preparation area to attempt rescue of a victim while firewater was unavailable. The small amount of water available on the fire engine was enough to protect the rescuers from the radiant heat from the fire, but was no protection against metal fragments. While the fire brigade did not respect the potential hazards of this incident, further investigation is needed to determine if it is a training problem or if the emergency response plan is deficient in this area. *(Emergency planning and response)*

The presence of the contractor (working in the instrument house) was not known to unit personnel. The contractor works in the area routinely, sometimes in the instrument house and sometimes in the rack. Because the instrument house is a noncode area, a permit is not required for the routine equipment calibration. The contractor did not check in with the control-room if he was going to work in the instrument house, but did check in if a permit was required. Interviews with the other contractors who work this job confirm that this was the standard practice. None of the unit personnel had complained to the contractors about not checking in for the routine work in the instrument house in the 2 years the contractors had been doing this job. *(Contractors, permitting)*

There was no procedure to evacuate nonessential personnel from the area in the case of a high LEL alarm or a power failure. Because the contractors and some other workers in the area do not have radios to monitor unit communications, they would not know to leave the area unless the evacuation alarm was sounded. *(Emergency planning and response)*

The LEL detectors have frequent false high alarms which make unit personnel less responsive to the alarms going off. Further investigation is needed to determine the source of the LEL detector failures. *(Mechanical integrity)*

The procedure that states that the operator goes directly to the area of a high LEL alarm puts the operator in harms way. Production personnel need better data to decide on the appropriate alarm response, but the risk to personnel to get that data needs to be reduced. *(Emergency planning and response)*

The plant shutdown (which distracted the operators and made the electric fire water pump unavailable) was caused by a thunderstorm that tripped the primary feeder and no backup feeder was installed. The unit should be designed to shutdown safely on loss of power. *(Process hazards analysis)*

The process hazards analysis for the raw material storage, catalyst preparation, catalyst storage areas was up for renewal this year. The prior PHA was not as thorough as expected by today's standards. The corporation has now established criteria for PHA leaders and has an approved list of resources.

Recommendations

The team looked at the structure of the trees and the bottom events on the trees to develop the following list of recommendations and timing. They also assigned each action to the appropriate individual in the plant. The due dates are shown in parentheses following the action.

1. Replace all corroded isopentane, catalyst mix, or fire damaged lines and equipment. For carbon steel lines without an inspection history, pull all insulation before inspection. *(Before startup)*
2. Review the rest of the mechanical integrity program to ensure all critical equipment, piping, and pumps have acceptable integrity and an established inspection program with guidelines for repair. Include inspection and repair of fireproof insulation in the program. *(By March 2002)*
3. Improve documentation of relief valve inspection and pop tests. Annual testing of relief valves is recommended until a valve has a history of good pop tests. Then the frequency can be slowly extended. *(Program established by March 2002)*
4. Review sloping of the catalyst mix kettle and catalyst storage tank dikes. Although the design intent was to slope the dikes to the sump, the dike collects liquid in some areas. Consider separating the dike for the catalyst mix kettle and catalyst storage tanks. *(Before startup)*
5. Establish a weekly fire pump start and check program to be sure that this equipment works as intended. Revise the procedure to run the diesel pumps for a minimum of 30 minutes to detect overheating problems. Establish a preventive maintenance program to oversee all the maintenance on all the fire water pumps. Establish a high priority (Priority 1) for repairs on the fire equipment. *(Before startup)*
6. Establish criteria for finding the cause of dead batteries on the diesel fire water pumps and for checking that recharged batteries retain the charge. Establish criteria for replacing impaired batteries. *(Before startup)*
7. Clearly define a process for delaying maintenance and capital work in the event of budget constraints. Critical repair work should not be delayed. Environmental, Safety, and Health reviews should be conducted on delay of scheduled mechanical integrity program inspections. *(Before startup)*

8. Conduct a thorough Process Hazards Analysis (PHA) for the following areas: raw material storage, raw material feed systems, catalyst preparation, and catalyst storage. The PHA leader must be on the approved corporate list. Ensure the following scenarios are considered:
   - Loss of utilities including electrical power, steam, cooling tower water, instrument air, and nitrogen.
   - Unit must be able to safely shutdown on loss of any critical utility.
   - Deadheaded pumps, especially those pumps carrying liquids with a low flash point.
   - Leaks on flammable or toxic material systems. Give special consideration to whether the LEL detectors are correctly located and whether they offer complete coverage.
   - Response to high alarms on unit LEL detectors.
9. Use Level of Protection Analysis to evaluate the reliability needed for safety instrumented functions. *(Before startup for all serious consequences identified in the process hazards analysis)*
10. Review the emergency procedures for the allowable time to diagnose and act and the required response time for the system to recover after corrective action. Do a human reliability analysis on the actions, including the time for an operator to walk to the remote location. For critical actions (high consequence potential) with a required short time period for diagnosis and action, automatic interlocks should be installed. Consider a fault tree analysis to determine the reliability of the interlock designs. *(By November 2002)*
11. Reenforce the management of change procedure with all personnel. Ensure that project leaders confirm that all parts of the change (such as alarm set point changes) are finished before the project is closed out. Ensure that operators follow the management of change procedure for disabling (silences alarms but they continue to log) or inhibiting alarms (prevents alarms from logging).
12. Establish a strong preventative maintenance program for the unit LEL detectors. Develop a good record keeping system for the testing program to aide in the diagnosis of problem detectors. *(Program established within 60 days after startup)*
13. Establish clear procedures for contractor and other non-operating personnel entry and check-in to production units. No one should be out in the operating area without the unit personnel knowing they are there (includes maintenance workers, engineers, and other workers who routinely enter the area). *(Before start-up)*
14. Establish criteria to pull the evacuation horn. Drill in evacuation once per quarter on each shift (one drill per quarter must be on

days). Consider a public address system to communicate with visitors, maintenance workers, contractors and others who may not have radio communication. *(By November 2001)*

15. Develop drills and talk-throughs for emergency procedures. Set priorities for emergency actions and have the personnel memorize and drill the most important actions. *(By November 2001)*
16. Improve training and drill for the fire brigade members to ensure that someone reports to the fire pump house. *(Before startup)*
17. Improve the emergency response procedures, training, and drills, to help the fire brigade members respect the potential hazards of an incident and avoid unnecessary exposure, particularly when fire fighting capabilities are below par. *(Plans complete by November 2001)*

Attachments

- Simplified Process Flow Diagram
- Plot Plan
- Sequence of Events
- Logic Trees

Criteria for Restart

1. All recommendations required for restart (labeled *before startup* in the above list) must be completed. The rest of the above-listed recommendations should be completed by the indicated dates.
2. All changes introduced during repair and installation of the recommendations must go through aprocess hazards analysis (PHA).
3. A walk-through safety, health, and environmental review must be completed after construction and before introduction of chemicals to ensure that repairs and additions have been made as intended.
4. Startup must be authorized by the signatures of the Operations Manager, Maintenance Manager, and Safety Supervisor (all three signatures required).

Signatures

____
Team Leader Date

____
Safety Supervisor Date

____
Operations Manager Date

____
Plant Manager Date

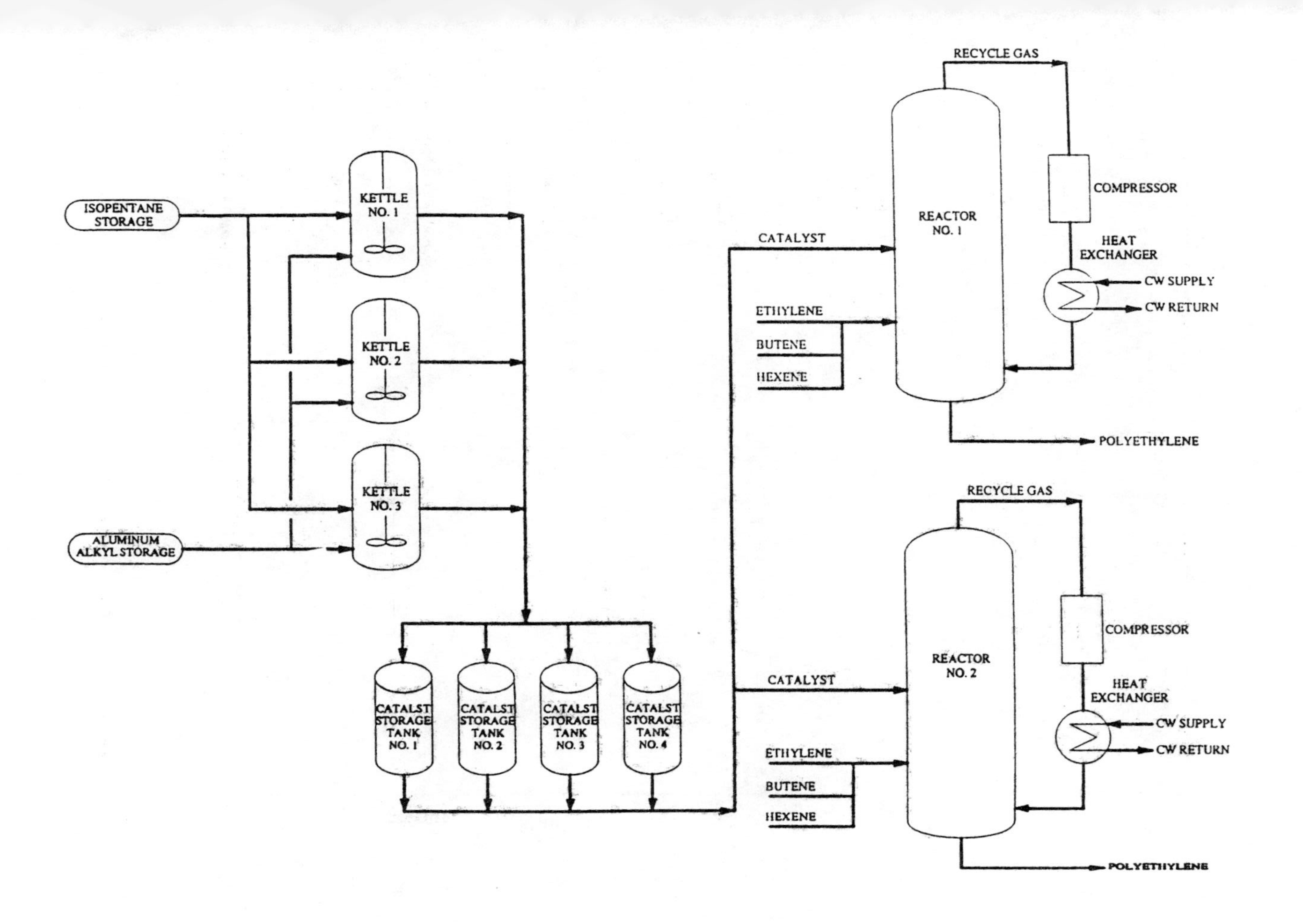
ISOPENTANE STORAGE
ALUMINUM ALKYL STORAGE
KETTLE NO. 1
KETTLE NO. 2
KETTLE NO. 3
CATALST STORAGE TANK NO. 1
CATALST STORAGE TANK NO. 2
CATALST STORAGE TANK NO. 3
CATALST STORAGE TANK NO. 4
RECYCLE GAS
REACTOR NO. 1
COMPRESSOR
HEAT EXCHANGER
CW SUPPLY
CW RETURN
CATALYST
ETHYLENE
BUTENE
HEXENE
POLYETHYLENE
RECYCLE GAS
REACTOR NO. 2
COMPRESSOR
HEAT EXCHANGER
CW SUPPLY
CW RETURN
CATALYST
ETHYLENE
BUTENE
HEXENE
POLYETHYLENE

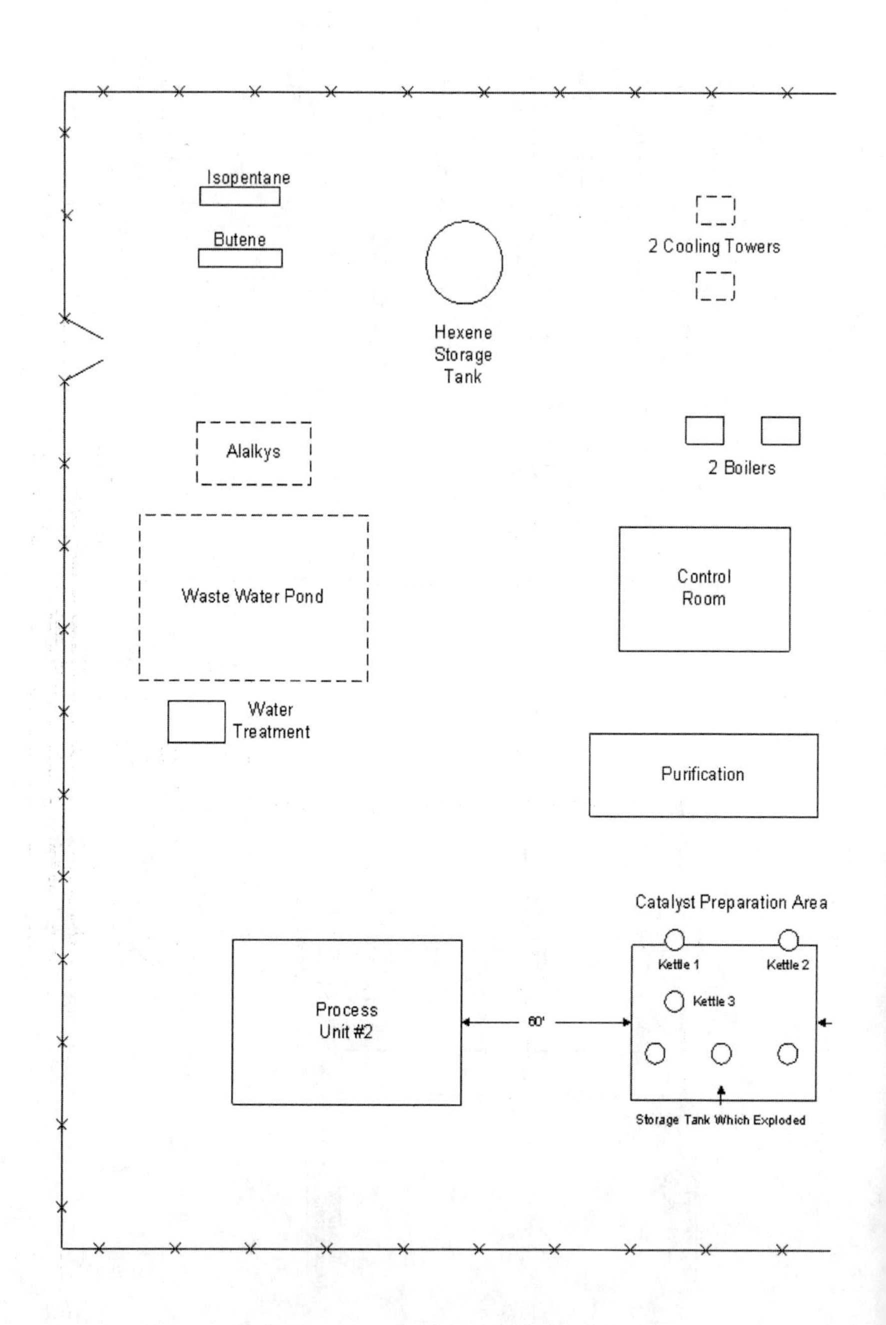
Isopentane
Butene
Hexene Storage Tank
2 Cooling Towers
Alalkys
2 Boilers
Waste Water Pond
Control Room
Water Treatment
Purification
Catalyst Preparation Area
Kettle 1
Kettle 2
Kettle 3
Process Unit #2
60'
Storage Tank Which Exploded

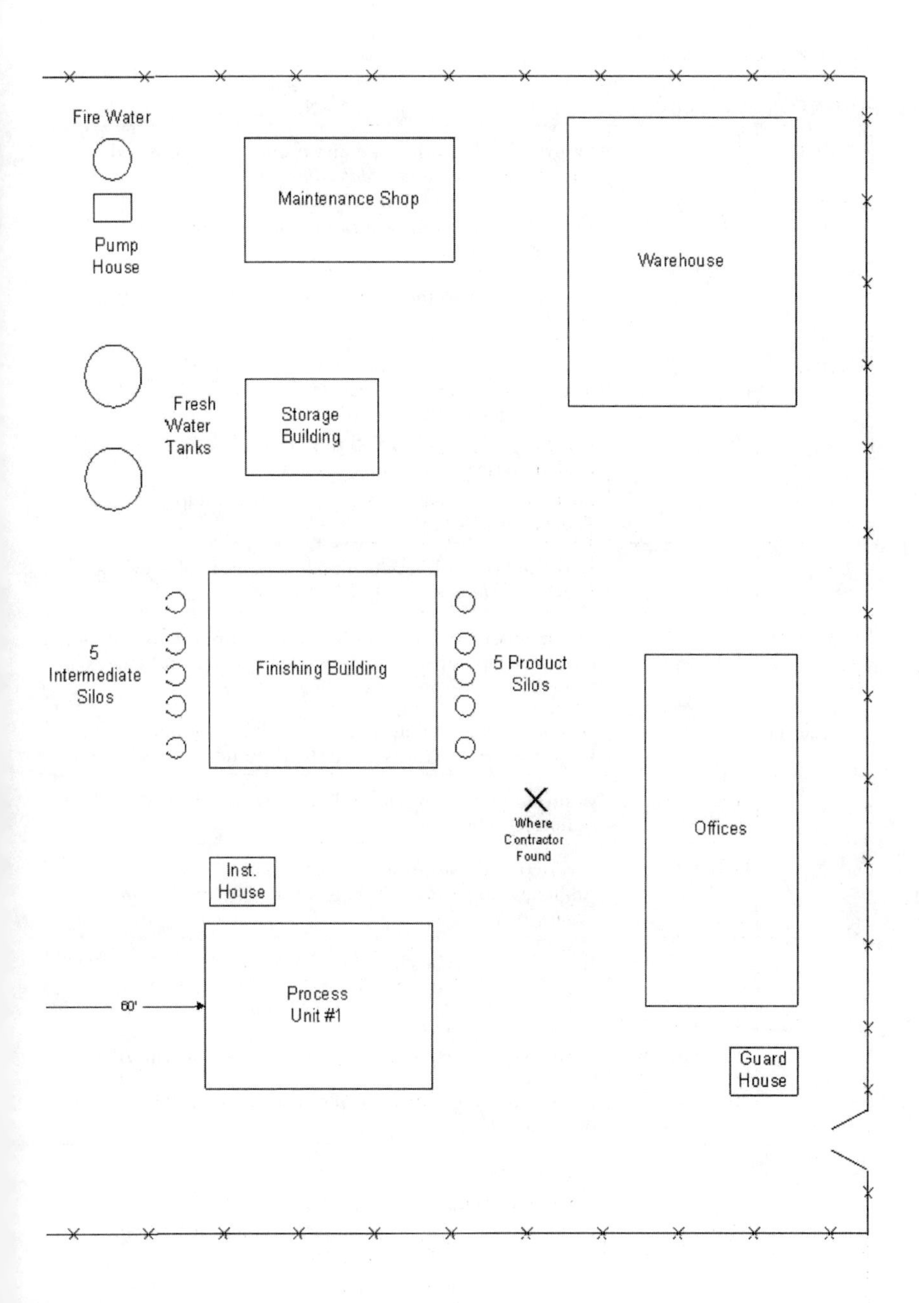
Fire Water
Pump House
Maintenance Shop
Warehouse
Fresh Water Tanks
Storage Building
5 Intermediate Silos
Finishing Building
5 Product Silos
Where Contractor Found
Offices
Inst. House
Process Unit #1
60'
Guard House

Sequence of Events

DATE	TIME	EVENT
7/1996		Last relief valve inspection and testing for Kettles and Catalyst Storage tanks. *(Maintenance records)*
2/2000		Corrosion control project proposed by maintenance superintendent.
3/2000		Last critical instrument check for No. 3 kettle *(Maintenance records)*
7/2001		No. 1 diesel fire pump taken out of service due to over heating during annual performance testing by outside agency. *(Maintenance records)* No. 2 diesel fire pump fails to start automatically due to weak batteries. *(Maintenance records)* Maintenance recharges No. 2 diesel fire pump batteries. *(Maintenance records)*
7/2001		Corrosion control work completed around Polyethylene Reactors. *(Maintenance records)*
7/3/2001		Last maintenance check of No. 2 diesel fire pump and the electric fire pumps. Test run of 5 minutes. (maintenance records)
8/1/2001	~ 10:30 A.M.	Service contractor enters area to calibrate equipment in the Polyethylene Reactor No. 1 instrument house. *(Interview)*
	10:30:33 A.M.	Control operator initiates filling of Kettle No. 3 (started remotely). *(DCS)*
	~11:00 A.M.	Severe thunderstorm starts. *(Interviews)* Ambient temperature 85°F, NW winds @ 3mph *(Plant weather station log)*
	11:00:47 A.M.	Kettle No. 3 reaches high 90% level *(DCS)* *Note:* High level alarm did not register in the DCS log. Later, the alarm is found to be inhibited.
	11:03:15 A.M.	Kettle No. 3 reaches 120 psig high pressure alarm *(DCS)*.
	11:03:45 A.M.	Kettle No. 3 high pressure alarm acknowledged by control operator *(DCS)*.
	11:05:03 A.M.	Plant wide electrical power outage. Isopentane supply trips off: no power. *(DCS)* Main control valve for isopentane storage tank fails closed (as designed) on electrical failure. *(DCS)* 120 psig pressure trapped in Kettle No. 3 and related piping *(Concluded from data)*. Outside operator goes to manually block in reactor feeds (*Part of emergency shutdown procedure)*.

DATE	TIME	EVENT
8/1/2001	~11:09 A.M.	Kettle No. 3 discharge piping cracks. Contents of kettle start dumping to the curb. Isopentane vapors spread as material flashes. *(Concluded from data).*
	11:09:30 A.M.	LEL detectors in Catalyst Prep area alarm. *(DCS)*
	After 11:09:30 A.M.	Control room operator requests by radio for the lead outside operator to visually inspect Kettle No. 3 due to high LEL alarm. Thunderstorm has passed and rain is diminishing.
	~11:10 A.M.	Whooshing" noise heard by many. *(Assumed to be fireball)* Contractor is just coming out of instrumentation house when he sees operator running towards catalyst prep area. Contactor sees fire flash throughout catalyst prep area. He remembers trying to get away from the heat.
	11:10:21 A.M.	Heat detector alarms for catalyst preparation area (Kettle No. 3 area) annunciate in control room. *(DCS)*
	After heat detector alarms	Control room operator tries to reach lead outside operator by radio, but there is no response. (Believe lead outside operator is critically burned.)
	~11:11 A.M.	Catalyst preparation supervisor activates plant fire brigade using plant fire alarm and notifies plant dispatch by radio *(Plant dispatch log)*
	~11:12 A.M.	Catalyst preparation supervisor notifies local volunteer fire department by telephone. *(Local fire department log)*
	~11:15 A.M.	Plant fire brigade reaches the emergency location. They: • see fire engulfing catalyst prep area (automatic deluge sprinkler had actuated, but no water was available) • see the lead outside operator down about 40 feet from the catalyst prep area, • find the injured (unidentified at that time) service contractor about 120 feet away, Plant fire brigade then: • tries to activate a fixed monitor, but no water flows, • sends one brigade member to fire pump house to check pump status.
	~11:18 A.M.	Fire brigade member reaches pump house and finds: • electric fire pump inoperable due to power failure, • one diesel fire pump inoperable due to known mechanical problems, • second diesel would not start due to dead batteries; calls for maintenance help. Several maintenance employees dispatched to repair diesel fire pump No. 2.

DATE	TIME	EVENT
8/1/2001	~ 11:20 A.M.	Fire brigade uses limited water supply on engine to shield two members of team and attempts rescue of lead operator. Another explosion occurs and four fire brigade members are injured by metal fragments.
	~11:22 A.M.	Local fire department arrives.
	After 11:22 A.M.	Spread of fire is slowed using water from fire department trucks.
	~11:30 A.M.	Maintenance completes move of batteries from No. 1 diesel fire pump to No. 2 diesel fire pump. No. 2 diesel fire pump is started.
	After 11:30 A.M.	Automatic deluge sprinkler system found to be severely damaged by fire/explosions and is now valved into OFF position. Three fixed fire monitors directed on fire at full flow. Two hose streams from hydrants directed on fire also.
	~11:58 A.M.	Fire deemed under control.
	~12:10 A.M.	Final extinguishment of fire.
8/2/2001		Lead operator dies from burn complications.
8/15/2001		No. 1 diesel fire pump repaired.

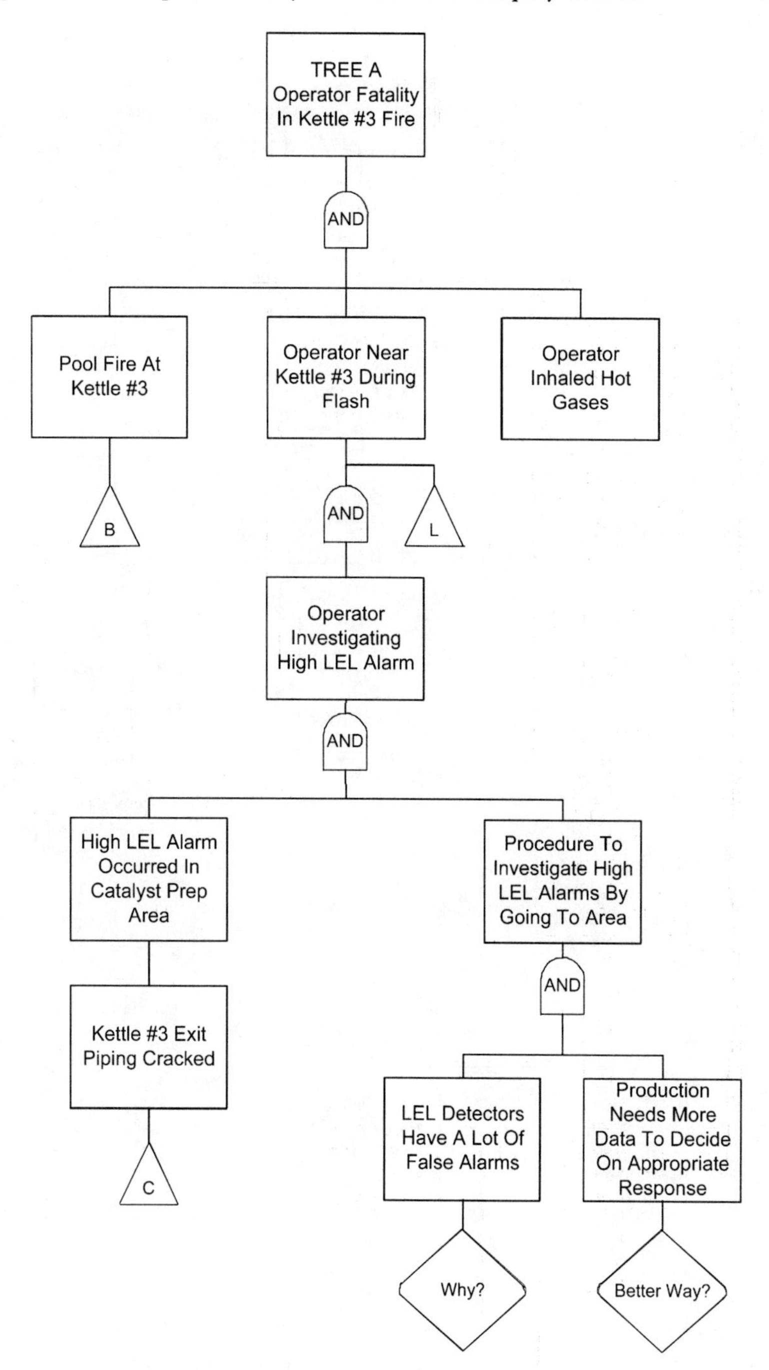
TREE A
Operator Fatality
In Kettle #3 Fire
AND
Pool Fire At
Kettle #3
Operator Near
Kettle #3 During
Flash
Operator
Inhaled Hot
Gases
B
AND
L
Operator
Investigating
High LEL Alarm
AND
High LEL Alarm
Occurred In
Catalyst Prep
Area
Procedure To
Investigate High
LEL Alarms By
Going To Area
AND
Kettle #3 Exit
Piping Cracked
C
LEL Detectors
Have A Lot Of
False Alarms
Production
Needs More
Data To Decide
On Appropriate
Response
Why?
Better Way?

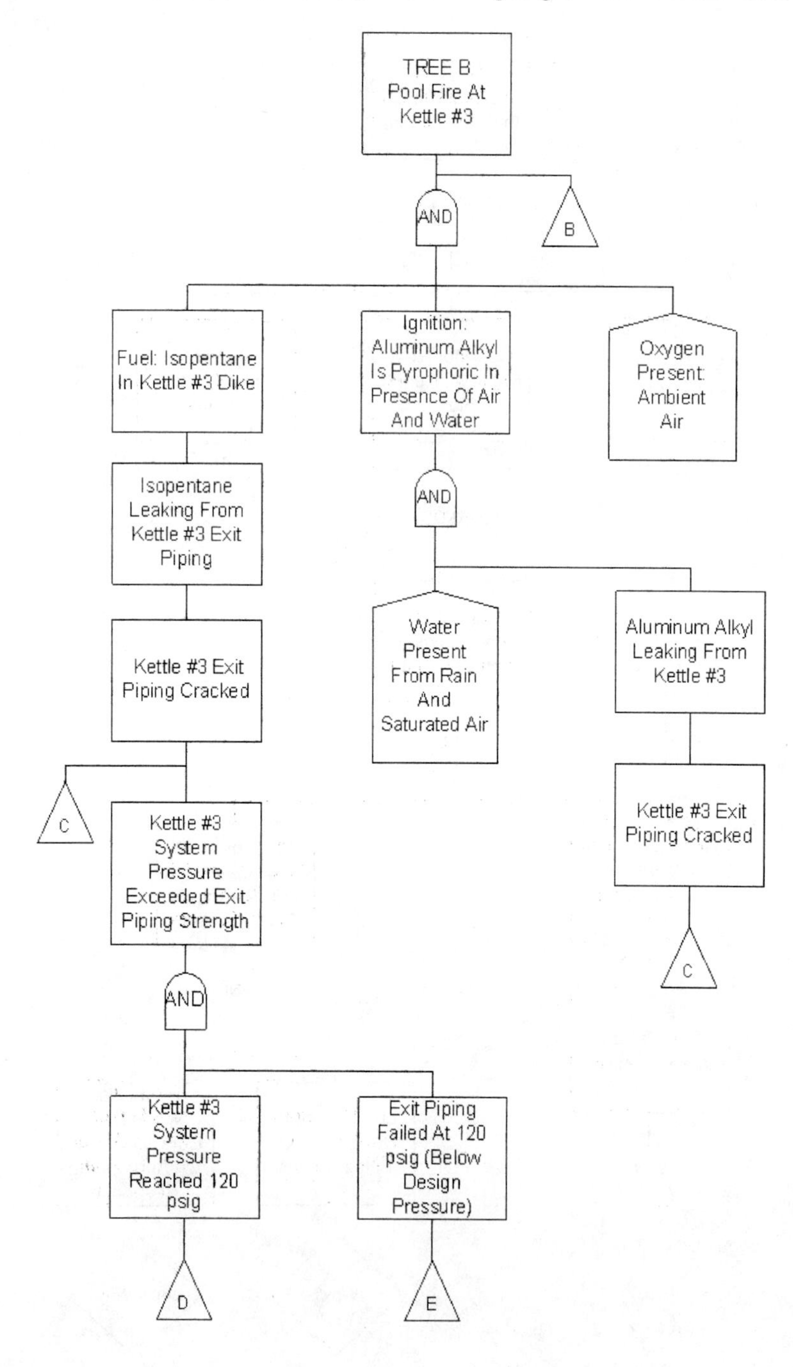
TREE B
Pool Fire At
Kettle #3
AND
B
Fuel: Isopentane
In Kettle #3 Dike
Ignition:
Aluminum Alkyl
Is Pyrophoric In
Presence Of Air
And Water
Oxygen
Present:
Ambient
Air
Isopentane
Leaking From
Kettle #3 Exit
Piping
AND
Kettle #3 Exit
Piping Cracked
Water
Present
From Rain
And
Saturated Air
Aluminum Alkyl
Leaking From
Kettle #3
C
Kettle #3
System
Pressure
Exceeded Exit
Piping Strength
Kettle #3 Exit
Piping Cracked
C
AND
Kettle #3
System
Pressure
Reached 120
psig
Exit Piping
Failed At 120
psig (Below
Design
Pressure)
D
E

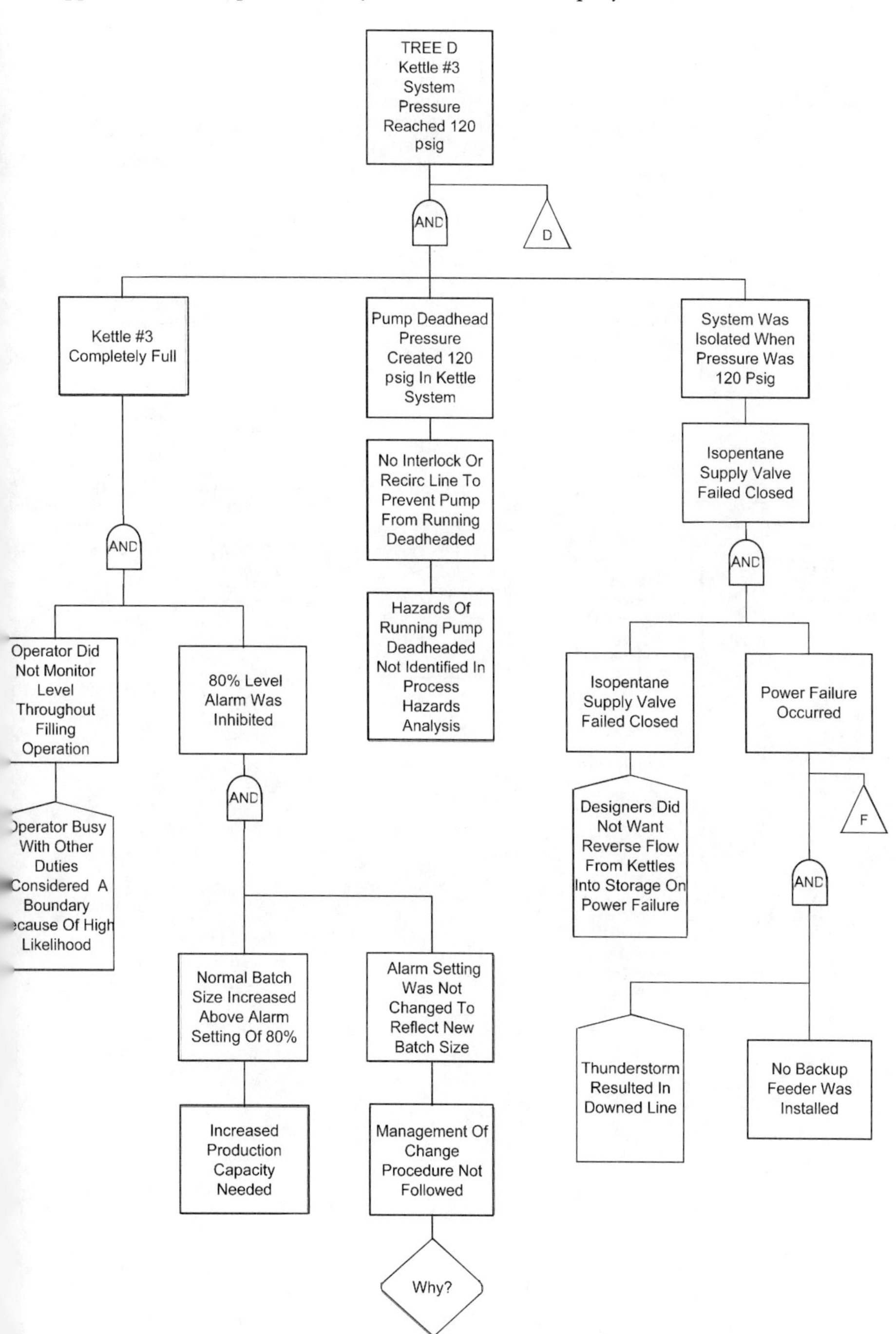

TREE D Kettle #3 System Pressure Reached 120 psig
AND
D
Kettle #3 Completely Full
Pump Deadhead Pressure Created 120 psig In Kettle System
System Was Isolated When Pressure Was 120 Psig
No Interlock Or Recirc Line To Prevent Pump From Running Deadheaded
Hazards Of Running Pump Deadheaded Not Identified In Process Hazards Analysis
Isopentane Supply Valve Failed Closed
AND
AND
Operator Did Not Monitor Level Throughout Filling Operation
80% Level Alarm Was Inhibited
Isopentane Supply Valve Failed Closed
Power Failure Occurred
AND
F
Designers Did Not Want Reverse Flow From Kettles Into Storage On Power Failure
AND
Normal Batch Size Increased Above Alarm Setting Of 80%
Alarm Setting Was Not Changed To Reflect New Batch Size
Thunderstorm Resulted In Downed Line
No Backup Feeder Was Installed
Increased Production Capacity Needed
Management Of Change Procedure Not Followed
Why?

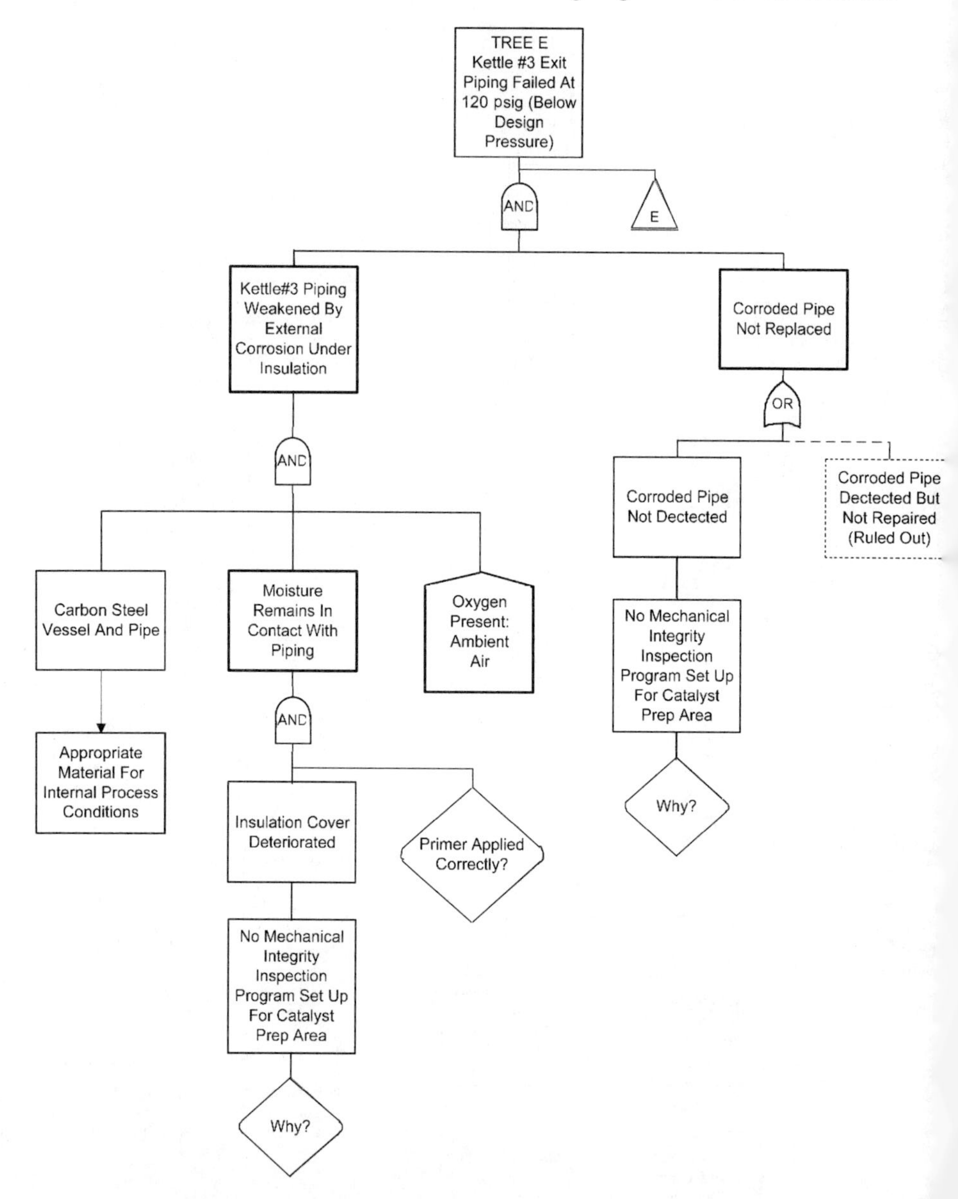
TREE E
Kettle #3 Exit Piping Failed At 120 psig (Below Design Pressure)
AND
E
Kettle#3 Piping Weakened By External Corrosion Under Insulation
Corroded Pipe Not Replaced
OR
AND
Corroded Pipe Not Dectected
Corroded Pipe Dectected But Not Repaired (Ruled Out)
Carbon Steel Vessel And Pipe
Moisture Remains In Contact With Piping
Oxygen Present: Ambient Air
No Mechanical Integrity Inspection Program Set Up For Catalyst Prep Area
AND
Appropriate Material For Internal Process Conditions
Insulation Cover Deteriorated
Primer Applied Correctly?
Why?
No Mechanical Integrity Inspection Program Set Up For Catalyst Prep Area
Why?

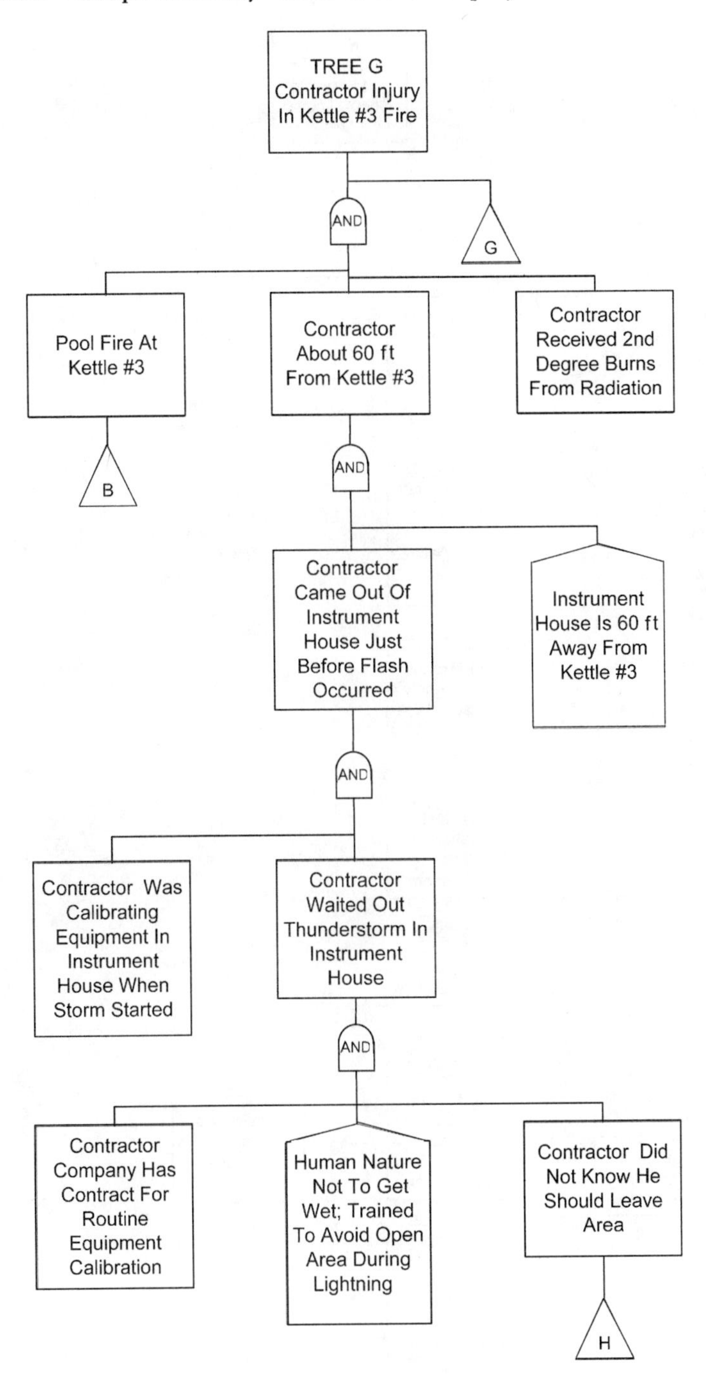
TREE G
Contractor Injury In Kettle #3 Fire
AND
G
Pool Fire At Kettle #3
B
Contractor About 60 ft From Kettle #3
Contractor Received 2nd Degree Burns From Radiation
AND
Contractor Came Out Of Instrument House Just Before Flash Occurred
Instrument House Is 60 ft Away From Kettle #3
AND
Contractor Was Calibrating Equipment In Instrument House When Storm Started
Contractor Waited Out Thunderstorm In Instrument House
AND
Contractor Company Has Contract For Routine Equipment Calibration
Human Nature Not To Get Wet; Trained To Avoid Open Area During Lightning
Contractor Did Not Know He Should Leave Area
H

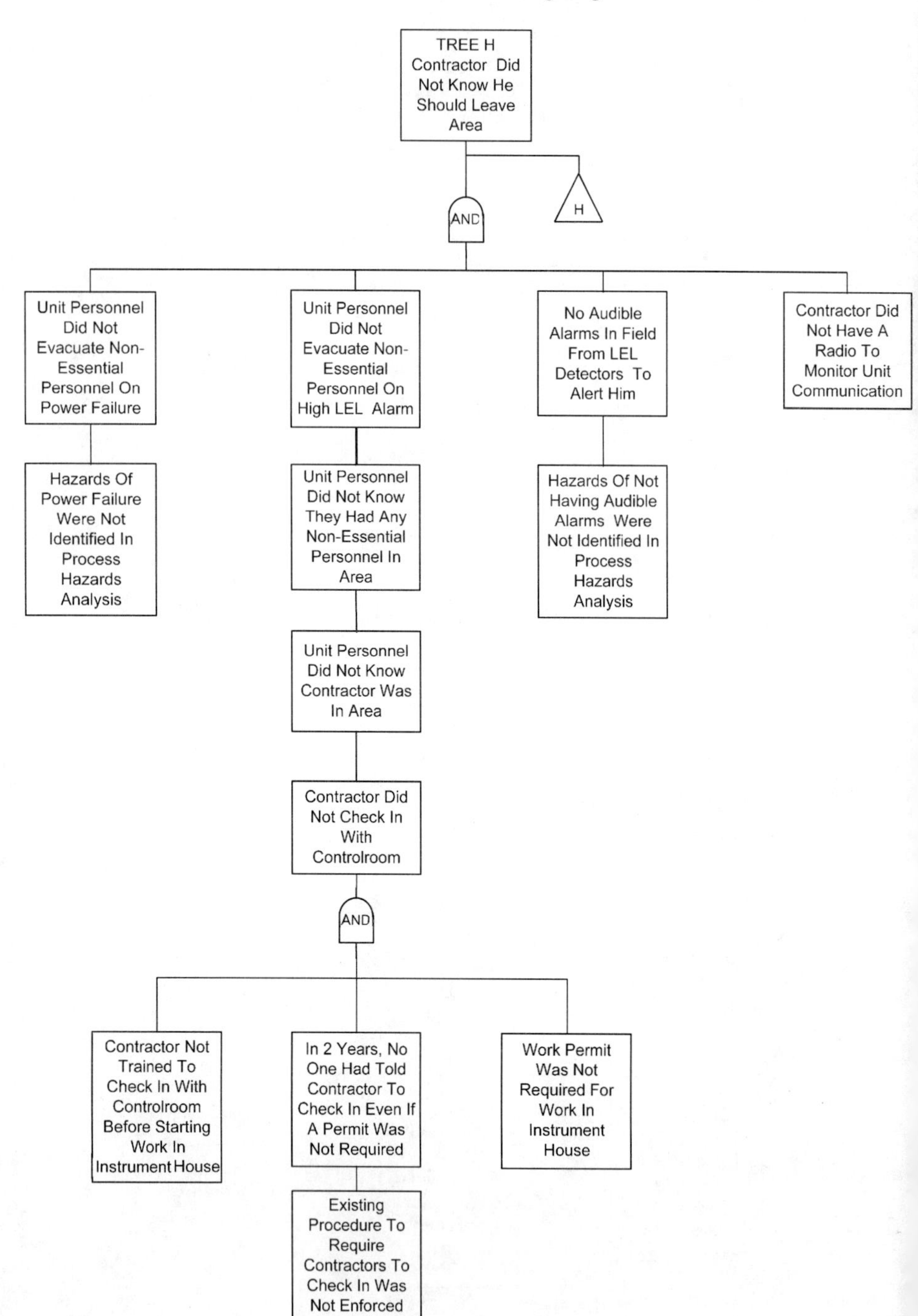
TREE H
Contractor Did Not Know He Should Leave Area
AND
H
Unit Personnel Did Not Evacuate Non-Essential Personnel On Power Failure
Unit Personnel Did Not Evacuate Non-Essential Personnel On High LEL Alarm
No Audible Alarms In Field From LEL Detectors To Alert Him
Contractor Did Not Have A Radio To Monitor Unit Communication
Hazards Of Power Failure Were Not Identified In Process Hazards Analysis
Unit Personnel Did Not Know They Had Any Non-Essential Personnel In Area
Hazards Of Not Having Audible Alarms Were Not Identified In Process Hazards Analysis
Unit Personnel Did Not Know Contractor Was In Area
Contractor Did Not Check In With Controlroom
AND
Contractor Not Trained To Check In With Controlroom Before Starting Work In Instrument House
In 2 Years, No One Had Told Contractor To Check In Even If A Permit Was Not Required
Work Permit Was Not Required For Work In Instrument House
Existing Procedure To Require Contractors To Check In Was Not Enforced

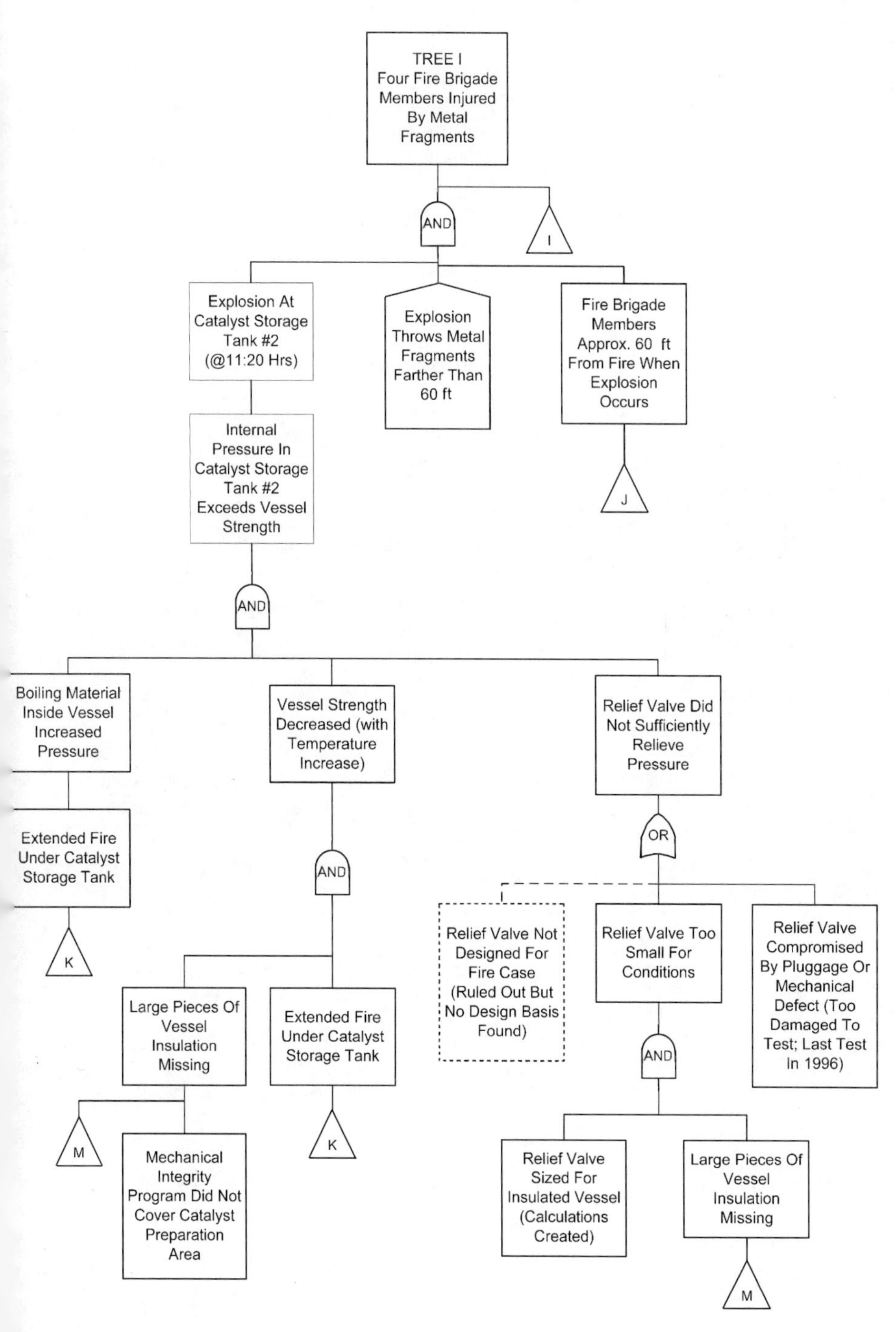
TREE I
Four Fire Brigade Members Injured By Metal Fragments
AND
I
Explosion At Catalyst Storage Tank #2 (@11:20 Hrs)
Explosion Throws Metal Fragments Farther Than 60 ft
Fire Brigade Members Approx. 60 ft From Fire When Explosion Occurs
Internal Pressure In Catalyst Storage Tank #2 Exceeds Vessel Strength
J
AND
Boiling Material Inside Vessel Increased Pressure
Vessel Strength Decreased (with Temperature Increase)
Relief Valve Did Not Sufficiently Relieve Pressure
Extended Fire Under Catalyst Storage Tank
OR
AND
K
Relief Valve Not Designed For Fire Case (Ruled Out But No Design Basis Found)
Relief Valve Too Small For Conditions
Relief Valve Compromised By Pluggage Or Mechanical Defect (Too Damaged To Test; Last Test In 1996)
Large Pieces Of Vessel Insulation Missing
Extended Fire Under Catalyst Storage Tank
AND
M
Mechanical Integrity Program Did Not Cover Catalyst Preparation Area
K
Relief Valve Sized For Insulated Vessel (Calculations Created)
Large Pieces Of Vessel Insulation Missing
M

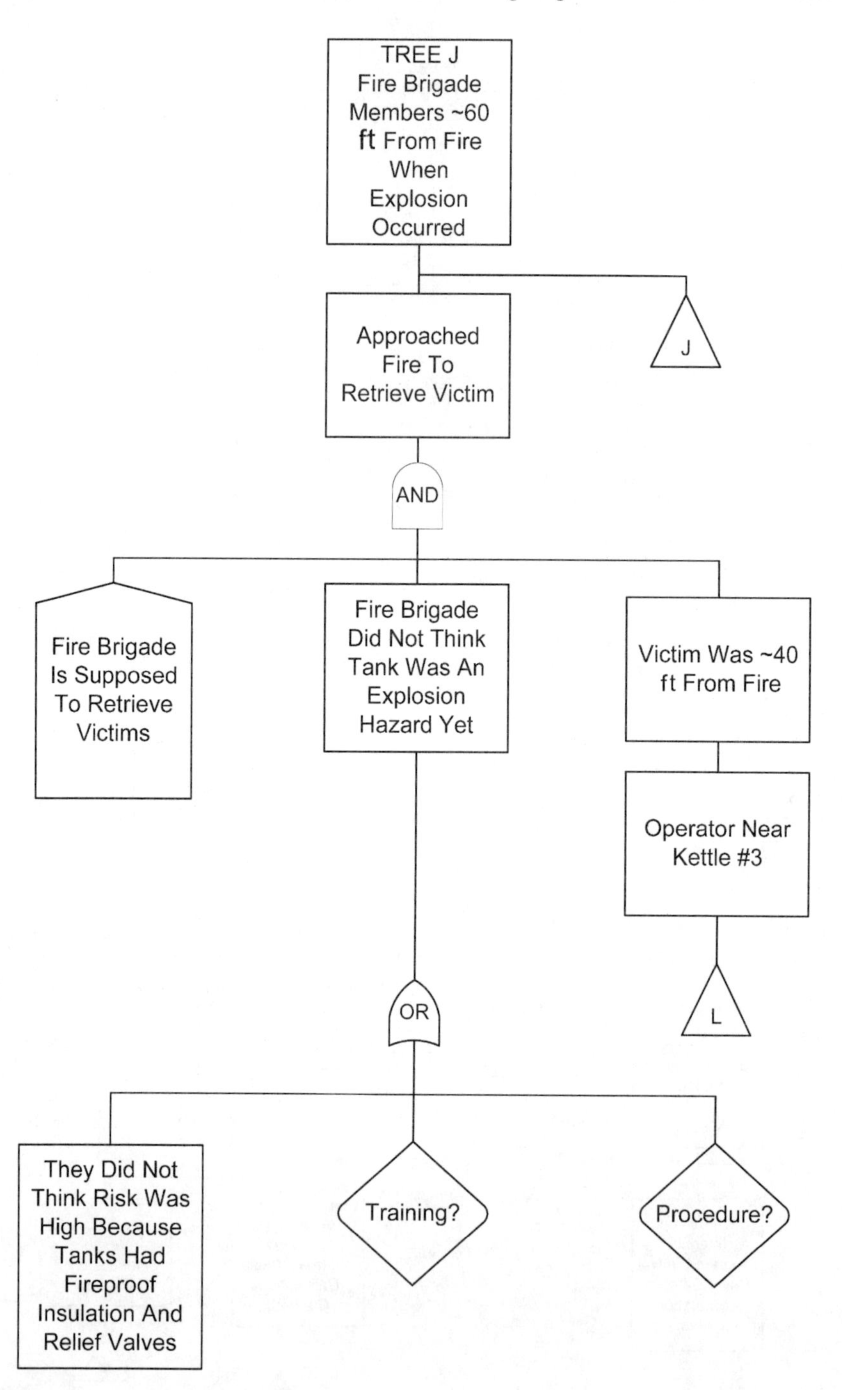
TREE J
Fire Brigade Members ~60 ft From Fire When Explosion Occurred
Approached Fire To Retrieve Victim
J
AND
Fire Brigade Is Supposed To Retrieve Victims
Fire Brigade Did Not Think Tank Was An Explosion Hazard Yet
Victim Was ~40 ft From Fire
Operator Near Kettle #3
L
OR
They Did Not Think Risk Was High Because Tanks Had Fireproof Insulation And Relief Valves
Training?
Procedure?

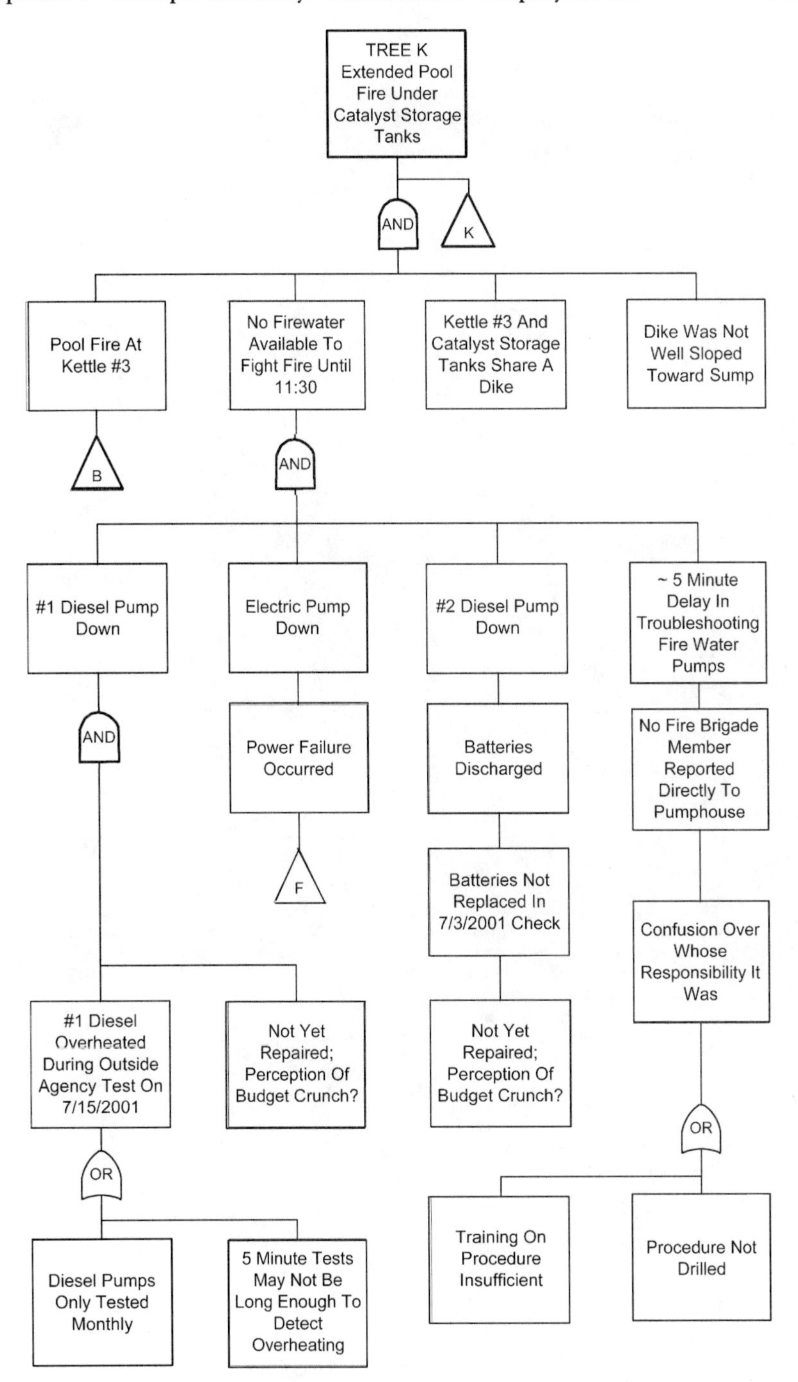
TREE K
Extended Pool Fire Under Catalyst Storage Tanks
AND
K
Pool Fire At Kettle #3
B
No Firewater Available To Fight Fire Until 11:30
AND
Kettle #3 And Catalyst Storage Tanks Share A Dike
Dike Was Not Well Sloped Toward Sump
#1 Diesel Pump Down
AND
#1 Diesel Overheated During Outside Agency Test On 7/15/2001
OR
Diesel Pumps Only Tested Monthly
5 Minute Tests May Not Be Long Enough To Detect Overheating
Not Yet Repaired; Perception Of Budget Crunch?
Electric Pump Down
Power Failure Occurred
F
#2 Diesel Pump Down
Batteries Discharged
Batteries Not Replaced In 7/3/2001 Check
Not Yet Repaired; Perception Of Budget Crunch?
~ 5 Minute Delay In Troubleshooting Fire Water Pumps
No Fire Brigade Member Reported Directly To Pumphouse
Confusion Over Whose Responsibility It Was
OR
Training On Procedure Insufficient
Procedure Not Drilled

Appendix E

Example Case Study—More Bang for the Buck: Getting the Most from Accident Investigations

S. E. Anderson and R. W. Skloss
Rohm and Haas Texas Incorporated
Deer Park, Texas

Presented at the
25th Annual Loss Prevention Symposium, August 18-22, 1991
Session: Case Histories

ABSTRACT
An incident investigation system based on a combination of fault tree analysis logic and the Deming principles of systems and quality was developed during a three-year period. The system was given an excellent test in the investigation of the explosion of a tank car filled with methacrylic acid. In this case, the event tree constructed during the investigation was easily converted into a fault tree. The fault tree so constructed was then used to evaluate the effectiveness of proposed corrective changes to the production, loading, and analytical systems before the changes were actually implemented. By making use of these tools to guide the inclusion of feedback loops (AND-gates in fault tree terminology), the intrinsic safety of the methacrylic acid production and shipping system was greatly enhanced at a relatively low cost.

Introduction

Obtaining the best results from supervisors' investigations of incidents that occur in the plant environment has been a goal of many safety professionals for many years, and the goal has often been quite elusive. In 1983, Plant Management became convinced of the merits of the accident causation theory of Dan Petersen[1] and decided to create a task group to attempt once more to develop an accident investigation technique that would achieve the following improvements:

- Improve quality of investigations
  - Force the investigator to go beneath the surface to the underlying causes
  - Foster attempts to find as many of the causes as practicable
  - Improve documentation of the investigations
  - Increase utility for training and information sharing
- Increase uniformity of investigations
- Improve utility of recommended corrective actions

Toward this end, the task group (called the Accident Investigation Committee, or AIC) was put together from a group of carefully chosen individuals to bring together the desired mix of skills and expertise. The chosen individuals contributed a broad range of experiences:

1. Safety professionals
2. A unit manager
3. A day foreman (second line manager)
4. An experienced process engineer skilled in hazard analysis

During about three years of development and testing, the AIC tried and screened a number of different approaches. The one that met most of the performance requirements was based on the marriage of fault tree analysis technology and Deming's concepts of quality and systems; we called it the Multiple-Cause, Systems-Oriented Incident Investigation (MCSOII) technique. From fault tree analysis we obtained a semirigorous technique that forces the participants to look more deeply into the layers of accident causation than they otherwise would, and from Deming came the idea that a failure in the plant "system" can produce a failure in "safety," and that all such failures are amenable to the same kinds of systems analysis and correction.

As a result of the AIC's efforts, we now have a process for investigating accidents in which we construct an event tree for each incident. The tree is quite similar to a fault tree from the quantitative risk analysis discipline, except that in the investigations we often sacrifice some structural rigor to get the most results in a reasonable time. Basically, the process uses a team to reconstruct the chronology of the incident and to construct the event tree. We try to include those who are most familiar with what actually happened, including the injured person(s) if any. We use the same basic method to investigate process failures, spills, injuries, or any other system failures. Emphasizing the system aspects of the failure removes much of the confrontational aspects of such proceedings, and facilitates achieving comprehensive results. The event trees, which are obtained, contain much information, communicate it in an easily understood form, are very useful in training, and can easily be converted to fault trees if desired.

After the investigation process described above had been in use for several years, an incident occurred that afforded a unique opportunity to use the system to its fullest and demonstrate the great potential of the process.

The Incident

Chronology

At 4:30 A.M. on July 21, 1988, a plant protection officer making rounds saw and heard vapors emitting from the relief valve on tank car UTLX 647014. This report indicated that the contents of the car, technical methacrylic acid (TMAA) were reacting, and that we had a serious situation. The car was in a marshalling yard awaiting transfer to a terminal. Cars filled with hazardous materials, which were near the reacting car, were removed, and empty cars were moved into position on its south side and west end. Remote fire monitors were placed into position on the north side and directed at the relief valve and the dome in an effort to control vapor emissions during the remainder of the reaction. The east end of the car could not be reached. Fortunately, the car at that end was empty. Personnel were kept away from the car as much as possible from the time vapor emissions were noticed. At about 12:25 A.M. on July 22 (about 20 hours after the problem became known), the car ruptured.

Effects

The forces released when the tank car skin ruptured were quite impressive. Parts of the car were found 250–300 yards (228–274 meters) away from the skin. The wheels of the car were driven into the ground about 2 feet (61 cm) while the wheels were still on the rails. The skin of the car was essentially flattened, with the impressions of the two rails clearly seen through the skin. The sound of the explosion was heard about 10 miles away. The overhead 138,000-volt electrical lines were severed, and the arcs resulted in two small grass fires. Plants connected to these electrical lines were shut down. There was no fire from the polymerization. In addition to the car that was destroyed, nearby cars were also damaged. Foam-like polymer covered an area about 200 by 50 yards (182 by 46 meters). Because of the precautions we had taken, there were no injuries.

The Investigation

On-the-spot photographs and videotapes were taken as soon as light permitted. Samples of polymer from different areas of the car were also collected. Records of loading and lab results for production and loading were also assembled. File samples were retrieved for testing. A 13-person team was formed to investigate the incident. The team was made up of persons from the AIC as well as from each group involved in producing, loading, analyzing, and shipping TMAA.

The Facts and Findings

- The car had been loaded on July 11, 1988, and the temperature had been in the 90 to 98°F range (30–33°C.) between loading and the explosion.
- The car was one of six that were being accumulated for combination in a ship's deep tank for subsequent export.
- These cars were stainless steel. This was the first time TMAA was loaded in cars that were not internally coated.
- File samples from the destroyed car indicated low levels of inhibitor (at the process rundown levels, not shipping levels). Other cars were sampled and found to be normal.
- Some of the polymer was found to contain about 300 ppm iron.
- The polymer was dehydrated to a significant degree. There was also almost no odor of MAA in the area covered with the polymer, indicating virtually complete conversion.

The Event Tree

The team developed the event tree from the facts in hand. It was clear that the car had exploded because of a polymerization. The first layer of significant causation we reached for the polymerization was as follows (see also Figure 1):

- The MAA in the car had ineffective inhibitor. Either the inhibitor had not been added or it was ineffective.
- The car had been held at moderately-high temperature for several days.
- The polymer showed evidence of iron contamination at levels high enough to promote polymerization. (Iron is a known promoter of polymerization.)

Each of these causes was followed to an end point by repeatedly asking "why" something happened. Branches are ended when no more questions

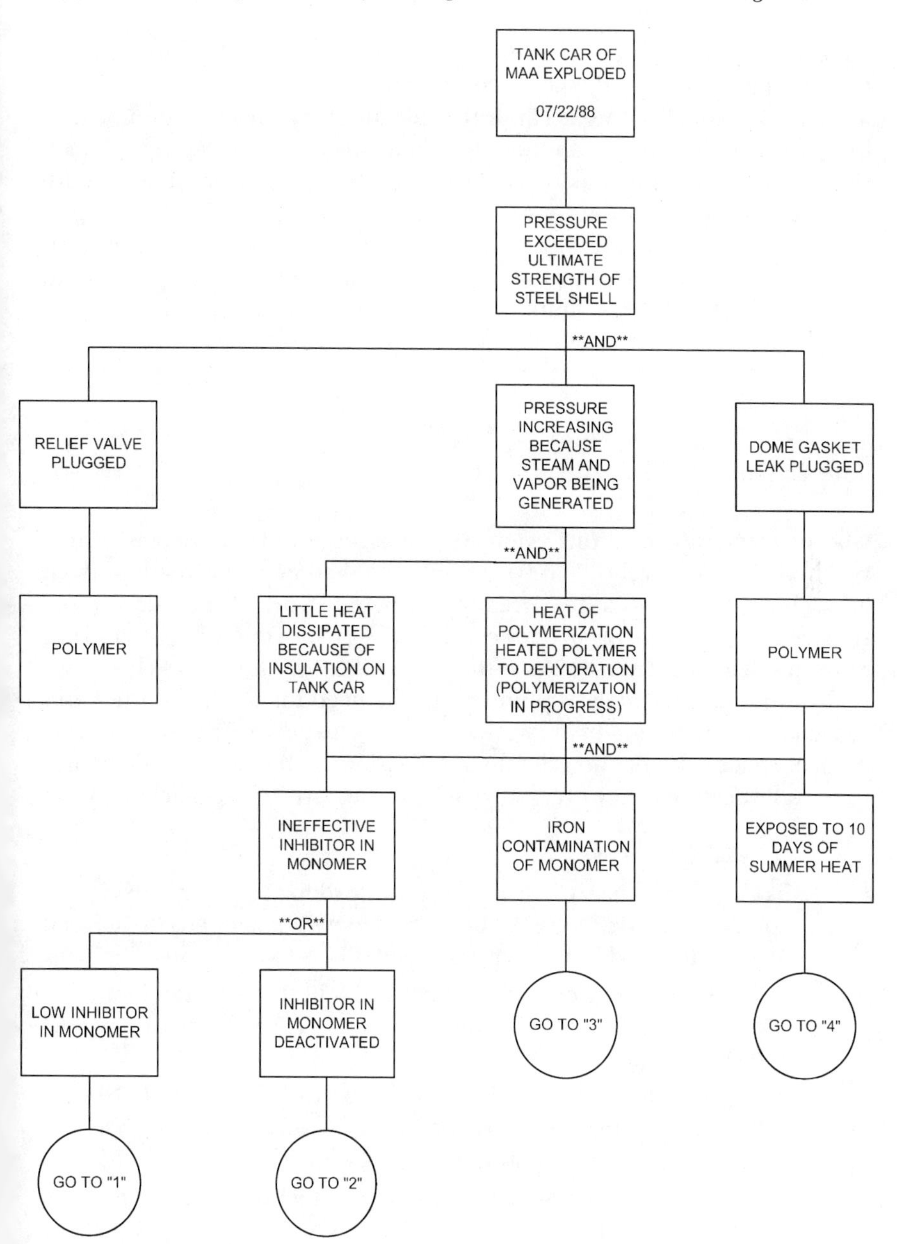

FIGURE 1. **Event tree for an MAA tank car explosion.**

about "why" something happened can be asked, or when we reach a reason like "It is naturally hot in Southeast Texas in the summer." These are called "primal events" in fault tree terminology. In our system, possible causes are followed even though we may not think they apply to a particular incident. This makes the incident investigation event tree more valuable for future applications, such as in training, because more than one scenario is included in the result. Thus we followed the branch dealing with contamination causing inhibitor deactivation to its conclusion; even though we did not think contamination was a cause in this case, it could have been. The detailed trees are shown as Figures 2–5.

The Causes Identified

The primary causes of this event were a direct result of systems, which resulted in a low level of TMAA stability. We had enjoyed freedom from accidents for years because a great many people worked very hard to see that everything was in order; however, there were not enough systems with built-in safeguards to ensure that the probability of failure was as low as we really wanted. This analysis pointed out the need for systems studies and subsequent improvements if we were to be satisfied with future performance. Space will not allow detailed discussion of all problems found and their ramifications; however, key deficiencies are listed below without comment:

1. Technical methacrylic acid was not a product we sold outside the company. It had largely been considered analogous to Glacial MAA, but it had important differences.
    - TMAA had a layer of dilute sulfuric acid that separated out upon standing. This dilute acid layer was very corrosive to ferrous materials.
    - No product code had been assigned to this material. No unique specifications for stability and inhibitor content had been established. (Again, handled by analogy.)
2. GMAA was routinely shipped in stainless steel cars; when lined cars became scarce, TMAA was loaded into stainless steel cars without hesitation or review.
    - The corrosion of the stainless steel by the acid layer was thought to have *promoted* the polymerization, but the *cause* was the low level of inhibitor; if the inhibitor level had been correct, the presence of the iron would not have mattered.

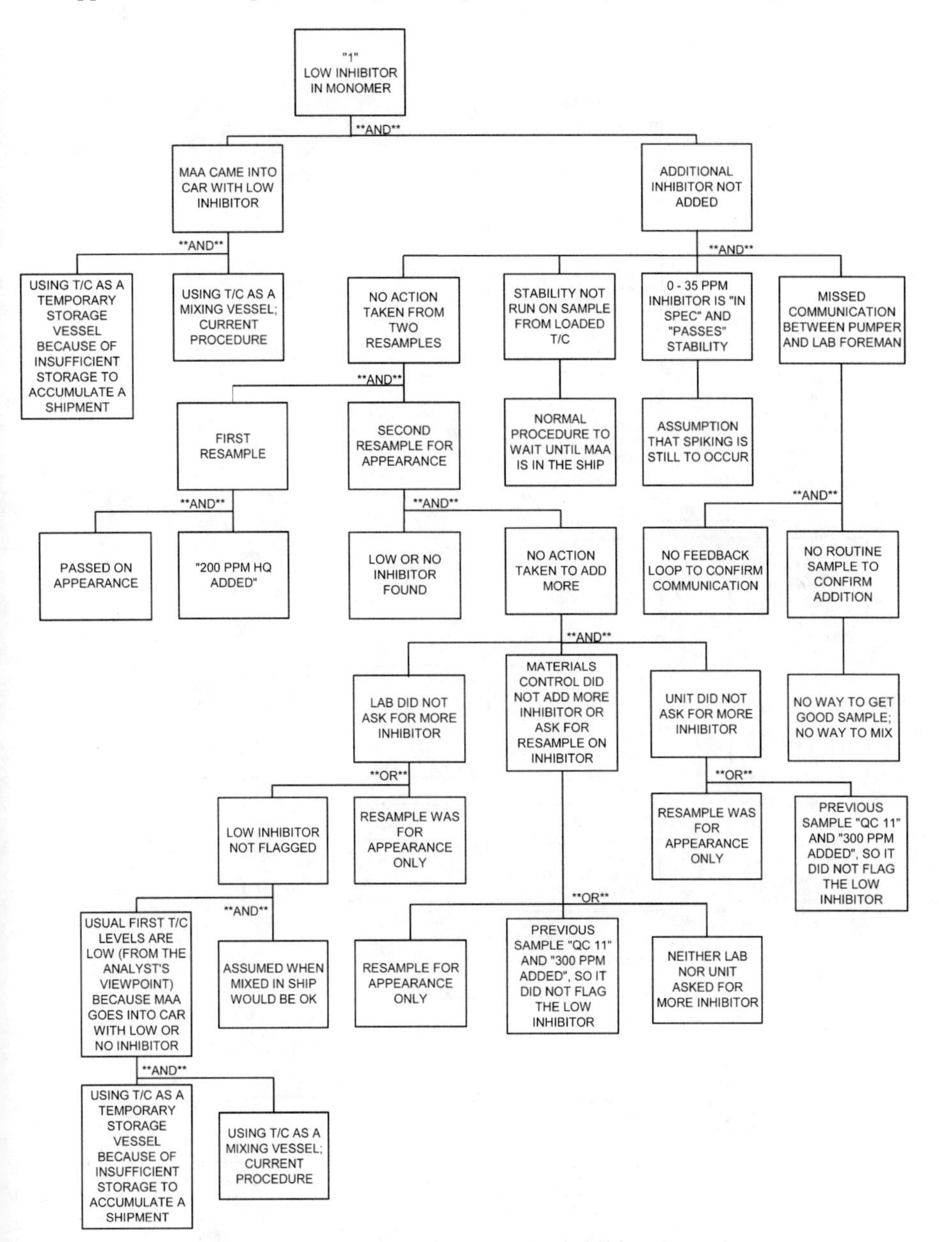

FIGURE 2. **Branch 1 of event tree (inhibitor branch).**

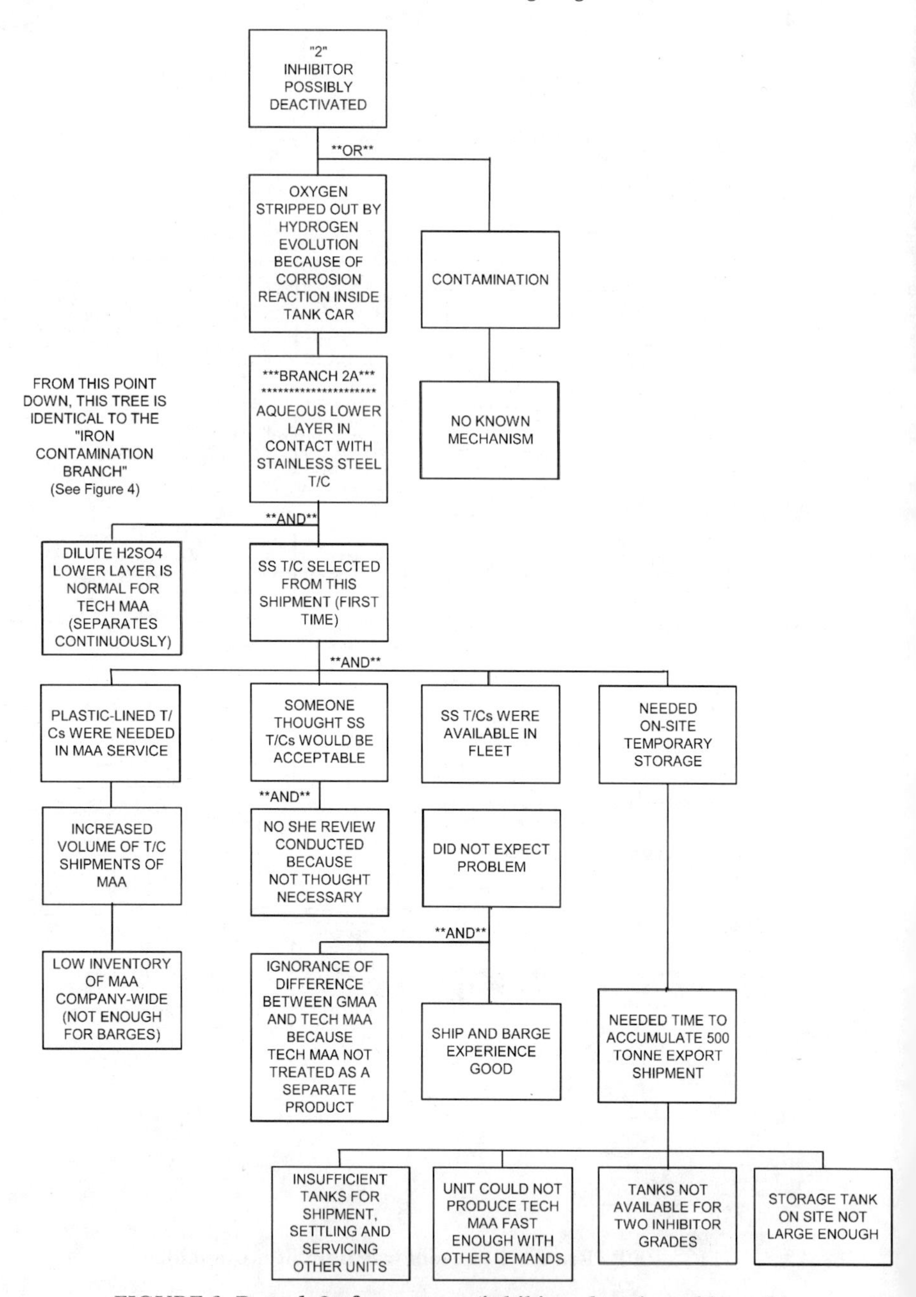

FIGURE 3. **Branch 2 of event tree (inhibitor deactivated branch).**

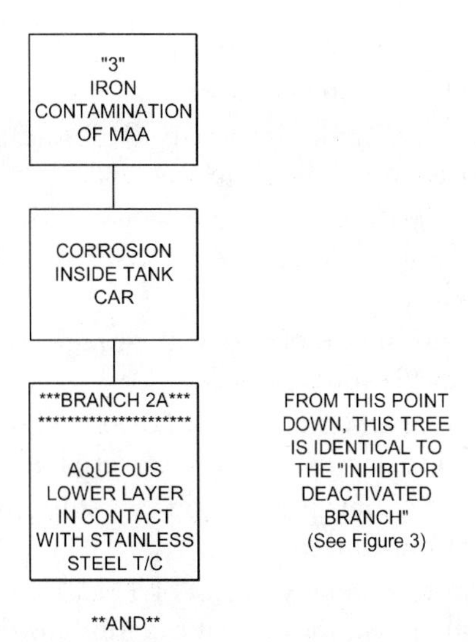

FIGURE 4. **Branch 3 of event tree (iron contamination branch).**

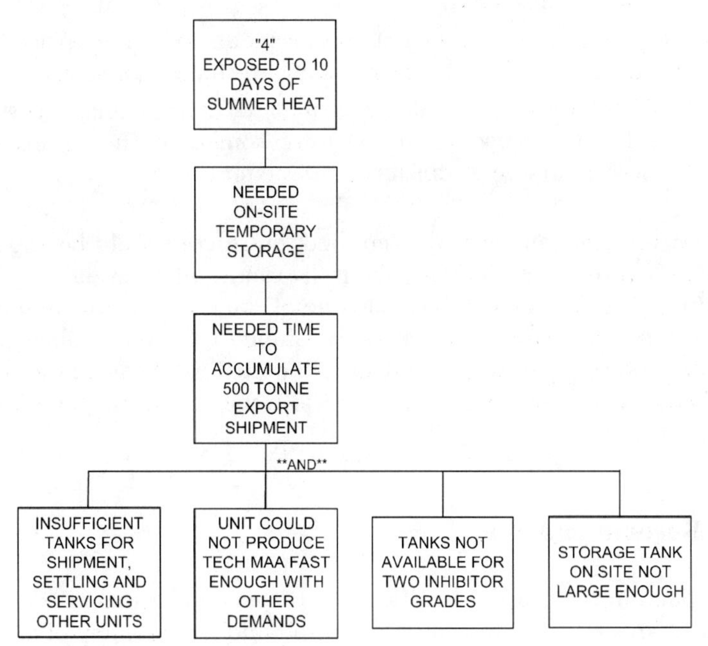

FIGURE 5. **Branch 4 of event tree (warm storage branch)**

3. The MAA was loaded into tank cars containing only the rundown inhibitor levels from the process.
   - There were no tanks in which inhibitor levels could be adjusted prior to loading the material.
4. After the inhibitor was added to the loaded tank car, there was no way to obtain analytical feedback to verify that the correct level had been achieved.
   - The car could not be mixed well enough to obtain a representative sample for feedback.
5. Responsibility for and ownership of the tank into which product was run down, and from which tank cars were loaded, were not clear.
   - Handing off from production to shipping allowed for missed communications.
6. In the laboratory system, samples of TMAA were not tested for stability unless analysis showed that the inhibitor was in specification. Neither was stability run on material for which no inhibitor results were available.
   - These procedures removed a possible feedback loop from the overall system.
7. Ship schedules and overseas requirements sometimes placed unrealistic demands on the production and shipping systems.
   - This meant that results of all analyses (in particular, the stability test took 24 hours) were sometimes not available before cars were transferred to the terminal.

Other shortcomings were identified, but these should be enough to show the nature of the findings about the causes of the event.

It was highly unlikely that only one of the causes above would have resulted in a polymerization. Properly inhibited TMAA is stable in stainless steel for long periods. On the other hand, TMAA having the amount of inhibitor contained in this tank car is also stable for long periods if iron is not present.

The Recommendations

As implied in the causes identified, the recommendations were to take a serious look at all the systems we used to handle monomers of all kinds. We deemed it especially critical to ensure that our systems had mechanisms to ensure feedback on important variables, such as inhibitor concentration, and that the results of analyses were reliable. (One cause for inaction after receipt of a bad inhibitor analysis was that the reputation of the analysis was so bad.) The investigation team recommended the formation of a team to work these systems issues in depth, and to make recommendations for sweeping changes in our operations where needed.

The Implementation of Recommendations

The shipping stability team (SST) was formed to implement the recommendations of the investigation. Our mission was as follows:

- To recommend changes that increase the reliability of the system that stores, prepares, loads, and approves monomers for shipment.
- To develop a quantitative measurement system of improvement over the current system.

The desired outcomes of the team's efforts were:

1. To have an understanding of the systems we have now
2. To reduce the risk of a polymerization during shipment
3. To have an assessment of the risk to the public
4. To implement longer term system changes to manage the public risk at acceptable levels

The SST was empowered by plant management and had members from every discipline involved in the shipping and handling of monomers at the plant: shipping, pumping, production, quality assurance, process chemistry, chemical engineering, and safety/risk analysis. Consequently, it was not difficult to obtain cooperation when requests were made of various entities.

We started the work by analyzing the event tree from the investigation (Figures 1–5) and pinpointed the areas in the system where the most benefit could be obtained with the least effort. To fulfill the first part of the mission, we essentially produced, or had produced, top-down flow charts (TDFCs) for all operations in the system. Only after the systems were completely defined could we confidently look for the most beneficial improvements.

To provide a baseline, a vehicle to monitor the effects of proposed changes, and to be able to fulfill the second part of the mission, we took the event trees and refined them into more rigorous (but still imperfect) fault trees that were capable of being analyzed using existing computer programs. The first generation fault tree corresponding most closely to the original event tree included all of the "challenges" (items identified by the AIC as having potential adverse effects on TMAA stability, such as "iron contamination," "exposed to 10 days of summer heat," and "inhibitor possibly deactivated").

However, this tree was unwieldy, and the SST had reached consensus early in its investigations that none of the visualized "challenges" would have caused the polymerization if there had been sufficient inhibitor in the material to start with. In order to expedite tree analysis, the subcommittee decided to concentrate only on the inhibitor system branch of the tree. This also allowed us to concentrate on the effects of changes in the system without the unnecessary complications of second-order effects that were not well quantified and which could not be controlled. (The weather in Houston is going to be hot in July, and there isn't much we can do about it.) The top event in all subsequent versions of the trees was "Insufficient inhibitor to meet stability specifications."

A subcommittee of the SST met to agree on the probabilities to assign to the various primal events. It was recognized that these probabilities were not necessarily correct; however, the same probabilities would be used in all versions of the trees, or changes would be assigned by the same people. This meant that there would be internal consistency, and comparisons between systems to evaluate the effects of incorporated improvements would be valid. This is an important point; no fault trees are perfect, and nobody should expect them to be. Nevertheless, the value of fault trees for comparisons remains excellent. The inhibitor branch of the original tree was designated "Old Inhibitor System—No Challenges." The complete trees are quite large and are not amenable to reproduction in this paper in readable format. Therefore, the trees have been reduced to include only the primal events from the Minimal Cut Sets. The reduction of "Old Inhibitor System—No Challenges" is shown as Figure 6. The calculated probability indicated that 4 out of 10 cars would not meet inhibitor specifications! The overriding primal events were loading the car with low inhibitor in the monomeric acid (always true) analytical difficulties, and inhibitor addition mechanics. The first improvement to the system was that the unit took on the responsibility for ensuring that the material was properly inhibited before releasing it for loading and shipping. More accurate scales were also added at the same time. This improved the accuracy of weighing inhibitor and added one feedback loop to the system, but no other changes were made at that time. These changes reduced the probability of error in weighing the inhibitor and added one "And" gate to the tree. ("And" gates are the best way to reduce the probability of an undesirable event.) The tree for that system was designated "New Inhibitor System—No Challenges" (Figure 7). In this system, probability calculations from the fault tree showed that only 8 cars out of 1000 would now fail specs. This is a significant improvement (almost three orders of magnitude) as a result of these relatively simple and inexpensive changes.

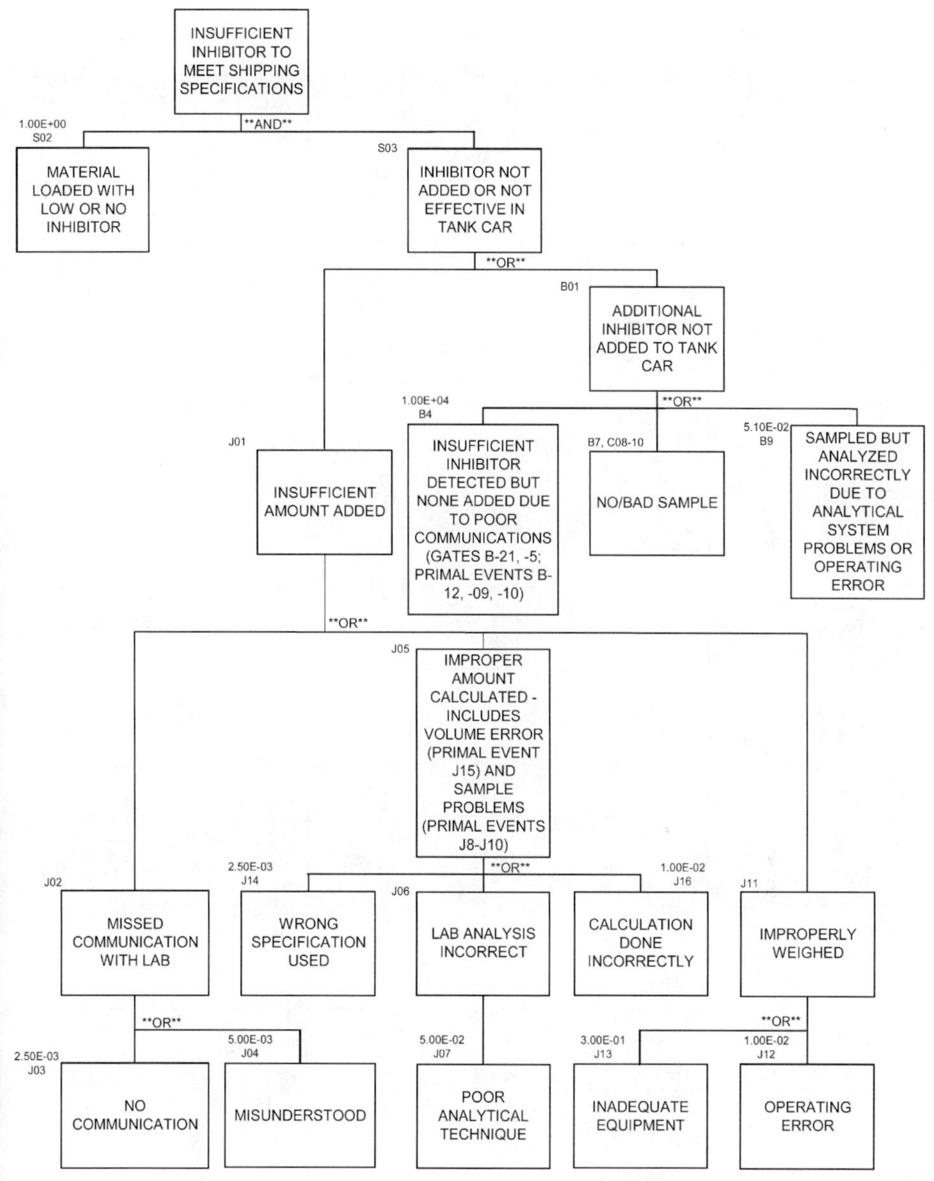

FIGURE 6. **Old Inhibitor System—No Challenges.**

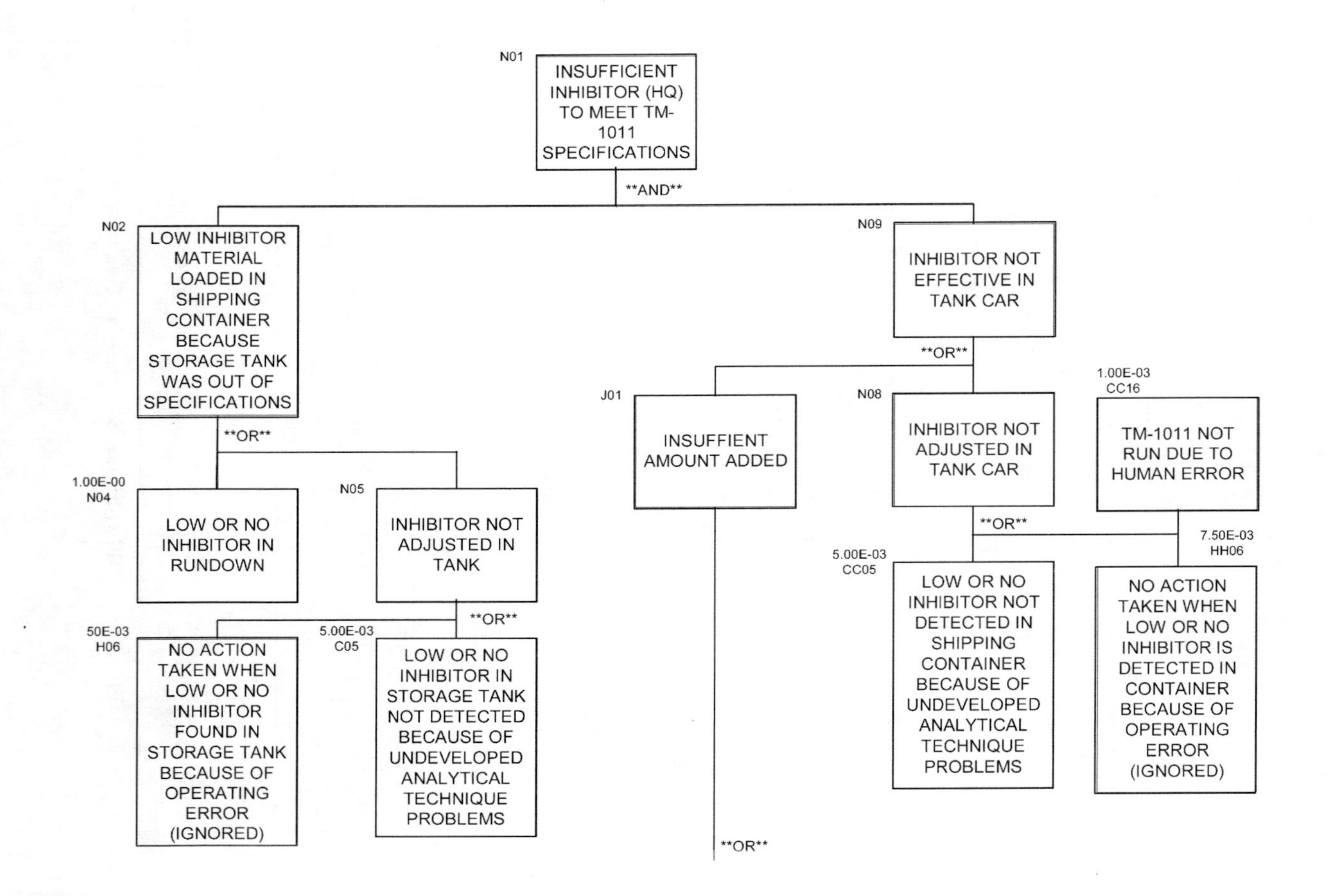
N01
INSUFFICIENT INHIBITOR (HQ) TO MEET TM-1011 SPECIFICATIONS
AND
N02
LOW INHIBITOR MATERIAL LOADED IN SHIPPING CONTAINER BECAUSE STORAGE TANK WAS OUT OF SPECIFICATIONS
OR
1.00E-00
N04
LOW OR NO INHIBITOR IN RUNDOWN
N05
INHIBITOR NOT ADJUSTED IN TANK
OR
50E-03
H06
NO ACTION TAKEN WHEN LOW OR NO INHIBITOR FOUND IN STORAGE TANK BECAUSE OF OPERATING ERROR (IGNORED)
5.00E-03
C05
LOW OR NO INHIBITOR IN STORAGE TANK NOT DETECTED BECAUSE OF UNDEVELOPED ANALYTICAL TECHNIQUE PROBLEMS
N09
INHIBITOR NOT EFFECTIVE IN TANK CAR
OR
J01
INSUFFIENT AMOUNT ADDED
OR
N08
INHIBITOR NOT ADJUSTED IN TANK CAR
OR
1.00E-03
CC16
TM-1011 NOT RUN DUE TO HUMAN ERROR
5.00E-03
CC05
LOW OR NO INHIBITOR NOT DETECTED IN SHIPPING CONTAINER BECAUSE OF UNDEVELOPED ANALYTICAL TECHNIQUE PROBLEMS
7.50E-03
HH06
NO ACTION TAKEN WHEN LOW OR NO INHIBITOR IS DETECTED IN CONTAINER BECAUSE OF OPERATING ERROR (IGNORED)

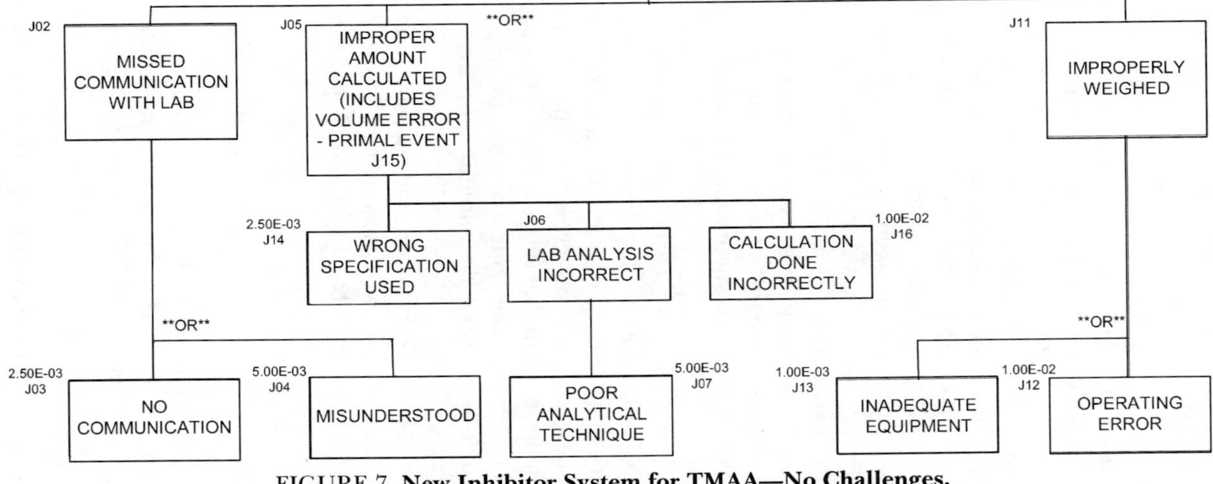

FIGURE 7. **New Inhibitor System for TMAA—No Challenges.**

Extensive work by the SST and all groups related to the TMAA system resulted in several additional basic systems changes that were implemented in the final version of the tree that was analyzed. These included the following:

1. The method for analyzing for inhibitor in TMAA was improved dramatically. This reduced the probability of error in that area and also increased the likelihood of appropriate action based on analytical results.
2. The inhibitor content of the rundown was increased to a level that was sufficient to ensure that the product could pass stability even if no additional inhibitor were added to the tank. This, in effect, added another "And" gate to the system.
3. Lab procedures were changed to assure that stability was always run on every sample regardless of inhibitor content. This added still more "And" gates to the tree.
4. Procedures for releasing shipments were improved, and a feedback loop was added to ensure that the surveyors did not accept tank cars for export without proper documentation. (Another "And" gate.)

The tree for this system was designated "Improved Inhibitor System—No Challenges" (Figure 8). Probability calculated from this fault tree indicated that fewer than 3 out of 10,000,000 cars would fail with this system. Again, these changes were relatively simple and cost very little to implement. I should mention again that we do not claim that these probabilities are **accurate**, but because the fault trees and the input data are consistent the comparisons remain meaningful and the results are dramatic.

The next generation system used an on-line analyzer to monitor the inhibitor concentration in the rundown continuously. This system would virtually have a vanishingly small probability of out of spec inhibitor.

The Results

1. This investigation/systems analysis/fault tree analysis of this incident provided an excellent opportunity to demonstrate the power of the investigation method and the ease with which it may be adapted to a systems analysis for facilitating improvements and corrective actions.
2. The power of systems-oriented thinking was clearly shown by the dramatic improvements, which were obtained by relatively simple and inexpensive changes.

3. A thorough understanding of systems is necessary if basic changes are to be made. This concept is the foundation of the Rohm and Haas Texas system for process safety management: "Know what you want to do." But that is another story.[2]
4. The SST work has been the basis for a complete revision of the monomer handling systems in the plant. In particular, we have looked at all of them to make sure that there were enough feedback loops ("And" gates in fault tree terminology) to ensure adequate inhibitor levels. The more loops there are, the greater the assurance. Examples of increasing integrity are given below:
   1. Sample well-mixed car contents and receive results before releasing.
   2. AND load only from tank containing correct material (that has been verified by sample results).
   3. AND run down from the process at a safe inhibitor level. Verify by sampling.
   4. AND use on-line inhibitor analyzer to verify performance of inhibitor addition systems.
   5. Guidelines for monomer systems were developed and are in place (Table 1).
   6. Once a polymerization incident is in progress, there are no known effective mitigation procedures.

TABLE 1
Monomer Handling Guidelines Recommended by the SST

1. No dry inhibitor should be added to any shipping container unless the container is gas-free
2. No inhibitor adjustment should be made without confirmatory feedback
3. Never assume that stability will pass
4. Analytical techniques must be statistically capable to be reliable
5. All monomers should be run down containing enough inhibitor to pass stability test
6. All monomers should be loaded at the inhibitor shipping specification to ensure a feedback loop
7. A tracking system is needed to account for all loaded containers of monomer

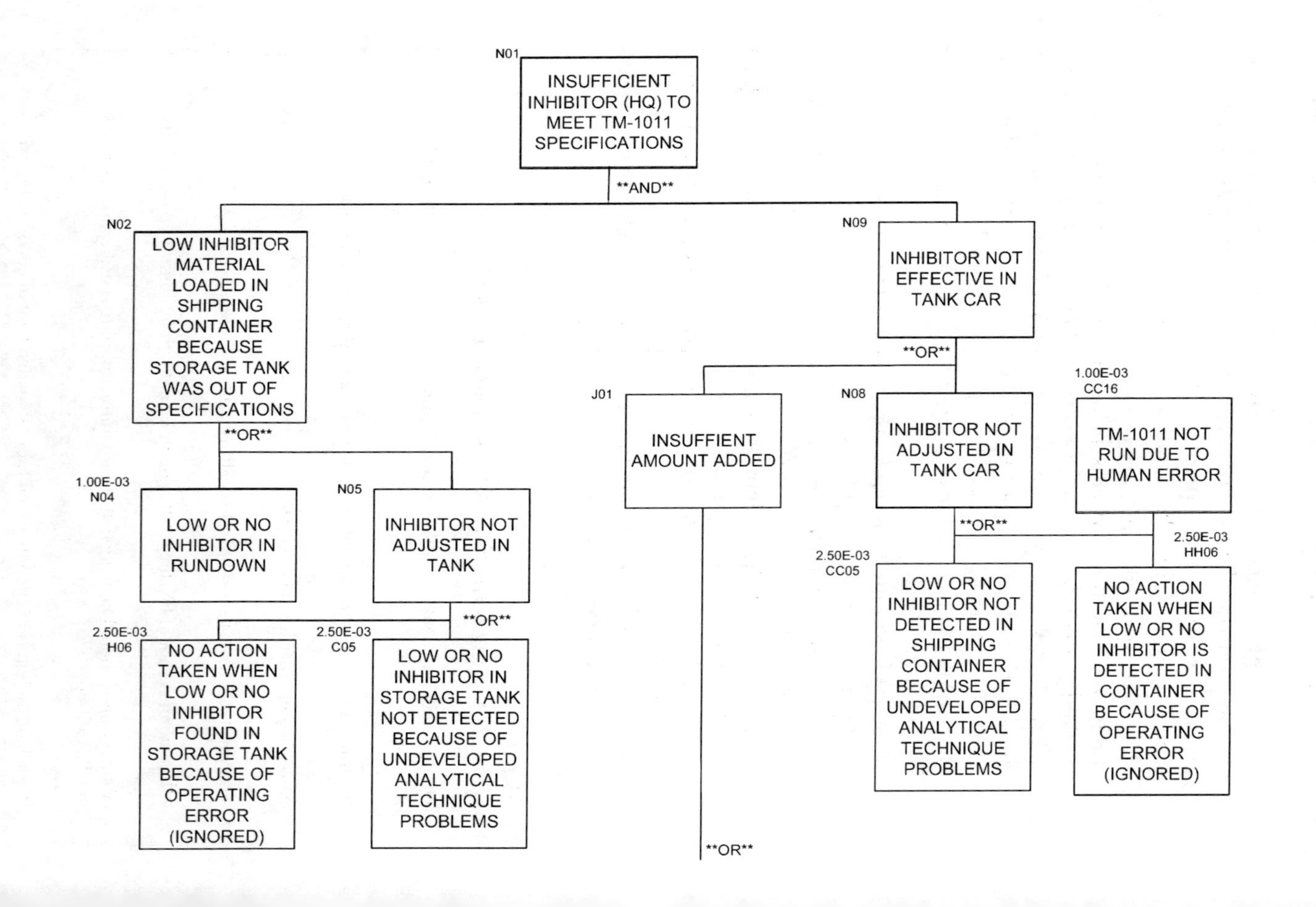
N01
INSUFFICIENT INHIBITOR (HQ) TO MEET TM-1011 SPECIFICATIONS
AND
N02
LOW INHIBITOR MATERIAL LOADED IN SHIPPING CONTAINER BECAUSE STORAGE TANK WAS OUT OF SPECIFICATIONS
OR
1.00E-03
N04
LOW OR NO INHIBITOR IN RUNDOWN
N05
INHIBITOR NOT ADJUSTED IN TANK
OR
2.50E-03
H06
NO ACTION TAKEN WHEN LOW OR NO INHIBITOR FOUND IN STORAGE TANK BECAUSE OF OPERATING ERROR (IGNORED)
2.50E-03
C05
LOW OR NO INHIBITOR IN STORAGE TANK NOT DETECTED BECAUSE OF UNDEVELOPED ANALYTICAL TECHNIQUE PROBLEMS
N09
INHIBITOR NOT EFFECTIVE IN TANK CAR
OR
J01
INSUFFIENT AMOUNT ADDED
OR
N08
INHIBITOR NOT ADJUSTED IN TANK CAR
OR
2.50E-03
CC05
LOW OR NO INHIBITOR NOT DETECTED IN SHIPPING CONTAINER BECAUSE OF UNDEVELOPED ANALYTICAL TECHNIQUE PROBLEMS
1.00E-03
CC16
TM-1011 NOT RUN DUE TO HUMAN ERROR
2.50E-03
HH06
NO ACTION TAKEN WHEN LOW OR NO INHIBITOR IS DETECTED IN CONTAINER BECAUSE OF OPERATING ERROR (IGNORED)

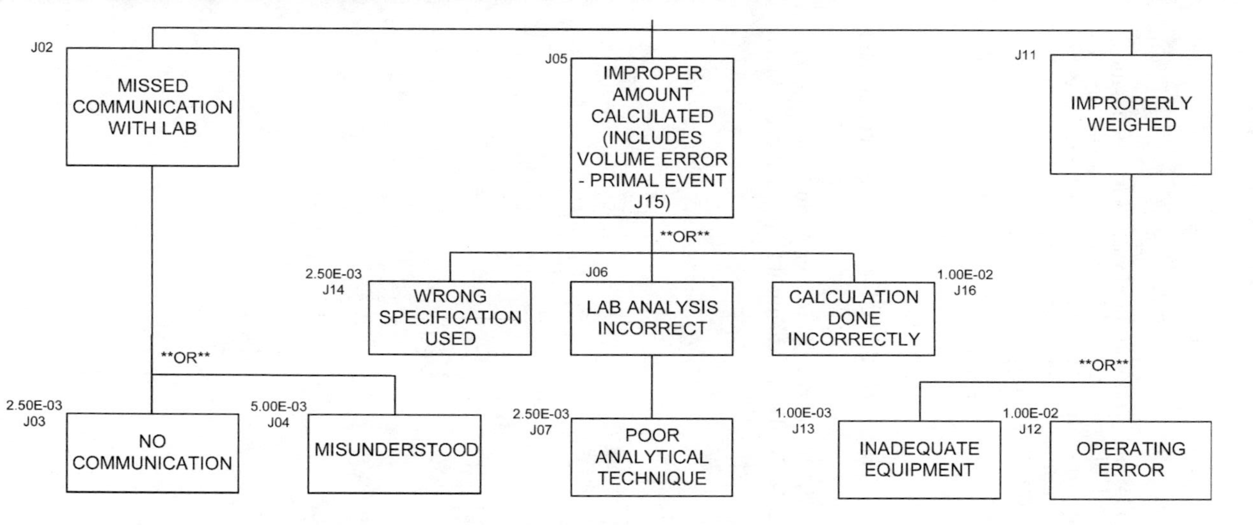

FIGURE 8. **Improved Inhibitor System for TMAA—No Challenges.**

Acknowledgments

The authors are grateful to the following groups and individuals for their candor, expertise, and dedication to the task of investigating this accident, analyzing the systems, and implementing the appropriate changes: A. M. Dowell; members of the AIC and the SST; plant and unit management; and the terminal personnel and surveyors who work in the shipping system.

References

1. Petersen, D. *Techniques of Safety Management*, 2nd ed. New York: McGraw-Hill, 1978.
2. Anderson, S. E. Dowell, A. M. III, and Martin, D. K. "An Audit System for Process Safety." Paper presented at the Texas Chemical Council Process Safety Seminar, June 4, 1990.

Appendix F

Selected OSHA and EPA Incident Investigation Regulations

OSHA Process Safety Management of Highly Hazardous Chemicals

29 CFR 1910.119

- 1910.119(m) Incident investigation.
- 1910.119(m) (1) The employer shall investigate each incident which resulted in, or could reasonably have resulted in a catastrophic release of highly hazardous chemical in the workplace.
- 1910.119(m) (2) An incident investigation shall be initiated as promptly as possible, but not later than 48 hours following the incident.
- 1910.119(m)(3) An incident investigation team shall be established and consist of at least one person knowledgeable in the process involved, including a contract employee if the incident involved work of the contractor, and other persons with appropriate knowledge and experience to thoroughly investigate and analyze the incident.
- 1910.119(m) (4) A report shall be prepared at the conclusion of the investigation which includes at a minimum:
- 1910.119(m) (4) (i) Date of incident;
- 1910.119(m) (4)(ii) Date investigation began;
- 1910.119(m)(4)(iii) A description of the incident;
- 1910.119(m)(4)(iv) The factors that contributed to the incident; and,
- 1910.119(m)(4)(v) Any recommendations resulting from the investigation.
- 1910.119(m)(5) The employer shall establish a system to promptly address and resolve the incident report findings and recommendations. Resolutions and corrective actions shall be documented.

- 1910.119(m)(6) The report shall be reviewed with all affected personnel whose job tasks are relevant to the incident findings including contract employees where applicable.
- 1910.119(m)(7) Incident investigation reports shall be retained for five years.

Accidental Release Prevention Requirements: Risk Management Programs Under Clean Air Act Section 112(r)(7)

40 CFR Part 68

Subpart C – Program 2 Prevention Program

Sec. 68.60 Incident investigation.

(a) The owner or operator shall investigate each incident which resulted in, or could reasonably have resulted in a catastrophic release.

(b) An incident investigation shall be initiated as promptly as possible, but not later than 48 hours following the incident.

(c) A summary shall be prepared at the conclusion of the investigation which includes at a minimum:

(1) Date of incident;
(2) Date investigation began;
(3) A description of the incident;
(4) The factors that contributed to the incident; and,
(5) Any recommendations resulting from the investigation.

(d) The owner or operator shall promptly address and resolve the investigation findings and recommendations. Resolutions and corrective actions shall be documented.

(e) The findings shall be reviewed with all affected personnel whose job tasks are affected by the findings.

(f) Investigation summaries shall be retained for five years.

Subpart D – Program 3 Prevention Program

Sec. 68.81 Incident investigation.

(a) The owner or operator shall investigate each incident which resulted in, or could reasonably have resulted in a catastrophic release of a regulated substance.

(b) An incident investigation shall be initiated as promptly as possible, but not later than 48 hours following the incident.

(c) An incident investigation team shall be established and consist of at least one person knowledgeable in the process involved, including a contract employee if the incident involved work of the contractor, and other persons with appropriate knowledge and experience to thoroughly investigate and analyze the incident.

(d) A report shall be prepared at the conclusion of the investigation which includes at a minimum:

(1) Date of incident;
(2) Date investigation began;
(3) A description of the incident;
(4) The factors that contributed to the incident; and,
(5) Any recommendations resulting from the investigation.

(e) The owner or operator shall establish a system to promptly address and resolve the incident report findings and recommendations. Resolutions and corrective actions shall be documented.

(f) The report shall be reviewed with all affected personnel whose job tasks are relevant to the incident findings including contract employees where applicable.

(g) Incident investigation reports shall be retained for five years.

Appendix G

Quick Checklist for Investigators

Even for experienced investigators, preparing to investigate a significant incident can be hectic in the mobilization period following the event. This checklist is intended to be used as a quick reminder of some key considerations for people on their way to an investigation. Incidents are unique and have unique requirements but the information included should be adequate reminders for most incidents.

Physical Items

Photographic Equipment

- 35mm camera
- Digital camera and cables
- Video camera and cables
- Film and batteries for cameras

Measurement Tools

- Tape measures
  - 100 feet
  - 10 feet
- 6 inch steel ruler
- Inside calipers
- Outside calipers

Documentation Aids

- Dictaphone
- Notebooks
- Clipboard
- Pens and pencils
- Laptop computer

Evidence Marking Aids

- Paint pens
- Grease pens
- Permanent markers

- Tags with wire or plastic connectors
- Orange flagging tape
- Evidence tag stickers

Evidence Collection Aids

- Self-closing plastic bags
- Tweezers
- Forceps
- Sample bottles

Personal Protective Equipment

- Hard hat
- Safety goggles
- Steel toed shoes
- Fire retardant coveralls
- Gloves
- Hearing protection

Other

- Cell phones
- Electric circuit tester
- Multi-purpose tool (pliers, knife, screwdriver, etc.)
- Compass
- Magnet
- Duct tape
- Mirror
- Small pocket mirror
- CCPS *Guidelines for Investigating Chemical Process Incidents* 2nd Edition
- Sticky notes
- Sticky flags
- Flashlights
- Magnifying glass
- Barricade tape

Action Reminders

- **Controlling the incident is first priority.** Until Incident Command has extinguished fires, evacuated injured personnel, completed a headcount, and contained spills/stopped releases, the control of the incident is first priority.
- **Secure the scene.** As soon as possible, protect the scene of the incident from disturbances. Work through operations, maintenance,

and emergency response personnel to ensure the scene is not disturbed. Establish a system to limit and control entry into the area.

- **Time sensitive evidence is a high priority.** Gathering evidence that might deteriorate with time should be a high priority.
  - Many electronic systems record data from operating units and then delete that data after a specified period of time, often 24 hours or less.
  - Some evidence such as burn char patterns, surface fractures, or volatile chemicals spills can degrade as a result of weather conditions (rain, wind, or sunlight)
- **Establish roles and expectations for the investigation team.** Roles and expectations need to be defined early so that there are no misunderstandings.
  - Is the incident investigation team also the primary point of contact with outside agencies such as OSHA, EPA, and CSB?
  - What expectations do local management and corporate management have for the investigation team for timing, interim reports, final reports, and defining requirements for startup of units or equipment?
  - What resources are available and just as important, what resources are not available?
- **Interviews need to be done promptly.** Memories fade with time and are influenced by discussions with other witnesses.
- **Interviewing techniques are important.**
  - Plan the interview. Do not do it haphazardly.
  - Interview one person at time and in a private comfortable setting. Use only one or two interviewers.
  - Set the interviewee at ease. One method is by asking questions about activities prior to the incident.
  - Be sensitive to the interviewee's emotional state.
  - Do not express opinions.
  - Do not lead the interviewee. Ask questions that allow the interviewee to describe the incident in their own words. Questions should be neutral, unbiased, and nonleading.
  - Do not interrupt the interviewee.
  - Use a plot plan to better understand
    « the location of interviewee
    « the location of people and activities the interviewee saw
    « movement of the interviewee
  - Ask what the interviewee saw, heard, felt, and smelled before, during, and after the incident.
  - Ask about timing/sequence of events to help develop the timeline.

- At the end of the interview, ask for opinions about the cause of the incident or if the interviewee has anything to add that was not already covered.

- **Gather information about the process early.** The investigation team will need information about the process. Gathering the information can sometimes be done while waiting to gain access to the unit for physical inspection and data gathering.
  - Plot plans
  - Process description
  - P&ID's
  - Information about the chemicals in the area
- **Follow established safety policies.** Incident investigation team members should lead by example by strictly following site safety policies.
- **Initial work is focused on "what" happened.** Determination of root causes is important to prevent recurrence of the incident, but the initial focus of the investigation team is to define "what" happened.
- **Capability of team members is critical.** A major investigation needs the best people available to represent each needed discipline. Frequently contractors/consultants will be needed for special expertise.
- **OSHA has specific requirements for the incident investigation teams.** OSHA 1910.119 (m) (3) states: *An incident investigation team shall be established and consist of at least one person knowledgeable in the process involved, including a contract employee if the incident involved work of the contractor, and other persons with appropriate knowledge and experience to thoroughly investigate and analyze the incident.*
- **Photograph the scene.** Photograph overall views and specific items.
  - Decide if still photography is adequate or if video photography might be more helpful.
  - Photographs should be taken to document as-found location, orientation, and condition of items deemed to be evidence.
  - If many photographs are taken, a log of each photograph with information such as item, location, orientation, and date may be helpful.
- **Establish a timeline.** The investigation of almost every significant incident will require the development of a time line to depict the sequence of events before, during, and after the incident.

- **Legal issues may be important too.** Remember that on a serious incident the company not only has an incident to investigate, but also must be prepared for significant and costly litigation.
- **Evidence that is removed from the scene needs to be secure.** Facilities need to be made available to limit access to physical evidence that is removed from the scene. A system needs to be put in place to control access to the evidence.
- **Documents used in the investigation need to be secure.** The investigation room needs to be secure. If investigation documents are stored in a room other than the investigation room, then that other room also needs to be secure.
- **Determine the method by which evidence will be gathered.**
  - Incidents with significant debris may require the establishment of a grid system to define the exact location of specific pieces.
  - Establish a method for documenting as found valve positions.
- **Develop a list of potential scenarios and remain open minded.** On complex incidents, it is sometimes helpful to develop a list of potential scenarios. Do not fall in the trap of only pursuing the initial obvious scenario. It is important to prove that the actual scenario did happen but it is also important to prove that other potential scenarios did not happen.

References

Center for Chemical Process Safety (CCPS), 1989. *Guidelines for Technical Management of Chemical Process Safety.* New York: American Institute of Chemical Engineers. Appendix C.

Baker, Q. A., Pierorazio, A. J., Ketchum, D. E., "Investigation of Explosion Accidents," Center for Chemical Process Safety International Conference and Workshop on Process Industry Incidents, October 2000, Orlando Florida. New York: AIChE, 2000.

Appendix H

Additional Resources

29 CFR 1904, *Recording and Reporting Occupational Injuries and Illnesses*. Effective January 1, 2002; The OSHA website for recordkeeping revisions is www.osha.gov/recordkeeping/index.html.

American Chemistry Council. *Responsible Care®, A Resource Guide for the Process Safety Code of Management Practices*. Washington, DC, 1990.

American National Standards Institute. *ANSI Z16.2 Method of Recording Basic Facts Relating to the Nature and Occurrence of Work Injuries*. New York.

American Society of Safety Engineers. *Dictionary of Terms Used in the Safety Profession*, 3rd ed. Des Plains, IA: American Society of Safety Engineers 1988.

Bea, Holdsworth, Smith. *Summary of Proceedings and Submitted Papers*. Workshop on Human Factors in Offshore Operations, 1996.

Bridges, W. G. "Get Near Misses Reported." Center for Chemical Process Safety International Conference and Workshop on Process Industry Incidents, October 2000, Orlando Florida. New York: AIChE, 2000.

Carper, K. *Forensic Engineering*. New York: Elsevier, 1989.

Center for Chemical Process Safety (CCPS). *Guidelines for Hazard Evaluation Procedures, Second Edition with Worked Examples*. New York: American Institute of Chemical Engineers, 1992.

Center for Chemical Process Safety (CCPS). *Guidelines for Technical Management of Chemical Process Safety*. New York: American Institute of Chemical Engineers, 1989.

Cook, R., Jones, A. "Establishing Criteria for Evaluating a Problem Solving System." AIAA Space 2001 Conference and Exposition, American Institute of Aeronautics and Astronautics, Inc., Houston, TX 2001.

Dowell, A. M., Philley, J., and Pearson, K. "Structured Root Cause Investigation." Training course presented at Texas Chemical Council Safety Seminar, 1999.

Feynman, R. P. *What Do You Care What Other People Think?* New York: Norton, 1988.

Gano, D. L. *Apollo Root Cause Analysis—A New Way of Thinking*. Apollonian Publications, September 1994.

Gilovich, T. *How We Know What Isn't So: The Fallibility of Human Reason in Everyday Life*. New York: Macmillian, 1991.

Greenwood, M., and Woods, H.M. "The Incidence of Industrial Accidents with Special Reference to Multiple Accidents," *Ind. Fatigue Res. Board*, Report 4, HMSO, London, England, 1919.

Hon. Lord Cullen. "The Public Inquiry into the Piper Alpha Disaster." London: UK Department of Energy, HMSO, 1990.

Jones, B. "A Process Quality Management Tool, Prevention of Unwanted Events, Training and Text," 1991.

Kepner, C. H. and Tregoe, B. B. *The Rational Manager*. 2nd ed. Princeton, NJ: Kepner-Tregoe, Inc. 1976.

Kletz, T. A. "Accident Investigation: How Far Should We Go?" Paper presented at the American Institute of Chemical Engineers Loss Prevention Symposium, 1983.

Kletz, T. A. "Organizations Have No Memory." Paper presented at the AIChE Loss Prevention Symposium, April 1979.

Kuhlman, R. *Professional Accident Investigation*. Loganville, GA: Institute Press, International Loss Control Institute.

Livingston, A. D., Jackson, and Priestly. *Root Causes Analysis: Literature Review*. WS Atkins Consultants Ltd. Birchwood, Warrington: WS Atkins House, 2001.

Loss Prevention Bulletin #061. Rugby, England: Institution of Chemical Engineers, February 1985.

Lucas, D. A. and Embrey, D. E. "Human Reliability Data Collection for Qualitative Modeling and Quantitative Assessment." In Colombari, V. (Ed.), *Reliability Data Collection and Use in Availability Assessment*. New York: Springer Verlag, 1989.

Norman, D. A. *The Design of Everyday Things*. New York: Basic Books, 1988.

Petersen, D. *Techniques of Safety Management*, 2nd ed. New York: McGraw-Hill, 1978.

Philley, J. "Root Cause Incident Investigation Can Be Tricky." Presented at International System Safety Conference, 2001.

Sanders, M. S. and McCormick. E. J. *Human Factors in Engineering and Design*, 6th ed. New York: McGraw-Hill, 1987.

Smith, K., and Franklyn, C. "Conquering Cultural Change in Incident Investigation." Center for Chemical Process Safety International Conference and Workshop on Process Industry Incidents, October 2000, Orlando Florida. New York: AIChE, 2000.

Stern, A., and Keller, R. "Human Error and Equipment Design in the Chemical Industry" *Professional Safety*. May 1991.

Vaughan, D. *The Challenger Launch Decision*. Chicago: University of Chicago Press, 1996.

Winsor, D. A. "Challenger: A Case of Failure to Communicate." *Chemtech*. September 1989.

Proceedings of the Center for Chemical Process Safety International Conference and Workshop: Process Industry Incidents—Investigation Protocols, Case Histories, Lessons Learned. New York: American Institute of Chemical Engineers, 2000.

Case Studies/Lessons Learned

Lessons Learned from an On-Plot Refinery Tank Explosion, K. Ann Paine

Case History of the Tosco Avon Refinery Investigation: January 1997 through November 1998, Dorian S. Conger

A Review of Past Accidents Occurring in Major Hazard Installations in Italy: Discussion of the Causes, Consequences, and Lessons Learned, Giancarlo Ludovisi and Fiorenzo Damiani

Lessons Learned from a Process Tank Explosion, Steven R. Marwitz, Randall P. Smith, and Rashid Hamsayeh

Management Systems

Incident Severity Rating and Investigation Guidelines, Dave Gaydos and Gary York

The Development of Approaches to Incident Selection for the Chemical Safety Board, Daniel E. Sliva and Jack Weaver

The Chemical Safety and Hazard Investigation Board's Process for Selecting Incident Investigations, Bill Hoyle, Shannon McCleary, and Isadore Rosenthal

Case Studies/Lessons Learned

A Structured Approach to Safe Design or Do the Safety Risks Outweigh the Environmental Benefit? Peter J. Hunt

Investigation of a Pesticide Explosion, Awilda Fuentes

A Compressor Failure That MOC May Not Have Caught! Alfred W. Bickum

Investigation Technologies

Premature Stopping Points for Determining the Root Cause of Human Error in Process Investigations, Jack Philley

Using Advanced Trending Techniques to Learn from Your Incident Statistics, Mark Paradies and Ed Skompski

Human Factors in Accident Investigations, Valerie E. Barnes

Case Studies/Lessons Learned

Lessons Learned from a Cold Weather Explosion and Fire in an Oil Refinery, Brian D. Kelly

Steam Line Rupture at Tennessee Eastman Division, Peter N. Lodal

Lessons Learned from the Longford Royal Commission Investigation into the Explosion and Fire on 25 September 1998 at the Esso Gas Processing Plant, Mark Boult Gary Kenney, Robin Pitblado

Risk Reduction by Learning from Incidents and Near-Misses, K. A. Ruppert and E. Meyer zu Riemsloh

Investigation Technologies

Root Cause Analysis-NOT What You Might Think, C. Robert Nelms

Investigation into the Root Causes of Repeated Incinerator Incidents, Donald K. Lorenzo

Investigation of Explosion Accidents, Quentin A. Baker, Adrian J. Pierorazio, Donald E. Ketchum

Case Studies/Lessons Learned

Central Collecting and Evaluating of Major Accidents and Near-Misses in the Federal Republic of Germany, Hans-Joachim Uth

Impact of Identifying Root Causes, Jack McCavit

A Reactive Chemical Incident: Morton International, Paterson, New Jersey, David Heller and William Hoyle

Management Systems

Safety Management through Learning from Experience in the Chemical Industry: Example of a New Incident Analysis Methodology, B. Wilpert, H. J. Uth, R. Miller, and E. Ninov

Using Process Tools, System Evaluation, and Accident Trends to Improve Operational Reliability, Ed Koshka

Quality Assurance in Incident Reporting and Investigation, Ujwal Ritwik

Management Systems

Organizational Unlearning: Detrimental Behaviors Present In Chemical Process Incident Investigation Teams, Robert K. Urian

A Case Study of the Use of Electronic Networking for Incident Notification, Response, Mitigation, and Sharing, David W. Owens

Legal Issues and Incident Investigations, Mark S. Dreux

Conquering Cultural Change in Incident Investigation, Kris Smith and Christy Franklin

Case Studies/Lessons Learnedr

Get Near Misses Reported, William G. Bridges

The Case of the Pressurized Drums (Unexpected Oxidation-Reduction Reaction), Larry G. Holloway

Acrylic Polymer Reactor Accident Investigation: Lessons Learned and Three Years Later, Michael Gromacki

Incident Data Workshop

Recent Developments in European Commission Tools to Manage Industrial Risks in Europe, Christian Kirchsteiger and Stuart Duffield

Establishment of an Industrial Accident Database in Korea, Hyuck-myun Kwon, Dae-ski Yim, Chang-kyu Lee, Soon Joong Kang, Hyung-suk Kim and Young-soon Lee

Chemical Accident Investigation Around the World Workshop

Chemical Accident Investigation and Management in Korea, Kyo-Shik Park and En Sup Yoon

The Loss and Claim Adjusting Process Workshop: Cost Estimation, Investigation Objectives, and the Settlement, Eric Lenoir, Straun Robertson, Michael Misurelli, Larry Collins, Berrin Tansel

Consequence Analysis of an Oil Refinery Explosion in Thailand, Pramuk Osiri, Berrin Tansel, Chalermchai Chaikittiporn, and Preecha Loosereewanich

Estimating Chemical Accident Costs in the United States: A New Analytical Approach, Larry Collins, Carmen D'Angelo, Craig Mattheissen, and Michael Perron

Poster Session

Insurance, Terrorism and the Risk Management Program, Dan'l Steward, Mike Duncan, and Ahmad Shafaghi

The Secret to Measuring Process Safety Performance: Combine Process Incident Data with Leading Indicators, Steve Arendt

Use of Computational Modeling to Identify the Cause of Vapor Cloud Explosion Incidents, J. Keith Clutter and Mark G. Whitney

Realistic Dispersion Modeling of Chlorine Release Incident. Al Waller

Appendix I

CD ROM Contents

Methodology Samples

- Apollo Root Cause Analysis (RCA) Sample
- Comprehensive List of Causes (CLC)
- Propane Tank Root Cause Analysis Example
  –Propane Tank Cause and Effect Chart (CEC)
- Root Cause Map™
- REASON® RCA Methodology
- Systematic Cause Analysis Technique (SCAT) Information
- TapRoot™ Example
- Why Tree Example

Electronic Files of Key Checklists and Examples

- Sample Program Endorsement Letter
- Quick Checklist for Investigators (electronic version of Appendix G)
- Investigation Plan Checklist
- Agency Inspection Preparation Checklist
- Checklist of Evidence Data to Be Collected
- Sample Questions for Witnesses
- Human Factors Checklist
- Sample Formal Report Outline
- Written Report Completeness Checklist
- Investigation Critique Checklist
- Investigation Follow-up Checklist
- PSM Investigation Compliance Checklist
- Investigation Improvement Checklist
- Case Study Analysis Checklist
- Example Investigation: Flashback from Waste Gas Incinerator into Air Supply Piping

Glossary

Accident—An event in which property damage, detrimental environmental impact, or human loss (either injury or death) occurs.

Accidental Chemical Release—An unintended, sudden release of chemical(s) from manufacturing, processing, handling, or on-site storage facilities to the air, water, or land.[1]

Action Tracking—A method of logging progress when implementing a task or set of tasks.

Ad Hoc Investigation—An incident investigation fashioned from the immediately available information and concerns. Typically, the ad hoc investigation is performed whenever there are no prior investigation procedures. A synonym to ad hoc is *unsystematic*.

Amelioration—Improvement of conditions immediately after an accident; treatment of injuries and conditions that endanger people and property.[2]

Anomaly—An unusual, abnormal, or irregular set of circumstances that, left unrecognized or uncorrected, may result in an incident.

Assumed Risk—A risk that has been identified, analyzed, and accepted at the appropriate management level, unanalyzed or unknown risks fall under *oversight* and *omissions by default*.

Audit Trail—The proof that systematic documentation of activities was performed in a way that allows an auditor to confirm compliance with required or desired organizational behavior.

Catastrophic—A loss of extraordinary magnitude in physical harm to people, with damage and destruction to property, or to the environment.[2]

Catastrophic incidents—Incidents that have major consequences with unacceptable lasting effects, usually involving loss of human life, severe off-site impacts, and/or loss of community trust with possible loss of franchise to operate.

Causal Factor—is a major unplanned, unintended contributor to the incident (a negative event or undesirable condition), that if eliminated would have either prevented the occurrence, or reduced its severity or frequency. (Also known as a critical causal factor or contributing cause.)

Cause—An event, situation, condition that results, or could result, directly or indirectly in an accident or incident.[2]

Chemical Process Quantitative Risk Assessment (CPQRA or QRA)— The quantitative evaluation of expected risk from potential incident scenarios. It examines both consequences and frequencies, and how they combine into an overall measure of risk. The CPQRA process is always preceded by a qualitative systematic identification of process hazards. The CPQRA results may be used to make decisions, particularly when mitigation of risk is considered.

Common Cause or Common Mode Failure—Failure, which is the result of one or more events, causing coincident failures in multiple systems or on two or more separate channels in a multiple channel system, leading to system failure. The source of the common cause failure may be either internal or external to the systems affected. Common cause failure can involve the initiating event and one or more safeguards, or the interaction of several safeguards.

Consequence—The cumulative, undesirable result of an incident, usually measured in health and safety effects, environmental impacts, loss of property, and business interruption costs.[1]

Consequence Analysis—The analysis of the expected effects of an incident, independent of its likelihood.[1]

Contributing Cause—Factors that facilitate the occurrence of an incident such as physical conditions and management practices. (Also known as *contributory factors*.)

Deductive Approach—Reasoning from the general to the specific. By postulating that a system or process has failed in a certain way, an attempt is made to determine what modes of system, component, operator, or organizational behavior contributed to the failure.

Enabling Event—An event that makes another event possible

Episodic Event—An event of limited duration, typically an incident. For example, release of hazardous materials, a spill, or an explosion.

Event—An occurrence involving the process caused by equipment performance, human action, or by an occurrence external to the risk control system. In Multilinear Event Sequencing (MES) an event is defined as one actor plus one action.

Evidence—Data on which the investigation team will rely for subsequent analysis, testing, reconstruction, corroboration, and conclusions.

Evidence gathering—the collection of data on which the investigation team will rely for subsequent analysis, testing, reconstruction, corroboration, and conclusions.

Failure—An unacceptable difference between expected and observed performance.(3)

Failure Mode and Effects Analysis (FMEA)—A hazard identification technique in which all known failure modes of components or features of a system are considered in turn and undesired outcomes are noted.(1)

Falsifiability—A concept where a specific effort is made to disprove a speculated hypothesis, in addition to the efforts made to prove the hypothesis.

Fault Tree—A method for representing the logical combinations of various system states that lead to a particular outcome (top event).(4)

Fault Tree Analysis—Estimation of the hazardous incident (top event) frequency from a logical model of the failure mechanisms of a system.(4)

Forensic Engineering—The art and science of professional practice of those qualified to serve as engineering experts in matters before the courts of law or in arbitration proceedings.(3)

Frequency—Number of occurrences of an event per unit time.

Hazard—A chemical, physical, or changing condition that has the potential for causing damage to human life, property, or the environment.(1)

Hazard and Operability Study (HAZOP)—A systematic qualitative technique to identify and evaluate process hazards and potential operating problems, using a series of guidewords to examine deviations from normal process conditions.(4)

Hazard Evaluation—The analysis of the significance of hazardous situations associated with a process or activity. It uses qualitative techniques to pinpoint weaknesses in the design and operation of facilities that could lead to accidents.

High Potential Incident—An event that, under different circumstances, might easily have resulted in a catastrophic loss.

Historic Incident Data—Data collected and recorded from past incidents.

Human Error—Any human action (or lack thereof) that exceeds some limit of acceptability (that is, an out-of-tolerance action) where the limits of human performance are defined by the system. Includes actions by designers, operators, or managers that may contribute to or result in accidents.

Human Factors—a discipline concerned with designing machines, operations, and work environments so that they match human capabilities, limitations, and needs. Includes any technical work (engineering, procedure writing, worker training, worker selection, and other items) related to the human factor in operator–machine systems.

Human Reliability Analysis—A method by which the probability of a person successfully performing a task is estimated.

Impact—The ultimate potential result of a hazardous event. Impact may be expressed in terms of numbers of injuries or fatalities, environmental or property damage, or business interruption.

Incident—An unusual or unexpected event, which either resulted in, or had the potential to result in serious injury to personnel, significant damage to property, adverse environmental impact, or a major interruption of process operations.

Incident Investigation—The management process by which underlying causes of undesirable events are uncovered and steps are taken to prevent similar occurrences.[4]

Incident Investigation Management System—A written document that defines the roles, responsibilities, protocols, and specific activities to be carried out by personnel performing an incident investigation.

Incident Investigation Team—A group of qualified people that examine an incident in a manner that is timely, objective, systematic, and technically sound to determine that information pertaining to the event is documented, probable causes are ascertained, and complete technical understanding of such an event is achieved.[3]

Incident Stereotype—A fixed or general pattern of incident causation. From a review of historical incident data it can be possible to identify “classes of incidents,” each with certain features (or typical, repeated patterns) in common; that is, incident stereotypes are defined.

Inductive Approach—Reasoning from individual cases to a general conclusion by postulating that a system element has failed in a certain way. An attempt is then made to find out what happens to the whole system or process.

Initiating Event—The event that initiates the scenario leading to the undesired consequence.

Injury—Physical harm or damage to a person resulting from traumatic contact between the body and an outside agency or exposure to environmental factors.

Job Safety Analysis (JSA)—A procedure that systematically identifies: (1) job steps, (2) specific hazards associated with each job step, and (3) safe job procedures associated with each step to minimize accident potential. Also called job hazard analysis.

Kaizen—A quality system using lessons learned

Latent Failure—Failure in a component because of a hidden flaw.

Layer of Protection Analysis (LOPA)—A process (method, system) of evaluating the effectiveness of independent protection layer(s) in reducing the likelihood or severity of an undesirable event.

Lessons Learned—Applying knowledge gained from past incidents in current practices.

Likelihood—An estimate of the expected frequency or probability of the occurrence of an event.[4]

Limited impact incidents—Incidents deemed to be controllable with local resources and which have no lasting effects.

Lockout/Tagout—A safe work practice in which energy sources are positively blocked away from a segment of a process with a locking mechanism and visibly tagged as such to help ensure worker safety during maintenance and some operations tasks.

Management of Change (MOC)—A mechanism to require safety analysis of a proposed change.

Management System—An administrative system that governs essential business activities.

Medical Treatment—As defined by OSHA, treatment (other than first aid) administered by a physician or by registered professional personnel under the standing orders of a physician.

Methodology—The use of a combination of two or more incident investigation tools to analyze the evidence and determine the root causes of the incident.

Minor incidents—Infrequent occurrence of these accidents or near misses have acceptable consequences but recurring events of this magnitude may warrant an investigation.

Mitigation—The act of causing a consequence to be less severe.

Morphological Approach—A structured analysis of an incident directed by insights from historic case studies but not as rigorous as a formal hazard analysis.

Near Miss—An event in which an accident (that is, property damage, environmental impact, or human loss) or an operational interruption could have plausibly resulted if circumstances had been slightly different.

Occupational Incident—An incident involving injury to workers.

Operational Interruption—is an event in which production rates or product quality is seriously impacted.

Organizational Error—A latent management system problem that can result in human error.

OSHA Recordable Cases—Work-related deaths, injuries and illnesses (other than minor injuries requiring only first aid treatment) which involve medical treatment, loss of consciousness, restriction of work or motion, or transfer to another job.[4]

OSHA Reportable Event—An incident that causes any fatality or the hospitalization of five employees or more requires a notification report to the nearest OSHA office.

PFD—Probability of failure on demand. The probability that a system will fail to perform a specified function on demand.

PHA—Process hazard analysis. A hazard evaluation of broad scope that identifies and qualitatively analyzes the significance of hazardous situations associated with a process or activity.

Prevention—The act of causing an event not to happen.

Probability—The expression for the likelihood of occurrence of an event or an event sequence during an interval of time or the likelihood of the success or failure of an event on test or on demand. Probability is expressed as a dimensionless number ranging from 0 to 1.

Process Control System—A system that responds to input signals from the process and the operator to generate output signals, causing the process to operate in the desired manner.

Process Hazard Analysis—An organized effort to identify and evaluate hazards associated with chemical processes and operations to enable their control. This review normally involves the use of qualitative techniques to identify and assess the significance of hazards. Conclusions and appropriate recommendations are developed. Occasionally, quantitative methods are used to help prioritize risk reduction measures.[4]

Process Safety—A discipline that focuses on the prevention of fires, explosions, and accidental chemical releases at chemical process facilities. Excludes classic worker health and safety issues involving working surfaces, ladders, protective equipment, and other things[1]

Process Safety Management—A program or activity that involves the application of management principles and analytical techniques to ensure process safety in chemical facilities. The focus is on preventing major accidents rather than dealing with classic worker health and safety issues.[1]

Process-Related Incident—An incident with impact, or potential impact, on process, equipment, people, and the environment. The incident could be internal or external to the process. An occupational incident can result from a process related incident.

Protection Layer—A device, system, or action that is capable of preventing a scenario from proceeding to the undesired consequence. (Also known as barrier.)

Proximate Cause—The causal factor that directly produces the effect without the intervention of any other cause. The cause nearest to the effect in time and space.[4]

Risk—A measure of potential economic loss, human injury or environmental insult in terms of the frequency of the loss or injury occurring and the magnitude of the loss or injury if it occurs.

Risk Analysis—The development of a quantitative estimate of risk based on engineering evaluation and mathematical techniques for combining estimates of initiating event frequency and independent protection layers and consequences.

Risk Assessment—The process by which the results of an analysis are used to make decisions, either through relative ranking of risk reduction strategies or through comparison with risk targets.

Risk Management—Systematic application of management policies, procedures, and practices that analyze, assess, and control risk in order to protect employees, the public, the environment, and company assets while avoiding business interruptions. It includes decisions to use suitable engineering and administrative controls for reducing risk.[4]

Risk Ranking—A decision making aid that ranks items, such as scenarios or proposed recommendations, in order of their potential associated risk exposure.

Root Cause—A fundamental, underlying, system-related reason why an incident occurred that identifies a correctable failure(s) in management systems. There is typically more than one root cause for every process safety incident.

Safeguard—Any device, system, or action that would likely interrupt the chain of events following an initiating event or that would mitigate the consequences.

Safety—A term denoting an acceptable level of risk. A state of relative freedom from and low probability of harm.

Safety Critical Actions—Specific steps humans take that provide layers of protection to lower the risk category of a specific scenario or scenarios from "unacceptable" to "acceptable" as defined by organizational risk tolerance criteria. Sometimes called *administrative control.* Such steps that further reduce the risk below "acceptable" might not be designated as safety critical actions.

Safety Critical Equipment—Engineering controls that provide layers of protection to lower the risk category of a specific scenario or scenarios from "unacceptable" to "acceptable" as defined by organizational risk tolerance criteria. Engineering controls that further reduce the risk below "acceptable" might not be designated as safety critical equipment.

Scenario—An event or sequence of events that results in undesirable consequences.

Sensor—Field measurement system (instrumentation) capable of detecting the condition of a process (for example, pressure transmitters; level transmitters, and toxic gas detectors).

Serious Injury—The classification for an occupational injury which includes all disabling work injuries and non-disabling work injuries as follows eye injuries requiring treatment by a physician, fractures, injuries requiring hospitalization, loss of consciousness, injuries requiring treatment by a doctor and injuries requiring restriction of motion or work, or assignment to another job.[(4)]

Significant incidents—Incidents that have, or would have in the case of near misses, consequences requiring considerable resources to mitigate and usually involve human injuries and/or major interruptions to operations.

Software—The programs, procedures, and related documentation associated with a system design and system operation. The system can be a computer system or a management system.

Task Analysis—An analytical process for determining the specific behaviors required of the human components in a man-machine system. It involves determining the detailed performance required of people and equipment and the effects of environmental conditions, malfunctions, and other unexpected events on both. Within each task to be performed by people, behavioral steps are analyzed in terms of (i) the sensory signals and related perceptions, (ii) the decisions, memory storage, and other mental processes, and (iii) the required responses.

Taxonomy—A structure for classifying incidents.

Technique—The manner in which an incident investigation tool is developed or used.

Tool—A device or means used at a discrete stage of the incident investigation to facilitate understanding of event chronology, causal factors, or root causes.

Underlying Causes—*See* root causes.

Validation—The activity of demonstrating that the safety-instrumented system under consideration, after installation, meets in all respects the safety requirements specification for that safety-instrumented system.

Verification—The activity of demonstrating by analysis or test, that, for the specific inputs, the deliverables meet, in all respects, the objectives and requirements set forth by the functional specification.

Witness—A person who has direct or indirect information related to an incident.

References

1. American Chemistry Council. *Responsible Care®: A Resource Guide for the Process Safety Code of Management Practices*. Washington, DC: American Chemistry Council, 1990.
2. Kuhlman, R. *Professional Accident Investigation*. Loganville, GA: Institute Press, International Loss Control Institute, 1977.
3. Carper, K. 1989. *Forensic Engineering*. New York: Elsevier, 1989
4. Center for Chemical Process Safety. *Guidelines for Technical Management of Chemical Process Safety*. New York: American Institute of Chemical Engineers, 1989.

Index

D

E

K

L

O

P

Q

R

Publications Available from the CENTER FOR CHEMICAL PROCESS SAFETY of the AMERICAN INSTITUTE OF CHEMICAL ENGINEERS

3 Park Avenue, New York, NY 10016-5991

CCPS Guidelines Series

Guidelines for Investigating Chemical Process Incidents, Second Edition
Guidelines for Analyzing and Managing the Security Vulnerabilities of Fixed Chemical Sites
Guidelines for Process Safety in Outsourced Manufacturing Operations
Guidelines for Process Safety in Batch Reaction Systems
Guidelines for Chemical Process Quantitative Risk Analysis, Second Edition
Guidelines for Consequence Analysis of Chemical Releases
Guidelines for Pressure Relief and Effluent Handling Systems
Guidelines for Design Solutions for Process Equipment Failures
Guidelines for Safe Warehousing of Chemicals
Guidelines for Postrelease Mitigation in the Chemical Process Industry
Guidelines for Integrating Process Safety Management, Environment, Safety, Health, and Quality
Guidelines for Improving Plant Reliability through Data Collection and Analysis
Guidelines for Use of Vapor Cloud Dispersion Models, Second Edition
Guidelines for Evaluating Process Plant Buildings for External Explosions and Fires
Guidelines for Writing Effective Operations and Maintenance Procedures
Guidelines for Chemical Transportation Risk Analysis
Guidelines for Safe Storage and Handling of Reactive Materials
Guidelines for Technical Planning for On-Site Emergencies
Guidelines for Process Safety Documentation
Guidelines for Safe Process Operations and Maintenance
Guidelines for Process Safety Fundamentals in General Plant Operations
Guidelines for Chemical Reactivity Evaluation and Application to Process Design
Tools for Making Acute Risk Decisions with Chemical Process Safety Applications
Guidelines for Preventing Human Error in Process Safety
Guidelines for Evaluating the Characteristics of Vapor Cloud Explosions, Flash Fires, and BLEVEs
Guidelines for Implementing Process Safety Management Systems
Guidelines for Safe Automation of Chemical Processes
Guidelines for Engineering Design for Process Safety
Guidelines for Auditing Process Safety Management Systems
Guidelines for Hazard Evaluation Procedures, Second Edition with Worked Examples
Guidelines for Process Equipment Reliability Data with Data Tables
Guidelines for Safe Storage and Handling of High Toxic Hazard Materials

CCPS Concept Series

Understanding Explosions
Essential Practices for Managing Chemical Reactivity Hazards
Layer of Protection Analysis: Simplified Process Risk Assessment

Wind Flow and Vapor Cloud Dispersion at Industrial and Urban Sites
Deflagration and Detonation Flame Arresters
Making EHS an Integral Part of Process Design
Revalidating Process Hazard Analyses
Electrostatic Ignitions of Fires and Explosions
Evaluating Process Safety in the Chemical Industry
Avoiding Static Ignition Hazards in Chemical Operations
Estimating the Flammable Mass of a Vapor Cloud
RELEASE: A Model with Data to Predict Aerosol Rainout in Accidental Releases
Practical Compliance with the EPA Risk Management Program
Local Emergency Planning Committee Guidebook: Understanding the EPA Risk Management Program Rule
Inherently Safer Chemical Processes: A Life-Cycle Approach
Contractor and Client Relations to Assure Process Safety
Understanding Atmospheric Dispersion of Accidental Releases
Expert Systems in Process Safety
Concentration Fluctuations and Averaging Time in Vapor Clouds

Proceedings and Other Publications

Center for Chemical Process Safety 17th Annual International Conference and Workshop: Risk, Reliability, and Security, 2002
Center for Chemical Process Safety International Conference and Workshop: Making Process Safety Pay:The Business Case, 2001
Center for Chemical Process Safety International Conference and Workshop: Process Industry Incidents—Investigation Protocols, Case Histories, Lessons Learned, 2000
Proceedings of the International Conference and Workshop on Modeling the Consequences of Accidental Releases of Hazardous Materials, 1999
Proceedings of the International Conference and Workshop on Reliability and Risk Management, 1998
Proceedings of the International Conference and Workshop on Risk Analysis in Process Safety, 1997
Proceedings of the International Conference and Workshop on Process Safety Management and Inherently Safer Processes, 1996
Proceedings of the International Conference and Workshop on Modeling and Mitigating the Consequences of Accidental Releases of Hazardous Materials, 1995
Proceedings of the International Symposium and Workshop on Safe Chemical Process Automation, 1994
Proceedings of the International Process Safety Management Conference and Workshop, 1993
Proceedings of the International Conference on Hazard Identification and Risk Analysis, Human Factors, and Human Reliability in Process Safety, 1992
Proceedings of the International Conference and Workshop on Modeling and Mitigating the Consequences of Accidental Releases of Hazardous Materials, 1991

SUPPLEMENT TO

Guidelines for Investigating Chemical Process Incidents, Second Editic

Center for Chemical Process Safety
American Institute of Chemical Engineers
3 Park Avenue
New York, NY 10016-5991

Getting started:
Insert CD into CD-ROM drive
Open Windows Explorer and select the file you wish to open.

Assistance:
If you have difficulties contact CCPS at (212) 591-7319 from 9:00 AM to 4:30 PM Monday through Friday